Creating Theoretical Research Frameworks Using Multiple Methods

Insight from ICT4D Investigations

Creating Theoretical Research Frameworks Using Multiple Methods

Insight from ICT4D Investigations

Sergey V. Samoilenko
Kweku-Muata Osei-Bryson

CRC Press is an imprint of the
Taylor & Francis Group, an **informa** business

AN AUERBACH BOOK

CRC Press
Taylor & Francis Group
6000 Broken Sound Parkway NW, Suite 300
Boca Raton, FL 33487-2742

© 2018 by Taylor & Francis Group, LLC
CRC Press is an imprint of Taylor & Francis Group, an Informa business

No claim to original U.S. Government works

Printed on acid-free paper

International Standard Book Number-13: 978-1-4987-7995-1 (Hardback)

This book contains information obtained from authentic and highly regarded sources. Reasonable efforts have been made to publish reliable data and information, but the author and publisher cannot assume responsibility for the validity of all materials or the consequences of their use. The authors and publishers have attempted to trace the copyright holders of all material reproduced in this publication and apologize to copyright holders if permission to publish in this form has not been obtained. If any copyright material has not been acknowledged please write and let us know so we may rectify in any future reprint.

Except as permitted under U.S. Copyright Law, no part of this book may be reprinted, reproduced, transmitted, or utilized in any form by any electronic, mechanical, or other means, now known or hereafter invented, including photocopying, microfilming, and recording, or in any information storage or retrieval system, without written permission from the publishers.

For permission to photocopy or use material electronically from this work, please access www.copyright.com (http://www.copyright.com/) or contact the Copyright Clearance Center, Inc. (CCC), 222 Rosewood Drive, Danvers, MA 01923, 978-750-8400. CCC is a not-for-profit organization that provides licenses and registration for a variety of users. For organizations that have been granted a photocopy license by the CCC, a separate system of payment has been arranged.

Trademark Notice: Product or corporate names may be trademarks or registered trademarks, and are used only for identification and explanation without intent to infringe.

Library of Congress Cataloging-in-Publication Data

Names: Samoilenko, Sergey, 1966- author. | Osei-Bryson, Kweku-Muata, author.
Title: Creating theoretical research frameworks using multiple methods : insight from ICT4D investigations / Sergey Samoilenko and Kweku-Muata Osei-Bryson.
Description: Boca Raton, FL : CRC Press, [2017]
Identifiers: LCCN 2017017363| ISBN 9781498779951 (hb : alk. paper) | ISBN 9781315155128 (eISBN)
Subjects: LCSH: Information technology--Developing countries. | Telecommunication--Developing countries. | Digital divide--Developing countries. | Economic development--Developing countries.
Classification: LCC HC59.72.I55 S35 2017 | DDC 338.9/260721--dc23
LC record available at https://lccn.loc.gov/2017017363

Visit the Taylor & Francis Web site at
http://www.taylorandfrancis.com

and the CRC Press Web site at
http://www.crcpress.com

Contents

Preface xiii
Authors xix

1 **Overview of Research on Information and Communication Technologies for Development** 1
  Introduction 1
  Some Options for Configuring an ICT4D Research Project 2
    ICT Issues and Focus Areas 2
    Context—Countries and Communities 3
    Research Methodologies 3
    Theoretical Frameworks 3
    Data Analysis Methods 4
  Conclusion 4
  References 5

2 **Overview on Framework of Neoclassical Growth Accounting** 11
  Exogenous and Endogenous Growth Models 12
  Formulation of the Neoclassical Production Function 13
  Transcendental Logarithmic (Translog) Production Function 15
  Weaknesses and Criticisms of the Framework of Neoclassical Growth Accounting 17
  References 18

3 **Overview on Complex Systems Theory and Chaos Theory** 19
  Complex Systems Theory: An Overview 19
  Complex Systems: Structural Perspective 20
  Complex Systems: Functional Perspective 24
  Chaos Theory: An Overview 25
  References 27

4 **An Overview on Product Cycle Theory and the Product Life Cycle Model** 29
  Product Cycle Theory: An Overview 29

v

vi ■ Contents

Product Life Cycle Model: An Overview of the Stages and Common
Criticisms...33
    Stage 1—Introduction.. 34
    Stage 2—Growth 34
    Stage 3—Maturity 35
    Stage 4—Decline 35
    Criticisms and Applications of Product Life Cycle Model 36
References...37

5 Overview on Decision Tree Induction..39
Introduction 39
    Classification Tree 41
    Regression Tree 41
DT Generation Process..43
    Recursive Splitting 45
    Selection of the Splitting Method.. 46
    Prepruning and Postpruning 46
        Prepruning.. 46
        Postpruning.. 46
    Recursive Splitting Example...50
    Software Implementation of the DT Generation Process 51
References...53

6 Overview on Cluster Analysis 55
Introduction 55
Understanding the Output of Clustering...57
Clustering Algorithms 60
    Similarity Metrics 61
Evaluating the Output of Clustering Algorithms 61
    The Issue of Quality: Assessing Cluster Validity..................................62
    The Issue of Usefulness: Goals for Clustering.....................................62
    Overview of a Process Model for Clustering.......................................63
Software Implementation of the Clustering Process.................................65
    Cluster Node: Some Important Parameter Settings............................67
References...68

7 Overview on Data Envelopment Analysis 71
Introduction 71
DEA Model: Common Guidelines and Assumptions 72
DEA: General Approach and Types of Efficiency 73
DEA Models: Common Orientations..74
DEA: Types of Models...76
DEA: Malmquist Index 77
References...79

Contents ▪ vii

8   Overview on Structural Equation Modeling..81
   SEM: Reflective and Formative Measurement Models.............................83
   SEM: Model Specification 84
   SEM: Two Common Approaches 85
   SEM: Preliminary Data Analysis and Factor Analysis................................86
   SEM: Assessment of the Measurement Model 88
   SEM: Assessment of the Structural Model..89

9   Overview on Artificial Neural Networks..91
   Introduction 91
   NN Induction Process for Directed Learning...93
      Computing the Inputs to the Nodes of the Hidden
      and Output Layers 94
      Computing Outputs of the Nodes of the Hidden and Output Layers ...95
      Network Architectures 95
      Weight Adjustment for the Backpropagation Algorithm
      for Multilayer Perceptron 95
      Error Functions 97
   Software Implementation of the NN Induction Process 97
   References..104

**10 Information Systems Fitness and Risk in IS Development:
      Insights and Implications from Chaos and Complex Systems
      Theories 107**
   Introduction 107
   Conceptual Foundations..109
   CT and ISD..113
   Insights and Implications for ISD 114
   IS and ISD from the Perspective of CST...118
      Self-Organization of the CSs..118
      Fitness Landscapes 118
   Fitness Landscapes and ISD...120
      Scenario One: Fitness Landscapes and ISD Process..........................121
      Scenario Two: Fitness Landscapes and ISD Product.........................122
      Scenario Three: Fitness Landscapes and the Process and Product
      of ISD 122
   Conclusion...123
   Acknowledgment 125
   References..125

**11 Design of the Research Workbench for Investigations Relying
      on Multitheoretical Support..129**
   Performance Analysis for Complex Systems 130
   Cybernetics-Based Analytic Support System 133

Cybernetics: An Overview 134
Implications of the General Principles of Cybernetic Systems
for Designing the PAS 135
Structural Components of PAS 135
Conclusion 140
References 140

12 Investigation of Determinants of Total Factor Productivity: An Analysis of the Impact of Investments in Telecoms on Economic Growth in Productivity in the Context of Transition Economies 143

Introduction 143
Description of the Data 146
Research Methodology 146
    Phase 1 146
    Phase 2 148
    Phase 3 150
    Phase 4 151
Null Hypotheses of the Study 152
    Phase 1 152
    Phase 2 152
    Phase 3 153
    Phase 4 154
Results and Discussion 154
Conclusion 164
Acknowledgment 166
References 166

13 Human Development and Macroeconomic Returns within the Context of Investments in Telecoms: An Exploration of Transition Economies 169

Introduction 169
Background 172
Theoretical Framework and Research Questions of the Study 176
Overview of the Data 177
Results of the Data Analysis 178
Discussion and Conclusion 178
Acknowledgment 182
References 182

14 The Spillover Effects of Investments in Telecoms: Insights from Transition Economies 187

Introduction 187
Research Problem of the Study 191
Overview of the Data 195

Results of Data Analysis 199
    Preliminary Data Analysis..199
    PLS Analysis: Steps, Procedures, and Results 199
        Assessment of the Measurement Model 199
        PLS Analysis: Assessment of the Structural Model 203
Discussion of the Results 203
Limitations of the Study 209
Conclusion...210
Acknowledgment 211
References..211
Appendix 214

15 Investigating Factors Associated with the Spillover Effect of Investments in Telecoms: Do Some Transition Economies Pay Too Much for Too Little? 215
Introduction 215
Background of the Study 218
Theoretical Foundation and Research Questions of the Study 220
Overview of the Data...223
Research Methodology 224
Results of the Data Analysis..229
    RQ1: Steps 1, 2, and 3..229
    RQ2: Steps 4 and 5 231
    RQ3: Steps 6 and 7 231
Discussion of the Results 235
Limitations of the Study and Future Research 239
Conclusion...240
Acknowledgment 241
References..241
Appendix 245

16 Understanding the Human Capital Dimension of Information and Communication Technology and Economic Growth in Transition Economies 247
Introduction 247
Background 249
The Research Problem...250
Determining Appropriate Policy Options 251
Proposed Approach to Policy Analysis 253
    Step 1: Using MR to Determine the Presence of Complementarity.....253
    Step 2: Using DEA to Determine the Sources of Relative Inefficiency...253
The Sample and Panel Data 256

Data Analysis 257
    Analysis Procedure 257
        Step 1: Analysis of H01 257
        Step 2: Analysis of H02 258
    Discussion and Conclusion 260
    Acknowledgment 261
    References 262

17 An Exploration of the Effects of the Interaction between Information and Communication Technology and Labor Force on Economic Growth in Transition Economies 265

Introduction 265
Overview of the Data and Background of this Study 267
Theoretical Framework 269
    Theory of Complementarity and Translog Production Function 270
    Formal Definition of the Research Problem 272
Results of the Data Analysis 274
Summary and Conclusion 280
Acknowledgment 281
References 281
Appendix 284

18 Contributing Factors to Information Technology Investment Utilization in Transition Economies: An Empirical Investigation 287

Introduction 287
Theoretical Framework 290
    Growth Accounting 290
Overview of the Data 291
Methodology: Searching for the Determinants of the Efficiency
of Utilization of Investments in Telecoms 292
    Phase 1: DEA 292
        Data Used to Perform DEA 293
    Phase 2: CA 294
        Data Used to Perform CA 295
    Phase 3: DT 295
        Data Used to Perform DT 295
Results 296
    Results: DEA 296
    Results: CA 298
    Results: DT 299
Contribution of the Study 303
Summary and Conclusion 306
Acknowledgment 308

References 308
Appendix 311

19 Socioeconomic Impact of Information and Communication Technology Capabilities in Sub-Saharan Economies: Using Association Rules to Describe the Structure of Complex Systems 315
Overview 315
Introduction 316
Research Framework 318
Data 318
Methodology 320
  Phase 1: DEA 320
  Phase 2: Data Transformation for ARM 321
  Phase 3: Market Basket Analysis via ARM 322
Research Questions and Null Hypotheses of the Study 322
Results of the Data Analysis 323
  Phase: 1 DEA 323
  Phase 2 and Phase 3: Data Transformation and MBA 324
Discussion of the Results 324
Conclusion 332
Acknowledgment 333
References 333

20 Improving the Relative Efficiency of Revenue Generation from ICT in Transition Economies: A Product Life Cycle Approach 335
Introduction 335
Theoretical Framework 338
  PLC Model 338
Description of the Data and the Background of the Study 339
Overview of the Methods and Techniques Used in the Study 341
Methodological Approach 343
  Phase 1: CA 343
  Phase 2: Output-Oriented DEA 345
  Phase 3: Input-Oriented DEA 346
  Phase 4: NN Simulation 346
  Phase 5: Output-Oriented DEA of the Simulated Data 347
Discussion of Results of the Data Analysis 347
  Results of Phase 1 347
  Results of Phase 2 349
  Results of Phase 3 351
  Results of Phases 4 and 5 352
Conclusions 357
  Summary of Results 357

        Contribution 358
        Limitations 358
        Future Research 359
    Acknowledgment 359
    References 359
    Appendix 363

21 Determining Strategies for Telecoms to Improve Efficiency in the Production of Revenues 373
    Introduction 373
    Overview of the Data 377
    Previous Findings 378
    Methodology 380
        Phase 1: DT Induction 381
        Phase 2: NN Simulation 382
        Phase 3: DEA 382
    Results of the Data Analysis 383
        Decision Tree Analysis 383
        NN Simulation 384
        Results of DEA 384
    Interpretation of the Results 384
    Discussion and Conclusion 386
    Acknowledgment 388
    References 388
    Appendix 392

Index 393

Preface

By now, it is commonly accepted that investments in information and communication technologies (ICTs) can facilitate macroeconomic growth in developed countries (Batchelor, Scott, & Woolnough, 2005). Numerous studies have shown that such investments resulted in generous payoffs for the United States (Jorgenson, 2001; Jorgenson & Stiroh, 2000; Oliner & Sichel, 2000; Stiroh, 2002) and European Union economies (Colecchia & Schreyer, 2002; van Ark, Melka, Mulder, Timmer, & Ypma, 2002; Daveri, 2002; Jalava & Pohjola, 2002). There appears to be no reason why other types of countries, namely, developing, least developed, or transition economies (TEs),* would not profit from investments in ICT at the macroeconomic level. However, at this point, any conclusion regarding this matter is premature, since the existing empirical evidence is limited.

Research standards in the area of ICT for development (ICT4D) are high, and it is a basic expectation that a theoretically sound conceptual investigation should yield actionable results. An additional expectation is that an on-the-ground study conducted in a given setting should add to the common body of knowledge based on a theory. In other words, one is expected to make a connection between the world of concepts and the world of reality. Middle-range theories and frameworks could help connect the case studies with grand theories, by helping to create a theoretically sound and practically applicable *research architecture* of ICT4D.

General research in the area of ICT4D may, arguably, be placed in two main categories. The first category comprises studies testing a grand, long-range theory within a specific context via deductive research methods. A common criticism of such studies is that the findings are too high level and general to be practically applicable due to their lack of context-specific recommendations. There is no connection between the high-level conceptual picture provided by the investigation and the boots-on-the-ground needs of practitioners. The second category comprises fairly narrow, context-specific investigations conducted in order to gain important insights into ways of dealing with real-world practical problems, generally by using inductive research methods. A common criticism of such studies is that they fail to

---

* The term *transition economies* refers to countries in the process of transitioning from a centrally planned economy to a market-oriented economy.

incorporate their findings, or newly discovered constructs, into the existing grand theories. There is no connection between the new findings or constructs and the established nomological network of existing theories.

It may be suggested that the purpose of scientific investigations in ICT4D is to *offer theoretically sound and practically applicable solutions to context-specific problems.* Often, however, studies in the area of ICT4D are either theoretically sound or practically applicable. It is rigor *versus* relevance; rarely is it rigor *and* relevance. This is not surprising, since an investigator armed with deductive methods often lacks sufficient constructs relevant to the setting; a theory does not inform researchers of a context. But in order to conduct a scientific investigation that is rigorous *and* relevant, the ICT4D researcher must attempt to translate high-level conceptual suggestions into actionable context-specific recommendations. And in order to do so, the investigator must be capable of dealing with two problems: *the problem of complexity* and *the problem of context.*

Firms, industries, and economies are examples of dynamic complex systems that are capable of exhibiting chaotic behavior. In order to survive in ever-changing environments, such systems respond by changing their structures and behaviors. As a result, the organizational structures of firms and economies, and their models of transformation of inputs into outputs, must also change. The economic models and business processes that worked at one time must be modified and revised often. Consequently, a researcher investigating the impacts of inputs on the outputs of dynamic systems may not have the luxury of relying on established business and economic models. Instead, the researcher faces the task of dealing with a black box, the inner workings of which must be investigated anew.

Researchers working in the area of ICT4D face numerous challenges associated with the context of investigations. We can identify four important aspects of a research setting that may complicate the generalizability of the results of an investigation beyond its context. Only the first aspect, the geographical specificity of the environment, can be considered static. The other three aspects—political, sociocultural, and economic—are dynamic.

This dynamic nature of the context is one of the factors that complicate theory- and framework-building efforts in the area of ICT4D. The complication is associated with the issue of representation of the context. Theory building may be conceived of as a two-step process. In the first step, which we call context representation, the relevant constructs and their interrelationships are identified. The outcome of the first step is the model of a context in the form of a theory. In the second step, which we call the test of representation, the validity of the proposed model of a context is tested. The results of the tests of the hypotheses generated are the outcome of the second step.

In this book, we demonstrate how creative use of various data analysis methods [e.g., data mining (DM), data envelopment analysis (DEA), and structural equation modeling (SEM)] and conceptual frameworks (e.g., neoclassical growth accounting, chaos and complexity theories) may be utilized for inductive and deductive purposes to develop and to test, in step-by-step fashion, theoretically sound frameworks for a large subset of ICT4D research questions. Specifically, we aim to showcase the utilization of DM, DEA, and SEM for the following purposes:

1. Identification of the relevant context-specific constructs (inductive application)
2. Identification of the relationships between the constructs (inductive application)
3. Development of a framework incorporating the constructs and relationships discovered (inductive application)
4. Testing of the constructed framework (deductive application)

In our approach to research, we rely on a multitheoretical perspective. Chapter 1 provides an overview of ICT4D. In Chapters 2, 3, and 4, we offer overviews of the relevant frameworks and theories: the framework of neoclassical growth accounting and the theory of complementarity (Chapter 2), complex systems and chaos theories (Chapter 3), and the product life cycle (PLC) theory (Chapter 4).

Chapters 5–9 provide nontechnical overviews of the DM and data analytic methods that are utilized in the later chapters: decision tree (DT) induction (Chapter 5), cluster analysis (CA) (Chapter 6), DEA (Chapter 7), SEM (Chapter 8), and neural networks (NNs) (Chapter 9).

Chapter 10 presents the argument that an organized social environment, such as a firm, industry, or country, may be investigated through the lens of complex systems and chaos theories. This chapter lays the basis for Chapter 11, which treats the design of studies with the multitheoretical perspectives and multitool, multitechnique methodologies developed in later chapters.

Chapter 12 describes an application of a four-phase methodology supported by three data analytic tools [DEA, multivariate regression (MR), and ordinary least-squares regression (OLS)] to investigate the factors impacting growth in productivity. The framework of neoclassical growth accounting and theory of complementarity serve as a foundation for the research framework supporting the investigation.

Chapter 13 is dedicated to investigating the impact of human development, as measured by the Human Development Index (HDI), on macroeconomic outcomes and productivity. Multivariate regression (MR) and DEA are the data analytic tools used in the study. The conceptual underpinning of the research is a *push–pull* model of the macroeconomic impact of investments in telecoms, which is consistent with the framework of neoclassical growth accounting and the theory of complementarity.

Chapter 14 demonstrates a possible solution to the problem of endogeneity associated with the framework of neoclassical growth accounting, where growth in productivity is unexplained by capital and labor. The conceptual foundation of the proposed solution is based on an indirect relationship between capital, labor, and productivity. The methodology outlined in the chapter relies on SEM, DEA, and MR.

Chapter 15 continues the line of inquiry by investigating the factors associated with growth in productivity. The study relies on a seven-step methodology that utilizes SEM, DEA, and DT to investigate whether growth in productivity is primarily due to changes in technology or to changes in efficiency, and whether labor and capital impact the changes.

Chapter 16 concentrates on the human capital dimension of growth in productivity, evidence of which was uncovered in the previous chapter. The research approach of the study is consistent with the theory of complementarity and the framework of neoclassical growth accounting, and relies on DEA and MR to formulate human capital strategies that explicitly take into consideration the level of investments in ICT.

Chapter 17 investigates a possible complementarity between human capital and investment components of the Cobb–Douglass production function of the framework of neoclassical growth accounting. MR serves as a primary tool of the data analysis.

While Chapters 10–17 presented evidence that human capital and investment capital are complementary and are reliable sources of economic growth, Chapter 18 is dedicated to investigating whether the efficiency of utilization of resources contributes to economic growth, as well as what factors possibly impact the levels of efficiency. This investigation relies on a neoclassical framework of growth accounting and the theory of complementarity, and employs a three-phase methodology utilizing DEA, CA, and DT.

Chapter 19 demonstrates an application of concepts of complex systems theory to the investigation of a socioeconomic impact of ICT. DEA and association rule mining are the data analytic methods used in this investigation.

Chapter 20 concentrates on the capital investment component of the Cobb–Douglas production function and investigates strategies for increasing the relative efficiency of the production of revenues from information and telecommunications technologies. The framework of the study is consistent with the framework of neoclassical growth accounting and relies on the PLC model. The investigation demonstrates how a five-step methodology incorporating CA, DEA, and NN methods allows for determining the areas along the PLC curve where increased investments in ICT will likely have the most macroeconomic impact in a resource-efficient manner.

Chapter 21 proposes a four-step methodology enabling a policy maker to formulate context-specific strategies for increasing the relative efficiency of the production of revenue, as well as to pinpoint the appropriate implementation routes of the identified strategy. The conceptual foundation of this study is the framework of neoclassical growth accounting, and the methodology utilizes DEA, CA, and NN methods.

References

Batchelor, S., Scott, N., & Woolnough, D. (2005). Good practice paper on ICTs for economic growth and poverty reduction. *DAC Journal*, 6(3), 27–95.

Colecchia, A., & Schreyer, P. (2002). ICT investment and economic growth in the 1990s: Is the United States a unique case? A comparative study of nine OECD countries. *Review of Economic Dynamics*, 5(2), 408–442.

Daveri, F. (2002). The new economy in Europe, 1992–2001. *Oxford Review of Economic Policy*, 18(3), 345–362.

Jalava, J., & Pohjola, M. (2002). Economic growth in the new economy: Evidence from advanced economies. *Information Economics and Policy*, 14(2), 189–210.

Jorgenson, D. W. (2001). Information technology and the US economy. *American Economic Review*, 91(1), 1–32.

Jorgenson, D. W., & Stiroh, K. J. (2000). *US economic growth in the new millennium*. Brookings Papers on Economic Activity, 1, 125–211.

Oliner, S. D., & Sichel, D. E. (2000). The resurgence of growth in the late 1990s: Is information technology the story? *Journal of Economic Perspectives*, 14(4), 3–22.

Stiroh, K. (2002). Information technology and the U.S. productivity revival: What do the industry data say? *American Economic Review*, 92(5), 1559–1576.

van Ark, B., Melka, J., Mulder, N., Timmer, M., & Ypma, G. (2002). ICT investments and growth accounts for the European Union, 1980–2000 (Research Memorandum GD-56). Groningen: Groningen Growth and Development Centre.

Authors

Sergey V. Samoilenko, PhD, is an associate professor and the chair of the Department of Computer Information Systems & Computer Science at Averett University, Danville, Virginia.

Previously, he was an associate professor and the chair of the Department of Computer Information Systems & Computer Science at Virginia Union University, Richmond, Virginia.

He earned a PhD and an MS in Information Systems from Virginia Commonwealth University, and a BS in Industrial Engineering from the Institute of Trade Technology, St. Petersburg, Russia.

Sergey's research areas include information technology (IT) and productivity, IT for development, decision support systems, data mining, and information systems development. He has published in the *European Journal of Operational Research*, *Journal of Global Information Technology Management*, *International Journal of Production Economics*, *Expert Systems with Applications*, and *Information Systems Frontiers*, among other journals, as well as in numerous conference proceedings. Sergey has also published two books and contributed numerous chapters to edited books.

Kweku-Muata Osei-Bryson, PhD, is a professor of information systems at Virginia Commonwealth University in Richmond, Virginia, where he also served as the coordinator of the Information System PhD program during 2001–2003. He is also a visiting professor of computing at the University of the West Indies at Mona, Jamaica. Previously, he was professor of information systems and decision sciences at Howard University in Washington, DC. He has also worked as an information systems practitioner in industry and government. He earned a PhD in Applied Mathematics (Management Science & Information Systems) from the University of Maryland at College Park; an MS in Systems Engineering from Howard University; and a BSc in Natural Sciences from the University of the West Indies at Mona.

His research areas include data mining, decision support systems, knowledge management, IS security, e-commerce, information technology (IT) for development, database management, IS outsourcing, and multicriteria decision making.

He has published in various leading journals including *Decision Support Systems, Information Systems Journal, Expert Systems with Applications, European Journal of Information Systems, Information Systems Frontiers, Knowledge Management Research & Practice, Information Sciences, Information & Management, Journal of the Association for Information Systems, Journal of Information Technology for Development, Journal of Database Management, Computers & Operations Research, Journal of the Operational Research Society,* and *European Journal of Operational Research*. He serves as associate editor of the *INFORMS Journal on Computing* and the *Journal of Information Technology for Development*, as a member of the editorial board of the *Computers & Operations Research* journal, and as a member of the International Advisory Board of the *Journal of the Operational Research Society*. He is also coeditor of four books: *Advances in Research Methods for Information Systems Research: Data Mining, Data Envelopment Analysis, Value Focused Thinking*, Springer (2014); *Knowledge Management for Development: Domains, Strategies and Technologies for Developing Countries*, Springer (2014); *Knowledge Discovery Process and Methods to Enhance Organizational Performance*, CRC Press (2015); *Strategic Project Management: Contemporary Issues and Strategies for Developing Economies*, CRC Press (2015).

Chapter 1

Overview of Research on Information and Communication Technologies for Development

This chapter provides a brief overview of previous research on information and communication technology for development (ICT4D).

Introduction

In this chapter, we present an overview of research on information and communication technologies for development (ICT4D), which, as noted by Walsham (2017), "has a history going back some 30 years." For a more comprehensive exposition of this research field, the reader may consult studies such as those of Walsham (2017), Roztocki and Weistroffer (2015), Avgerou (2008), and Donner (2008).

To date, ICT4D research can be considered to fall into the broad categories of (1) the relationship between information and communication technologies (ICTs) and development at the macro or micro level and (2) adoption/diffusion/implementation of ICT in "developing" countries (including those considered to be transitional or emerging economies) without necessarily giving attention to the issue of development. With regard to the first category, a major aim is often to understand and/or demonstrate how ICT may be used to combat poverty and promote economic growth and sustainable development in developing countries, economies in transition, and least developed countries, rooted in the recognition, which has proliferated in recent years, that quickly accessed, properly adapted, and broadly shared information and knowledge are key drivers of economic growth and social opportunity. An underlying speculation behind such research involves examination of the validity of theories of modernization, which, as noted by Qureshi (2015), "are based on the notion that technological change developed in the West should be used to develop economies that are poor." Alternatively, ICT4D research can be considered to fall into three groups: (1) studies that explain or predict the adoption/diffusion/implementation of ICTs; (2) those that assess the impacts of ICTs on people, organizations, and economies; and (3) those that attempt to explore and explain the usage of various ICTs.

Some Options for Configuring an ICT4D Research Project

In this section, we briefly discuss previous ICT4D research in terms of the ICT issues and focus areas of previous studies; the countries and communities that provided the context; and the research methodologies, theoretical frameworks, and data analysis methods that were utilized. For each subsection, we list examples of relevant papers in order to, if necessary, sensitize the reader to the wide variety of options for configuring an ICT4D research project.

ICT Issues and Focus Areas

A variety of ICT issues have been explored in previous studies, including ICT adoption, diffusion, and usage (e.g., Avgerou et al. 2016; Wong et al. 2016; Zaremohzzabieh et al. 2016; Van der Schyff & Kraus 2014; Bernroider et al. 2011; Gryczka 2011; Bailey & Ngwenyama 2010; Pavon & Brown 2010; Donner & Tellez 2008; Brown et al. 2007; Brown & Licker 2003; Brown 2002), ICT implementation (e.g., Cavalheiro & Joia 2016; Lech 2012), ICT strategy (e.g., Maryska et al. 2012), impacts of ICT (e.g., Bankole et al. 2015; Samoilenko & Osei-Bryson 2011), ICT education (e.g., Maryska et al. 2012), e-learning (e.g., Charlton-Laing & Grant 2015), and ICT security (e.g., Barclay 2014; Barrett-Maitland et al. 2016; Gupta et al. 2017;). Roztocki and Weistroffer (2015) report that for ICT4D research that focused on the transition economies of Europe, ICT diffusion and ICT implementation were the two most explored issues. These two ICT issues have also attracted the attention of researchers focusing on other geographic regions, though with regard to less economically developed regions, there is also increased interest in exploring the impacts of ICTs on the people, organizations, and economies.

Focus areas previously studied include banking (e.g., Gupta et al. 2017; Borena & Negash 2016; Mansingh et al. 2015), cultural issues (e.g., Krauss 2013), education (e.g., Stanimirovic & Vintar 2015), employment (e.g., Pavon & Brown 2010), governance (e.g., Canares 2016; Prakash 2016), e-government (e.g., Cumbie & Kar 2016), health care (e.g., Prakash 2016; Stanimirovic & Vintar 2015; Purcarea et al. 2011), trade and commerce (e.g., Boateng 2016; Bankole et al. 2015), telecommunications (e.g., Wong et al. 2016; Kaba & Osei-Bryson 2013; Mimbi et al. 2011; Brown et al. 2004), community empowerment (e.g., Nemer 2016; Bailey & Ngwenyama 2010; Qureshi 1998), rural economic development (e.g., Alam & Wagner 2016; Wyche & Steinfield 2016), and small and medium enterprises (SMEs) and microenterprises (e.g., Xiong et al. 2015; Qureshi et al. 2010).

Context—Countries and Communities

The contexts of previous ICT4D research projects include Africa (e.g., Kibere 2016; Bankole et al. 2015), Asia (e.g., Gupta et al. 2017; Wong et al. 2016; Prakash 2016), Latin America (e.g., Ferreira et al. 2016; Nemer 2016), the Caribbean (e.g., Barrett-Maitland et al. 2016; Bailey & Ngwenyama 2013; Bailey 2009), transition economies of Europe (e.g., Samoilenko 2014; Samoilenko & Osei-Bryson 2011; Stanimirovic & Vintar 2015), and SMEs and enclaves of minority groups or marginalized communities located in the "developed world" (e.g., Cumbie & Kar 2016; Xiong et al. 2015). The contexts for these studies include both what is often referred to as the developing world (e.g., Africa, Latin America, most of Asia) and some countries and communities in Europe, North America, and Australasia.

Research Methodologies

Research methodologies include design science (e.g., Islam & Grönlund 2012), grounded theory (e.g., Potnis 2016), action research (e.g., Krauss 2013), qualitative behavioral science (e.g., Donner & Tellez 2008), ethnography (e.g., Kibere 2016; Ferreira et al. 2016; Bailey & Ngwenyama 2013), case studies (e.g., Boateng 2016), quantitative behavioral science (e.g., Kaba & Osei-Bryson 2013; Samoilenko & Osei-Bryson 2011), the Delphi method (e.g., Khan et al. 2010), hybrid data analytic methods (e.g., Samoilenko 2008; Samoilenko & Osei-Bryson 2014), and value-focused thinking (e.g., Barrett-Maitland et al. 2016).

Theoretical Frameworks

Various theoretical frameworks have been adopted and adapted, including absorptive capacity theory (e.g., Bernoider et al. 2011), actor network theory (e.g., Cavalheiro & Joia 2016), affordance theory (e.g., Wyche & Steinfield 2016), bureaucracy theory (e.g., Klun & Dečman 2006), complementarity theory (e.g., Bankole et al. 2015), contingency theory (e.g., Ifinedo & Singh 2011), diffusion of innovation theory (e.g.,

Ulmanis & Deniņš 2012), growth theory (e.g., Bauer et al. 2010), Hofstede's cultural dimensions framework (Kaba & Osei-Bryson 2013), theories of neoclassical growth accounting (e.g., Samoilenko & Osei-Bryson 2011), the resource-based view (Ifinedo & Singh 2011), social presence theory (e.g., Doern & Fey 2006), stakeholder theory (e.g., Soja 2011), the technology acceptance model (TAM) (e.g., Zaremohzzabieh et al. 2016; Brown 2002), the theory of reasoned action [applied in Guinea (Kaba et al. 2006)], and transaction cost theory (e.g., Doern & Fey 2006). For example, TAM (e.g., Brown 2002) and Hofstede's cultural dimensions framework (Kaba & Osei-Bryson 2013) have been used to explore and/or explain issues related to ICT usage in South Africa and Guinea, respectively. Similarly, complementarity theory and theories of neoclassical growth accounting have been used to explain the impacts of investments in ICT on various aspects of economic growth in South Africa (e.g., Bankole et al. 2015) and the transition economies of Europe (e.g., Samoilenko & Osei-Bryson 2011).

In several studies, multiple theoretical frameworks were used. For a more detailed exposition on theoretical frameworks used in previous ICT4D research, the reader may consult reviews such as that of Roztocki and Weistroffer (2015).

Data Analysis Methods

Again, a variety of methods have been used, with structural equation modeling (e.g., Borena & Negash 2016; Samoilenko 2014) probably being the most popular. However, recently, other methods that provide additional exploration capabilities have been utilized. These include ethnographic decision tree modeling (e.g., Bailey & Ngwenyama 2013), data envelopment analysis (e.g., Samoilenko & Osei-Bryson 2010), decision tree induction (e.g., Samoilenko & Osei-Bryson 2008), cluster analysis (e.g., Skaletsky et al. 2014; Samoilenko & Osei-Bryson 2010), neural networks (e.g., Samoilenko & Osei-Bryson 2010), and multivariate adaptive splines (e.g., Bankole et al. 2015). Some studies have also used combinations of these newly adopted methods with the more popular methods, such as factor analysis.

Conclusion

An ICT4D researcher is required to conduct a scientific investigation that is rigorous and relevant, and that offers a significant contribution to the body of knowledge. To the extent that a theory (Whetten 1989) can be considered to involve *whats* (i.e., *factors* that could logically be considered to be part of the explanation of the phenomenon of interest), *hows* (i.e., *causal relationships* between the factors), *whys* (i.e., the underlying dynamics that justify the selection of the factors and causal relationships), and *who/where/when* (i.e., temporal and contextual factors that indicate the range of the theory), the material discussed in section "Some Options for Configuring an ICT4D Research Project" plus that presented in relevant review studies (e.g., Walsham 2017; Roztocki & Weistroffer 2015; Donner 2008) could be useful in designing a value-adding ICT4D research project.

References

Alam, M. M. & Wagner, C. (2016). The relative importance of monetary and non-monetary drivers for information and communication technology acceptance in rural agribusiness. *Information Technology for Development*, 22(4), 654–671.

Aker, J. & Mbiti, I. (2010). Mobile phones and economic development in Africa. *Journal of Economic Perspectives,* 24(3), 207–232.

Avgerou, C. (2008). Information systems in developing countries: A critical research review. *Journal of information Technology,* 23(3), 133–146.

Avgerou, C., Hayes, N., & La Rovere, R. L. (2016). Growth in ICT uptake in developing countries: New users, new uses, new challenges.

Bailey, A. (2009). Issues affecting the social sustainability of telecentres in developing contexts: A field study of sixteen telecentres in Jamaica. *The Electronic Journal of Information Systems in Developing Countries*, 36.

Bailey, A. & Ngwenyama, O. (2010). Bridging the generation gap in ICT use: Interrogating identity, technology and interactions in community telecenters. *Information Technology for Development*, 16(1), 62–82.

Bailey, A. & Ngwenyama, O. (2013). Toward entrepreneurial behavior in underserved communities: An ethnographic decision tree model of telecenter usage. *Information Technology for Development*, 19(3), 230–248.

Bankole, F. O., Bankole, O. O., & Brown, I. (2011). Mobile banking adoption in Nigeria. *The Electronic Journal of Information Systems in Developing Countries*, 47.

Bankole, F. O., Osei-Bryson, K. M., & Brown, I. (2015). The impacts of telecommunications infrastructure and institutional quality on trade efficiency in Africa. *Information Technology for Development,* 21(1), 29–43.

Barclay, C. (2014). Using frugal innovations to support cybercrime legislations in small developing states: Introducing the cyber-legislation development and implementation process model (CyberLeg-DPM). *Information Technology for Development*, 20(2), 165–195.

Barrett-Maitland, N., Barclay, C., & Osei-Bryson, K. M. (2016). Security in social networking services: A value-focused thinking exploration in understanding users' privacy and security concerns. *Information Technology for Development*, 22(3), 464–486.

Bauer, O., Němcová, Z., & Dvořák, J. (2010). E-commerce and its impact on customer strategy. *Economics & Management*, 397–407.

Bernroider, E. W., Sudzina, F., & Pucihar, A. (2011). Contrasting ERP absorption between transition and developed economies from Central and Eastern Europe (CEE). *Information Systems Management*, 28(3), 240–257.

Boateng, R. (2016). Resources, electronic-commerce capabilities and electronic-commerce benefits: Conceptualizing the links. *Information Technology for Development*, 22(2), 242–264.

Borena, B. & Negash, S. (2016). IT infrastructure role in the success of a banking system: The case of limited broadband access. *Information Technology for Development*, 22(2), 265–278.

Brown, I., Collins, T., Malika, B., Morrison, D., Muganda, N., & Speight, H. (2007). Global diffusion of the Internet XI: Internet diffusion and its determinants in South Africa: The first decade of democracy (1994-2004) and beyond. *Communications of the Association for Information Systems*, 19(1), 9.

Brown, I., Hoppe, R., Mugera, P., Newman, P., & Stander, A. (2004). The impact of national environment on the adoption of Internet banking: Comparing Singapore and South Africa. *Journal of Global Information Management*, 12(2), 1–26.

Brown, I. & Licker, P. (2003). Exploring differences in Internet adoption and usage between historically advantaged and disadvantaged groups in South Africa. *Journal of Global Information Technology Management*, 6(4), 6–26.

Brown, I. T. (2002). Individual and technological factors affecting perceived ease of use of web-based learning technologies in a developing country. *The Electronic Journal of Information Systems in Developing Countries*, 9.

Canares, M. P. (2016). Creating the enabling environment for more transparent and better-resourced local governments: A case of e-taxation in the Philippines. *Information Technology for Development*, 22(sup1), 121–138.

Cavalheiro, G. M. D. C. & Joia, L. A. (2016). Examining the implementation of a European patent management system in Brazil from an actor-network theory perspective. *Information Technology for Development*, 22(2), 220–241.

Charlton-Laing, C. & Grant, G. (2015). Who should champion e-learning projects in educational institutions? Emergent roles of school leadership in a national e-learning project in a developing country. In *Strategic Project Management: Contemporary Issues and Strategies for Developing Economies*, eds. C. Barclay & K.-M. Osei-Byson, CRC Press, Boca Raton, FL, pp. 249–268.

Cumbie, B. A. & Kar, B. (2016). A study of local government website inclusiveness: The gap between e-government concept and practice. *Information Technology for Development*, 22(1), 15–35.

Doern, R. R. & Fey, C. F. (2006). E-commerce developments and strategies for value creation: The case of Russia. *Journal of World Business*, 41(4), 315–327.

Donner, J. (2008). Research approaches to mobile use in the developing world: A review of the literature. *The Information Society*, 24(3), 140–159.

Donner, J. & Tellez, C. A. (2008). Mobile banking and economic development: Linking adoption, impact, and use. *Asian Journal of Communication*, 18(4), 318–332.

Ferreira, S. M., Sayago, S., & Blat, J. (2016). Going beyond telecenters to foster the digital inclusion of older people in Brazil: Lessons learned from a rapid ethnographical study. *Information Technology for Development*, 22(sup1), 26–46.

Gupta, S., Yun, H., Xu, H., & Kim, H. W. (2017). An exploratory study on mobile banking adoption in Indian metropolitan and urban areas: A scenario-based experiment. *Information Technology for Development*, 23(1), 127–152. http://dx.doi.org/10.1080/02681102.2016.1233855

Gryczka, M. (2011). The effects of delays in the dissemination and usage of broadband Internet. *Folia Oeconomica Stetinensia*, 10(1), 186–201.

Ilavarasan, P. V. & Levy, M. R. (2012). ICT access and use by microentrepreneurs in Mumbai, India: A value chain model analysis. In *Proceedings of the Fifth International Conference on Information and Communication Technologies and Development*, ACM, New York, pp. 259–267.

Islam, M. S. & Grönlund, Å. (2012). Applying design science approach in ICT4D research. In *Practical Aspects of Design Science*, eds. M. Helfert & B. Donnellan, EDSS 2011. *Communications in Computer and Information Science*. Springer, Berlin, Germany, vol. 286, pp. 132–143.

Kaba, B., Diallo, A., Plaisent, M., Bernard, P., & N'Da, K. (2006). Explaining the factors influencing cellular phones use in Guinea. *The Electronic Journal of Information Systems in Developing Countries*, 28.

Kaba, B. & Osei-Bryson, K. M. (2013). Examining influence of national culture on individuals' attitude and use of information and communication technology: Assessment of moderating effect of culture through cross countries study. *International Journal of Information Management*, 33(3), 441–452

Khan, G. F., Moon, J., Rhee, C., & Rho, J. J. (2010). E-government skills identification and development: Toward a staged-based user-centric approach for developing countries. *Asia Pacific Journal of Information Systems*, 20(1), 1–31.

Kibere, F. N. (2016). The paradox of mobility in the Kenyan ICT ecosystem: An ethnographic case of how the youth in Kibera slum use and appropriate the mobile phone and the mobile Internet. *Information Technology for Development*, 22(suppl 1), 47–67.

Klun, M. & Dečman, M. (2006). E-public services: The case of e-taxation in Slovenia. *Financial Theory and Practice*, 30(3), 233–252.

Krauss, K. (2013). Collisions between the worldviews of international ICT policymakers and a deep rural community in South Africa: Assumptions, interpretation, implementation, and reality. *Information Technology for Development*, 19(4), 296–318.

Lech, P. (2012). Information gathering during enterprise system selection: Insight from practice. *Industrial Management & Data Systems*, 112(6), 964–981.

Mansingh, G., Rao, L., Osei-Bryson, K.-M., & Mills, A. (2015). Profiling Internet banking users: A knowledge discovery in data mining process model based approach. *Information Systems Frontiers*, 17(1), 193–215.

Maryska, M., Doucek, P., & Kunstova, R. (2012). The importance of ICT sector and ICT university education for the economic development. *Procedia-Social and Behavioral Sciences*, 55, 1060–1068.

Mimbi, L., Bankole, F. O., & Kyobe, M. (2011). Mobile phones and digital divide in East African countries. In *Proceedings of the South African Institute of Computer Scientists and Information Technologists Conference on Knowledge, Innovation and Leadership in a Diverse, Multidisciplinary Environment*. ACM, New York, pp. 318–321.

Nemer, D. (2016). Online favela: The use of social media by the marginalized in Brazil. *Information Technology for Development*, 22(3), 364–379.

Pavon, F. & Brown, I. (2010). Factors influencing the adoption of the World Wide Web for job-seeking in South Africa. *South African Journal of Information Management*, 12(1), 1–9.

Prakash, A. (2016). E-governance and public service delivery at the grassroots: A study of ICT use in health and nutrition programs in India. *Information Technology for Development*, 22(2), 306–319.

Potnis, D. (2016). Managing seven dimensions of ICT4D projects to address project challenges. In *Human Development and Interaction in the Age of Ubiquitous Technology*, ed. H. Rahman, IGI-Global Publications, Hershey, PA, pp. 227–249.

Purcarea, V. L., Petrescu, D. G., Gheorghe, I. R., & Petrescu, C. M. (2011). Optimizing the technological and informational relationship of the health care process and of the communication between physician and patient–Factors that have an impact on the process of diagnosis from the physician's and the patient's perspectives. *Journal of Medicine and Life*, 4(2), 198–206.

Qureshi, S. (1998). Fostering civil associations in Africa through GOVERNET: An administrative reform network. *Information Technology for Development,* 8(2), 121–136.

Qureshi, S., Kamal, M., & Wolcott, P. (2010). Information technology interventions for growth and competitiveness in micro-enterprises. In *Global Perspectives on Small and Medium Enterprises and Strategic Information Systems: International Approaches,* eds. P. Bharati, I. Lee, & A. Chaudhury, IGI-Global Publications, Hershey, PA, pp. 306–329.

Qureshi, S. (2015). Are we making a better world with information and communication technology for development (ICT4D) research? Findings from the field and theory building.

Roztocki, N. & Weistroffer, H. R. (2015). Information and communication technology in transition economies: An assessment of research trends. *Information Technology for Development,* 21(3), 330–364.

Samoilenko, S. (2008). Contributing factors to information technology investment utilization in transition economies: An empirical investigation. *Information Technology for Development,* 14(1), 52–75.

Samoilenko, S. & Osei-Bryson, K.-M. (2008). Increasing the discriminatory power of DEA in the presence of the sample heterogeneity with cluster analysis and decision trees. *Expert Systems with Applications,* 34(2), 1568–1581.

Samoilenko, S. & Osei-Bryson, K. M. (2010). Determining sources of relative inefficiency in heterogeneous samples: Methodology using cluster analysis, DEA and neural networks. *European Journal of Operational Research,* 206(2), 479–487.

Samoilenko, S. & Osei-Bryson, K. M. (2011). The spillover effects of investments in telecoms: Insights from transition economies. *Information Technology for Development,* 17(3), 213–231.

Samoilenko, S. V. (2014). Investigating the impact of investments in telecoms on microeconomic outcomes: Conceptual framework and empirical investigation in the context of transition economies. *Information Technology for Development,* 20(3), 251–273.

Sen, A. (2001). *Development as Freedom.* Oxford University Press, Oxford.

Skaletsky, M., Soremekun, O., & Galliers, R. D. (2014). The changing—And unchanging—face of the digital divide: An application of Kohonen self-organizing maps. *Information Technology for Development,* 20(3), 218–250.

Soja, P. (2011). Examining determinants of enterprise system adoptions in transition economies: Insights from Polish adopters. *Information Systems Management,* 28(3), 192–210.

Stanimirovic, D. & Vintar, M. (2015). The role of information and communication technology in the transformation of the healthcare business model: A case study of Slovenia. *Health Information Management Journal,* 44(2), 20–32.

Summer, A. & Tribe, M. (2008). *International Development Studies: Theories and Methods in Research and Practice.* Sage, Los Angeles, p. 176.

Ulmanis, J. & Deniņš, A. (2012). A management model of ICT adoption in Latvia. *Procedia-Social and Behavioral Sciences,* 41, 251–264.

Van der Schyff, K. & Krauss, K. E. (2014). Higher education cloud computing in South Africa: Towards understanding trust and adoption issues. *South African Computer Journal,* 55(1), 40–55.

Walsham, G. (2017). ICT4D research: Reflections on history and future agenda. *Information Technology for Development,* 23(1), 18–41.

Whetten, D. (1989). What constitutes a theoretical contribution? *Academy of Management Review,* 14(4), 490–495.

Wong, C. Y., Chandran, V. G. R., & Ng, B. K. (2016). Technology diffusion in the telecommunications services industry of Malaysia. *Information Technology for Development,* 22(4), 562–583.

Wyche, S. & Steinfield, C. (2016). Why don't farmers use cell phones to access market prices? Technology affordances and barriers to market information services adoption in rural Kenya. *Information Technology for Development,* 22(2), 320–333.

Xiong, J., Qureshi, S., & Lamsam, T. T. (2015). A framing analysis of African-American and Native-American owned micro-enterprises: How can information and communication technology support their development? In *System Sciences (HICSS), 2015 48th Hawaii International Conference on System Sciences.* IEEE, New York, pp. 4315–4324.

Zaremohzzabieh, Z., Samah, B. A., Muhammad, M., Omar, S. Z., Bolong, J., Hassan, S. B. H., & Mohamed Shaffril, H. A. (2016). Information and communications technology acceptance by youth entrepreneurs in rural Malaysian communities: The mediating effects of attitude and entrepreneurial intention. *Information Technology for Development,* 22(4), 606–629.

Chapter 2

Overview on Framework of Neoclassical Growth Accounting

It is hard to argue against benefits brought to a society by economic growth, and it is hard to find an example of a society that rejects the prosperity delivered by its growing economy. The verdict of the international community is out—economic growth is a good thing, and the more of it you have, the better of you are. The question then becomes how to bring economic growth about, and the first step in answering this question is explaining the economic growth that took place previously. The approach to growing an economy is fairly simple—if we can explain the past, then we can use the obtained insights to impact the future. However, an explanation depends on the perspective on the nature of economic growth. There are many models that can help in explaining the factors impacting economic growth, but in this chapter, we concentrate on the framework of neoclassical growth accounting.

Originated from the work of Solow (1957), the neoclassical framework of growth accounting has since been widely used by researchers. The objective of the framework is to decompose, using neoclassical production function, the rate of growth of the economy into contributions from different inputs. In simple terms, the production function tells us that if economic growth took place, then it could be explained by growth in the levels of capital, labor, and everything else that cannot be explained by labor and capital. The framework does not make any claims that the production function offers a theoretically true representation of economic growth, but rather, uses the function as a vehicle of interpretation of the economic growth that took place. Consequently, it is not a grand theory in the sense that it could be used to make predictions about future economic growth, but rather, it is a method of partially explaining an economic growth that took place. First and foremost, the framework of neoclassical growth accounting is an empirical tool.

While some of the inputs, such as labor and capital, could be explained as drivers of economic growth, other inputs cannot be explained so easily—we simply do not know what they are. As a result, the framework breaks down observed economic growth into two types of components, known factors and unknown factors, that, taken together, serve as a residual (commonly referred to as the *Solow residual*), as something else of unclear provenance that helps to grow an economy. This "something else" is commonly referred to as multifactor productivity or total factor productivity (TFP).

As a result, according to the framework of growth accounting, economic growth could be obtained via two venues—by increasing levels of known inputs (e.g., capital and labor) and by increasing the contribution from the unknown inputs (e.g., TFP). It is important to point out that the residual—the unknown inputs positively impacting economic growth—must stay unexplained. This does not mean that investigators are prohibited from getting insights into the nature of the black box that is TFP; it simply means that according to the framework, a part of the economic growth is, well, unexplained. As time goes by and investigators are able to explain economic growth to a greater extent using more observable factors, the residual portion represented by TFP may shrink but will never disappear completely.

Exogenous and Endogenous Growth Models

The framework of neoclassical growth accounting is considered to be an *exogenous* model of economic growth, because the primary means of growing an economy, according to the framework, is via TFP—the term exogenous to the model. While endogenous factors such as labor and capital can bring economic growth, those *endogenous* factors, eventually, would be negatively affected by the law of diminishing returns. Thus, it is a technological progress, expressed as the Solow residual, that is a consistent and reliable vehicle of economic growth. And this is one of the commonly noted shortcomings of the framework—the exogenous model explains economic growth by simply stating the presence of the technological progress as a major factor of growth, without providing any explanation regarding its nature.

Unlike the exogenous model, endogenous models of economic growth aim to bring a representation of technological progress within the model—make it one of the independent variables in a model. For example, the *AK model* of endogenous growth relies on the representation of human knowledge as the driver of economic growth. According to this approach, which amounts to a production function $Y = A*K$ (hence the name *AK*), a long-term economic growth is possible via capital accumulation because the knowledge component offsets the impact of the diminishing returns to scale that are unavoidable in the exogenous model.

Another endogenous approach, the *Uzawa–Lucas model*, identifies human capital growth via education to be the engine of economic growth. According to this perspective, economic growth is driven by growth in human capital, which is dependent on time spent on education and a measure of efficiency with which education increases the impact of human capital.

Yet another endogenous model of economic growth explains technological progress via research and development (R & D)–related investments and activities. According to this perspective, R & D will result in the increased specialization of the workforce, which then would develop new innovative products that would contribute to growth in productivity.

One of the important impacts of the endogenous-versus-exogenous difference in perspective is that on policy making. In the case of the exogenous model of economic growth, any impact of economic policy would be limited—any intervention would result in temporary economic growth, after which an economy is bound to return to its steady state and continue growing via productivity.

The endogenous perspective, however, would allow for a long-term impact of economic policy on growth in productivity—simply because a part of the economic growth is explained within the model.

Formulation of the Neoclassical Production Function

The neoclassical production function relates output and inputs in the following manner:

$$Y = f(A, K, L)$$

Where
- Y = output [often in the form of gross domestic product (GDP)]
- A = level of technology/total factor productivity
- K = capital stock
- L = quantity of labor/labor force

Based on the provided function, growth accounting uses Cobb–Douglas production function:

$$Y = A * K^\alpha * L^\beta$$

Where α and β are constants determined by technology.

If $\alpha + \beta > 1$, returns are increasing to scale, and if $\alpha + \beta < 1$, returns are decreasing to scale. In the case of constant returns to scale, $\alpha + \beta = 1$; thus, $\beta = 1 - \alpha$, which gives following formulation:

$$Y = A * K^\alpha * L^{1-\alpha}$$

Based on this formulation, it is easy to see that any changes in the level of output ΔY could only be explained by changes in the levels of capital ΔK or labor ΔL, or changes in TFP ΔA.

Let us consider an example where levels of A, K, and L have changed over the same period of time t.

If the original level of capital K changed over the period of time by ΔK, then the proportional impact of ΔK on increase ΔY from the original level of Y could be expressed as

$$\alpha * \Delta K / K = \Delta Y / Y$$

If α is 0.7 and the proportional change in capital $\Delta K/K$ is 5%, then the proportional corresponding change in output could be calculated as

$$\Delta Y / Y = \alpha * \Delta K / K = 0.7 * 5\% = 3.5\%$$

Similarly, if changes in labor ΔL took place over a period of time from the original level of L, then the proportional change in output could be expressed as

$$\beta * \Delta L / L = \Delta Y / Y, \text{ or}$$
$$(1 - \alpha) * \Delta L / L = \Delta Y / Y$$

In the case of $\alpha = 0.7$ and under the assumption of constant returns to scale,

$$\beta = 1 - \alpha = 1 - 0.7 = 0.3$$

If the proportional change $\Delta L/L$ is 2%, then we could calculate proportional change in output as follows:

$$\Delta Y / Y = (1 - \alpha) * \Delta L / L = 0.3 * 2\% = 0.6\%.$$

Finally, a proportional increase in TFP $\Delta A/A$, let us say 4%, would produce a proportional impact $\Delta Y/Y$ of the same magnitude.

$$\Delta A / A = \Delta Y / Y = 0.4\%$$

Overall, then, the proportional change in output could be expressed as

$$\Delta Y / Y = \alpha * \Delta K / K + (1 - \alpha) * \Delta L / L + \Delta A / A$$

It is important to note that this function does not necessarily represent the true relationships between the inputs and the output; rather, its purpose is simply to serve as a vehicle of the exploration and interpretation of the macroeconomic growth.

Out of three inputs used by growth accounting, only capital K and labor L could be observed in the data, while TFP would serve as a residual (Solow residual) term capturing that contribution to Y (GDP) that is left unexplained by the inputs of capital and labor. Thus, an important difference between the three inputs used by the Cobb–Douglas production function is that two inputs (K and L) are provided by the data, while the third input (A) is computationally derived. Therefore, the computed value of A would include not only all improvements that are derived from the utilization of the capital and labor, but also any errors that have been made from estimation of the inputs of K and L as well.

One of the endemic shortcomings of the neoclassical production function is that it does not allow investigating the presence of complementarities between the known factors. Initially introduced in economics by Edgeworth (1881), the concept of *complementarity* refers to the notion that the increase in one factor could result in an increased benefit received from its complementary factors. The theory of complementarity was applied, for example, to argue that if the benefits of the investments in information and communication technologies (ICTs) are to be reaped successfully at the macroeconomic level, then such investments could not be made in isolation from the investments in other areas. Thus, if the two investments, one in ICT and another one elsewhere, are more effective when taken jointly, rather than separately, such investments are considered to be complementary.

However, even if the complementarity of the factors exists within a given production function, it cannot be identified through the formulation offered by the Cobb–Douglas production function. Complementarity of the investments could only be discerned if the formulation allows for the presence of the interaction term between the specified endogenous factors. Thus, we turn our attention to the transcendental logarithmic production function, a brief overview of which is offered next.

Transcendental Logarithmic (Translog) Production Function

The standard Cobb–Douglas production function, as mentioned before, has a form of

$$Y = A * K^{\alpha} * L^{\beta}$$

By taking the logarithm, the following formulation can be obtained:

$$\log Y = \log A + \alpha * \log K + \beta * \log L$$

An extension to the previously given formulation of the Cobb–Douglas production function, which has a more general and flexible functional form, was proposed by Christensen et al. (1973). The formulation of this extension, called the transcendental logarithmic (translog) production function, is provided:

$$\log Y = \alpha_0 + \alpha_k * \log K + \alpha_L * \log L + 1/2\alpha_{kk} * (\log K)^2$$
$$+ 1/2\alpha_{LL} * (\log L)^2 + \alpha_{kL} * \log L * \log K$$

or

$$\ln Y = \alpha_k \ln K + \alpha_L \ln L + \beta_{kk}(\ln K)^2 + \beta_{LL}(\ln L)^2 + \beta_{kL} \ln L * \ln K$$

Thus, given $Y = f(A, K, L)$ and considering that A is a residual that could be expressed as an error term e, we can offer the following general forms for the Cobb–Douglas production function and translog production function:

Cobb–Douglas production function:

$$\log Y = \beta_0 + \beta_1 \cdot \log K + \beta_2 \cdot \log L + e$$

Translog production function:

$$\log Y = \beta_0 + \beta_1 \cdot \log K + \beta_2 \cdot \log L + \beta_3 \cdot \log K^2 + \beta_4 \cdot \log L^2 + \beta_5 \cdot \log K * \log L + e$$

It is easy to see that the Cobb–Douglas function is "nested" in translog function, and a test of the hypothesis that both functions describe the production process equally well would entail testing the following null hypothesis:

$$H0: \beta_3 = \beta_4 = \beta_5 = 0$$

The translog production function is more flexible than the Cobb–Douglas function in the sense that it allows testing for the presence of the interactions between the variables. For example, let us say that we are interested in investigating the following production function:

$$Y = f(A, K_{ICT}, K_{non\text{-}ICT}, L)$$

where
 Y = output (GDP)
 A = level of technology/total factor productivity
 K_{ICT} = investments in ICT
 $K_{non\text{-}ICT}$ = investments in non-ICT
 L = quantity of labor/size of labor force

Then we could test for the presence of the statistically discernible interaction between the two types of the investments (ICT and non-ICT) using the translog function as follows (again, keeping in mind that A is a residual expressed as error term e):

$$\log Y = \beta_0 + \beta_1 {*} \log K_{ICT} + \beta_2 {*} K_{non\text{-}ICT} + \beta_3 {*} \log L + \beta_4 {*} \log K_{ICT}^2$$
$$+ \beta_5 {*} \log K_{non\text{-}ICT}^2 + \beta_6 {*} \log L^2 + \beta_7 {*} \log K_{ICT} * \log K_{non\text{-}ICT}$$
$$+ \beta_8 {*} \log K_{ICT} {*} \log L + \beta_9 {*} \log K_{non\text{-}ICT} {*} \log L + e$$

Then the test for the presence of the interaction would involve testing of the following hypothesis:

H0: β_7 is not statistically discernible from 0 at a given level of α.

And again, in the case of $\beta_4 = \beta_5 = \beta_6 = \beta_7 = \beta_8 = \beta_9 = 0$, the translog production function would end up being formulated as a corresponding Cobb–Douglas production function:

$$\log Y = \beta_0 + \beta_1 {*} \log K_{ICT} + \beta_2 {*} K_{non\text{-}ICT} + \beta_3 {*} \log L + e$$

But if the interaction term between investments in ICT and non-ICT is significant (i.e., we reject the null hypothesis of $\beta_7 = 0$), then we have a reason to assume that such investments are complementary.

Weaknesses and Criticisms of the Framework of Neoclassical Growth Accounting

One of the appeals of using the neoclassical growth accounting framework lies in its simplicity; after all, only two factors, the TFP growth and the rate of increase in inputs, are used to explain the growth rate of the output. As a result, this relationship reflects the fundamental assumptions of the framework, namely, the presence of technological progress and the growth of labor. However, the flip side of the simplicity is the somewhat limited explanatory capability of economic growth. For example, while assuming the presence of technological progress, the framework neither explains the sources of the progress or the factors that impact the progress, nor accounts for any possible interactions between the technological progress and capital growth. In reality, though, capital investments would be affected by the technological progress, for the progress in information technologies has fueled capital investments in the economies of the United States and the other developed countries.

Also, according to another assumption of the growth accounting framework, that the capital is subject to the law of diminishing returns, convergence of the poor and wealthy economies must take place. The reality, however, reflects that the gap between poor and rich countries of the world is widening.

Nevertheless, the use of the framework of neoclassical growth accounting for the purposes of researching contributions of ICT investments to macroeconomic growth appears to be warranted, for this analytical framework has been used widely to estimate the contribution of ICT to economic growth in the context of developed and developing countries.

It was noted, however, that while the reliance of the most influential studies on ICT productivity and growth on the neoclassical framework allows us to take advantage of its theoretical assumptions, in the ICT-related and knowledge-intensive sectors of the economy, such assumptions are not easy to justify empirically. Nevertheless, despite the presence of the argument that in the strict sense, the common interpretation of the rapid growth in TFP as being ICT driven is not in complete agreement with the neoclassical framework's exogenous nature of TFP, there seems to be an agreement that it is natural to expect that ICTs would reveal their productivity impact on the overall economic efficiency and become visible in TFP. And, indeed, studies have demonstrated that the extensive investment in ICTs is associated with the improvements in TFP.

References

Christensen, R., Jorgenson, D., & Lau, L. (1973). Transcendental logarithmic production frontiers. *The Review of Economics and Statistics*, 55(1), 28–45.

Edgeworth, F. Y. (1881). *Mathematical Psychics: An Essay on the Applications to Mathematical Statistics*. Kegan Paul, London.

Solow, R. (1957). Technical change and the aggregate production function. *Review of Economics and Statistics*, 39(3), 312–320.

Chapter 3

Overview on Complex Systems Theory and Chaos Theory

In this chapter, we provide a general overview of complex systems theory and chaos theory. The two perspectives offer an investigator valuable vantage points allowing for inquiry into an internal structure as well as an external behavior of such complex entities as firms, industries, and economies. One of the aims of the overview is to propose that complex systems theory and chaos theory could complement each other and, thus, serve as a broader foundation for a scientific inquiry than if the theories were considered in isolation.

Complex Systems Theory: An Overview

The goal of the first part of the chapter is to offer a general, nontechnical overview of *complex systems theory* (CST). The purpose of CST is to offer an analytic lens allowing for investigating the structure and behaviors of *complex systems (CSs)*—those systems that consist of many parts connected by nonlinear relationships and are characterized by emergent behavior.

To accomplish our goal, we proceed by following a two-step approach. First, we take a look at CSs from a structural perspective—we describe what CSs are composed of. Then, we discuss the behavior of CSs—we outline some of the common patterns of actions that characterize them.

Complex Systems: Structural Perspective

An undertaking of defining what CS is could be accomplished in two steps—first, offering a definition of a *system*, and then introducing the term *complexity*. First and foremost, it is important to keep in mind that the term *system* refers to a conceptualization of something of interest—anything could be viewed as a system with varying, according to our needs, degrees of complexity.

A common definition of a system is that of a *collection of interrelated components that function together for a given purpose*. Based on this definition, a mob of people is not a system, while a military unit is. It is the interrelationships that exist between the components of the system that allow those components to function in accord. Thus, a common representation of a system is *(k, n)*, where *k* designates the number of components comprising the system and *n* refers to the number of the relationships between *k* components. In essence, a system is *a collection of components and relationships between the components*.

A component, an element, or a part of a system (we use the terms interchangeably here) is a conceptual designation that does not carry any restrictions or prescriptions regarding the nature or size of a component. This gives rise to a circular definition of a system being a collection of components, and a component being a part of a system. This concept also lends itself easily to *nesting*—where an element of a system could be viewed as a system itself, and a system could be viewed as component of a larger system.

There is also no restriction on the nature and number of connections between the components—elements comprising the system do not have to be physically bound together (e.g., human body, mechanical watch, etc.) but could be connected by virtual or information channels (e.g., business department, online store, etc.). Consequently, a collection of physical and physically connected parts comprising an automotive engine is conceptually similar to a collection of virtual entities connected by information channels comprising a system of a business-to-business (B2B) marketplace in e-commerce.

The way in which elements of the system are interconnected is context specific and not prescribed by CS—therefore, two firms, one with a rigid hierarchical organization relying on a predefined reporting structure and another being a matrix-based organization utilizing an open-door policy, could both be viewed as systems.

Elements of a system could be bound together by *linear* or *nonlinear* relationships. If the relationships between the components are linear, then the resulting system is *additive, output proportional*, and *predictable*. *Additivity* implies that the system is characterized by, and is equal to, the collection of its parts. Bricks stacked together on a pallet comprise an additive linear system that consists of all the bricks and a pallet. *Output proportionality* stands for changes in the level of inputs resulting in proportional changes in outputs. Every layer of bricks added to the pallet adds a certain number of pounds in weight and a certain number of inches in height to the baseline presented by the pallet. *Predictability* of a system refers to the ability to predict the level and the type of outputs based on the level and the type of inputs. If we want to stack 10 layers of bricks on a pallet, then we can easily predict the resulting height and weight of the completed pallet.

Let us consider an example of using a concrete mixer at a construction site. If a concrete mixer could produce 3 volume-units of concrete using 2 units of cement and 1 unit of water per hour then by adding a second mixer, we could expect to produce 6 volume-units in an hour using 4 units of cement and 2 units of water. If the system of concrete mixers is linear, then we know that this system is nothing else other than a collection of two mixers (additive) that is capable of producing $n*3$ units of concrete from $n*2$ units of cement and $n*1$ units of water (proportional for any n), which, over the period t *(no. of hours), will* produce $t*6$ units of concrete (predictable for any t).

In general, linear systems do not give us any surprises in the way they behave, and they are easy to analyze and troubleshoot. In the previous example, the system is decomposable to two concrete mixers, and inputs are decomposable to cement mix and water. However, while the perspective of *linearity* carries its own advantages (e.g., modularity, simplicity, ease of control), this point of view cannot account for most of the real-world scenarios. Systems characterized by linear relationships are not complex—they could be large and complicated if the number of components is great enough, but not complex. Linearity does not give rise to complexity.

While we can expect that the meaning of the term *system* would be understood almost universally, the term *complexity* is not easy to define due to the dual nature of its general and context-specific use. For the purposes of this chapter, we offer a definition of complexity as of an *assessable degree of difficulty*. We use the term *assessable degree* to indicate that the state of complexity is not of Boolean nature, where the differentiation is of *complex versus simple*, or *complex versus not complex* type. Instead, the state of complexity could be assessed using a quantitative (e.g., 1–10) or a qualitative (e.g., more, less) scale of some sort.

Commonly, the meaning of the term *complexity* (e.g., degree of difficulty of what, exactly?) is often dictated by the context—researchers in such various fields as mathematics and computer science, biology, economics, medicine, and so on have different measures of complexity that refer to different aspects of the domain of interest. The complexity of building a spaceship capable of going to Venus and back is a different sort of complexity from finding a cure for cancer or solving the problem of world hunger. Regardless of the multitude of fields using the term complexity, it is probably safe to state that they all refer to complexity as a degree of difficulty of creating, or describing, or predicting something relevant to the field. In the case of CST we are dealing with a structure and a behavior of systems; thus, we can define complexity in the context of CSs as a *degree of difficulty of describing or predicting a structure and a behavior of a system*. In the case of a linear system, this definition helps us to understand why such systems are not complex—it is not difficult to describe or predict a structure or a behavior of a linear system.

CSs are characterized by nonlinear relationships between their components. The concept of nonlinearity could be introduced in many different and sophisticated ways, but the simplest and the most effective explanation is that the relationships between the elements in the system are simply "not linear." For all intents and purposes, all the conceptualized systems in the world could be separated into two groups—one is a group of linear systems, and the other a group containing everything else. So, nonlinearity is an umbrella term for everything else that is not linear.

Let us consider our previous example, where the amount of concrete produced by a single mixer could be expressed as

$$YUnitsOfConcrete = (2UnitsOfCement + 1UnitOfWater)*XNumberOfHours, \text{ or } Y = 3X$$

If this equation holds for any X (e.g., being equal to 1, or 50, or 5000), then the system is linear. This means that knowing the initial condition of the system and the path that is described by the equation $Y = 3X$, we can predict the future state of the system. The system is deterministic (cement mix and water determine concrete) and predictable (the equation always holds).

In reality, however, we know that this representation would not work reliably—mixers slow down, parts wear down, mixers break, delays are introduced, and so on. This means that knowing the initial condition of the system and its path, we cannot predict the future state of the system—the system is deterministic but not predictable. So, the system that more accurately describes the real world context is not linear—it is nonlinear. But what causes this nonlinearity of a system?

This difficult question could be answered in a very simple way—nonlinearity is caused by the *instability* of the relationships between the system components. We could discover that new relationships form (e.g., as a result of developed complementarities and symbiosis between the elements) and that the strength of the existing relationships varies—becoming stronger in some cases and weaker in other cases. The internal changes in the system are a response of the system to the outside forces from the system's environment. Relationships between the system's components are a structural part of the system, and as the structure begins to change, so does the behavior of the system. And as the new structure emerges, it gives rise to *emergent behaviors* of the CS.

The ecosphere and the human immune system are among the most obvious examples offering an illustration of the emergent properties of CSs. Let us extend our previous example of a concrete mixer to a small construction crew to demonstrate concepts of emergent structure and emergent behavior as tools allowing a CS to adjust to the changes in the environment. We do so by introducing two new components of the system, an operator of the mixer and a carpenter, and a new condition, the mixer's production capacity can vary. In the case of a linear system, the addition of an operator does not impact the overall behavior of the system—it is simply a constant with a value of 1. However, in the case of a nonlinear system, the value of an operator lies in the ability to alter the proportion of the mix required for producing the concrete as well as capability to adjust the production volume. By manipulating the cement-to-water ratio, the operator would be able to adjust the system to produce concrete that is tailor-made to the specific requirement (e.g., compressive strength) of the environment, for we cannot expect that the same concrete would be used for decorative and structural purposes. Consequently, the weight of the resulting units of concrete would vary according to the mixture. Additionally, based on the demand, the operator could manipulate the mixer to produce various quantities of concrete per hour—more than the usual average if needed, or less if demand is not there. It is also clear that no construction crew could survive and thrive if the members do not help each other—consequently, it is expected that in the time of need, the carpenter will help the operator and vice versa.

This is another characteristic of CSs—the behavior of the system could not be inferred by knowing its components. In the case of our example, even if we know that there is water, there is cement mix, and there is a mixer and an operator, we do not know in what proportion they will be mixed and what would be the weight of the resultant units of concrete. Similarly, if we know the *average* amount of concrete that is produced in the period of an hour, we could not predict what amount would *actually* be produced. Finally, even if we know the membership of the crew, we cannot predict what exactly each member would be doing—the operator could be mixing concrete or helping the carpenter. The system will adapt depending on the requirements of the environment—it will react based on the received *feedback*. This ability to adapt is predicated on the presence of nonlinear relationships between components—a system with a great number of components and linear relationships could be large and complicated, but it could never be complex in the sense of producing emergent structures and behaviors.

One can easily see that any nonlinear system could be made more complex by simply adding new elements. Such addition increases the level of complexity of the resulting system—*the degree of difficulty of describing or predicting a structure and a behavior of a system* increases. This process could be reversed by reducing the size of the system or by replacing some of the nonlinear relationships with fixed linear ones.

It is important to reiterate that the systems perspective is just a conceptualization utilizing two types of building blocks—components and relationships between the components. We could make this conceptualization as simple, or as complex, as needed—while linear systems allow us to present a useful but very limited abstraction of the real world, complex systems perspective gives us a better tool for representation of reality. Consequently, while a stick figure could serve as an easy-to-recognize minimalistic representation of a human being that has its place on a street sign, there is no doubt that an endocrinologist would prefer a much more complex representation of a human being to suit his/her needs.

Complex Systems: Functional Perspective

At this point, we take a look at functional characteristics of CSs. The nonlinear nature of CSs allows them to *adapt* their structure and behavior according to the pressures of the environment—hence, it is often stated that the subject of CST is *complex adaptive systems* (CASs). The process of adaptation is achieved through self-organization of the system that results in new patterns of relationships between the system's components. It is important to note that the process of self-organization is internally controlled, and not directed from the outside of the system. The self-organization of the system is an internal response to the external pressures that is supported by the formation of emergent properties.

It is important to point out that the process of self-organization is driven by the global goal of preserving the system, and not the local goals of assuring the existence of any of the system's components. It is quite possible, for example, that a new product development team would contain accounting, marketing, business, and manufacturing professionals, and it is possible that every representative would have a primary goal of his/her own benefit, rather than the benefit of the whole team. However, only in the case where all the members of the team are considered to aim for the global goal of team success (whatever it could mean) are we justified to look at the team through the lens of CST.

The purpose of self-organizing behavior is to acquire and maintain a balance between the system and its environment—and as the environment changes, so does the system. CSs aim to preserve their existence within the environment—they are not trying to "one-up" or "get ahead" of the environment; the response is the reaction to the external stimuli. Consequently, viable CSs coevolve with their environment and adapt to its pressures. This continuing coevolution and adaptation are made possible by the *sensitivity* of the system, which reacts even to small events taking place in the environment by reorganizing itself. Thus, small environmental events could give rise to big changes in the structure of the system.

Additionally, the concept of sensitivity of a CS also applies to its own structure—CSs are very sensitive to their *initial conditions*, and two systems that differ very little in terms of structure would take different adaptation routes via generating different sets of emergent properties. As a result, two CSs with a very similar initial structure could end up, as a result of coevolution and adaptation, having significantly different structures and behaviors. Clearly, even in the case of identical twins, we do not expect them to remain identical in all aspects later on in life.

We can summarize functional characteristics of CSs as follows: CSs continuously evolve and adapt to the environment, and the adaptation is driven by a self-organizing behavior resulting in development of emergent properties that is enabled by the sensitivity of the system to its own structure (e.g., importance of initial conditions) and environmental changes (e.g., significance of small events).

There are many real-world examples that demonstrate the described patterns of the behavior of CSs—construction crews at the job site, agile software development teams, hospital medical units, and special operations military units are among the most obvious ones. Based on the introduction presented up to this point, we can see a utility in using CST as a perspective allowing for inquiring into the internal structure and internal behavior of real-world entities. Unfortunately, CST does not offer investigators an external perspective for studying behavior of CSs within its environment—this is the area where chaos theory (CT) may be a better guide.

Chaos Theory: An Overview

In this part of the chapter, we look at the behavior of systems through the lens of CT. Specifically, we offer a general nontechnical overview of the major concepts of the theory as it may apply to CSs. This clause is important to note, for a system does not have to be complex to exhibit a pattern of behavior that is characterized as chaotic.

One of the concepts important to CT is that of a *phase space*, which is an abstract space representing all possible states of a CS within its environment. By taking an external view to the behavior of a CS, we can perceive the movement of the system through a variety of states—charting its own path within the environment. At this point, we know that for CS, this movement, the point-in-phase-space to point-in-phase-space progression, is a result of nonlinear interactions of the components bringing about emergent properties of the system in response to external forces.

While, fundamentally, CST aims to answer questions regarding the internal structure and internal behavior of CSs, the purpose of CT is to inquire into where within its environment the system would end up. Given this distinction, we can use CST to investigate changes in internal structure and internal behavior of a firm under the pressure of Porter's five competitive forces, but we would utilize CT to inquire into where, within the competitive landscape, the firm might end up.

The practicality of predicting a point in phase space of a CS is intuitively important: for example, our ability to accurately forecast weather—a path of a hurricane—carries enormous tangible and intangible implications. Unsurprisingly, it is the prediction of the weather that originated CT—in 1963, Edward Lorenz discovered that small differences in the initial condition (due to rounding differences) result in very significant differences in the outcome. Interestingly, it took more than 10 years to associate Lorenz's discovery with the term *chaos theory*—the name was officially coined in 1975 by Li and Yorke (Li & Yorke 1975).

Intuitively, one would expect that small initial differences, when applied to the same formula, would result in small differences in the end result—if we round up the original values of the initial condition to the third versus the sixth digit, then the difference in the result would be exactly of the same magnitude—the difference of rounding up to the third versus the sixth digit. This turned out not to be the case, and the phenomenon of small original differences leading to significant differences in the results was later coined *butterfly effect* (Lorenz 1972).

After Lorenz concluded that it was impossible to predict the weather accurately, he became interested in whether smaller and simpler systems could also be sensitive to initial conditions. After scaling down from a 12-equation weather system to a 3-equation system and running the simulation, Lorenz plotted seemingly random results and discovered that the predicted output consistently stayed on a double spiral. This seemingly random yet ordered output differed from previously known steady state and periodic behaviors of a system.

A normal, undisturbed behavior of a system could be represented by the collection of its phase space points, which, taken together, form the system's *attractor*—the points of the attractor itself are pulled from the *basin of attraction*. Phase space points of a disturbed system will spread out into its basin of attraction, and as the system settles, its behavior will settle on its attractor.

If a system settles on a single point and does not deviate from it, then this forms a *fixed-point attractor*. A behavior of a system cycling through the same set of points in its phase space is characterized by a *limit-cycle attractor*. A *torus attractor* (commonly portrayed in the form of a doughnut), on the other hand, depicts the behavior of a system that travels between the phase points within the same area while never repeating the same path. The attractor discovered by Lorenz (fittingly known as the *Lorenz attractor*) is a type of *strange attractor* (Ruelle & Takens 1971).

Strange attractors could form an infinite number of trajectories within the same region of the phase space, but those trajectories never intersect and never repeat. Unlike in the case of fixed-point, limit-cycle, and torus attractors, where close points in phase space tend to stay close together over time, close points of strange attractors tend to diverge exponentially as time goes by. It is strange attractors that characterize the behavior of chaotic systems, where the sensitive dependence on initial conditions is the cause for divergence of the points in the phase space.

As demonstrated by the Lorenz attractor, a chaotic system does not have to be complex, for a chaotic behavior is a result of sensitivity to initial conditions applied to a reiteration of the rule describing the system. However, the complex phenomena and artifacts of this world are more fittingly perceived as nonlinear CSs. Consequently, it is of value to investigate the conditions under which behavior of CSs could become chaotic. It is important to emphasize that a chaotic system is a system that exhibits, or is capable of exhibiting, a chaotic behavior—a system may or may not be behaving chaotically at a given point in time. In general, there are three stages, of phases, of the behavior of a system—ordered, critical (edge of chaos), and chaotic. The process of the changing the behavior—as the system transitions between the stages—is referred to as *phase transition*.

Phase transitions take place as a response of the system to a *control parameter*, which is an external event, disturbance, or some sort of an input that causes a change in behavior—a common example is that of water transitioning through the states of being solid, liquid, and gas in response to changes in temperature. The process of phase transitioning that takes a system from order to chaos follows a path of *period doubling*, or *bifurcation*, the point where small changes in control parameters result in such changes in the behavior of the system that its period doubles. Thus, the path of a system from order to chaos could be seen as a path traversing a series of bifurcation points, where each consequent period doubling makes the system less and less stable.

In 1975, Mitchell Feigenbaum discovered that the phase transitioning that leads the system from order to chaos through the series of bifurcations is ordered and quantifiable. It turns out that as the system proceeds through the series of bifurcations on the path to chaos, each consequent periodic region is smaller than the previous region by 4.669 (Strogatz 1994). This constant, referred to as the *Feigenbaum number*, characterizes the behavior of all systems that traverse the path of period-doubling bifurcations on the way to chaos.

The system that transitions through the phases and becomes chaotic is characterized by the gradual loss of control—when the system enters the chaotic region, it cannot be controlled any longer, only disturbed. This process of phase transition is very important if applied to CSs. Let us consider a firm existing in a dynamic competitive environment to be conceptualized as CS. Such a firm would have to continuously utilize feedback coming from its external business landscape to develop its new emergent properties and adapt to the changes. This means that the firm would always be impacted by the changes in its control parameter (or multiple parameters)—this precludes the firm from ever being in an ordered state, for the firm would always be responding to perturbations and, as a result, transitioning through phases. At the same time, for obvious reasons, it is of utmost importance to not let the system phase transition all the way into the chaotic region.

This situation—"between a rock and a hard place"—provides valuable insights into the structure and the behavior of CSs. From one side, a dynamic environment would, consistently, manifest itself via changes in control parameters—an ordered state of a system means that the system is not sensitive enough to outside changes. At the same time, a system that is overly sensitive to the external environment may end up behaving chaotically. Thus, it is the goal of keeping a system on the edge of chaos, where it is not in order any longer but not yet in a chaotic state, that is worth pursuing.

References

Li, T.-Y. & Yorke, J. (1975). Period three implies chaos. *The American Mathematical Monthly*, 82(10), 985–992.

Lorenz, E. N. (1972). Predictability: Does the flap of a butterfly's wings in Brazil set off a tornado in Texas? 139th Annual Meeting of the American Association for the Advancement of Science (29 December 1972), in Essence of Chaos (1995), Appendix 1, 181.

Ruelle, D. & Takens, F. (1971). On the nature of turbulence. *Communications in Mathematical Physics*, 20, 167–192.

Strogatz, S. H. (1994). *Nonlinear Dynamics and Chaos*. Westview Press, Perseus, New York, USA.

Chapter 4

An Overview on Product Cycle Theory and the Product Life Cycle Model

In this chapter, we introduce to our reader two models—product cycle theory, which deals with patterns of international investment and trade, and the product life cycle model, which primarily refers to the patterns associated with sales and revenues of a product or service.

Product Cycle Theory: An Overview

The product cycle theory (PCT) originated with the work of Raymond Vernon (1966) in the context of international investments and international trade. Specifically, Vernon aimed to address reasons for why the empirical data coming from the United States did not conform to the Heckscher-Ohlin (H-O) model of international trade, according to which countries would export goods that use plentiful resources and are cheap to produce, and would import goods that use scarce resources and are expensive to produce.

Based on the H-O model, a country like the United States, where the labor is expensive and capital resources are abundant, would import labor-intensive and export capital-intensive products, for the comparative cost of producing and importing clearly suggests that the United States could get more "bang for a buck" from exporting products that require a lot of abundant capital than from exporting products that require a lot of scarce labor. Basically, the model predicts that a well-paid brain surgeon with a fully booked schedule would not do his/her own yard work and a low-level retail store clerk working part time would not pay someone to do his/her laundry.

While the predictions made by the H-O model seem to be intuitive, the relative cost of resources and labor within the model could not account for the *Leontief paradox*—Wassily Leontief discovered that in 1947, US imports were 30% more capital intensive than US exports.

Thus, noting the inadequacy of the concept of comparative cost in explaining changes in the patterns of international trade and investment, Vernon proposed that time is one of the important factors causing shifts in trade and investments. Consequently, PCT could be seen as an attempt to express changes in investments and trade associated with a product not as being influenced by gradual changes in comparative cost of the product, but as being impacted by stages in the lifetime of the product. For a product to exist, and eventually be traded, it must go through the cycle—first developed and introduced to the market, then produced in a cost-effective and cost-efficient manner via building economies of scale, and then maintained on the market. As a result of this progression, the relative cost of a product changes over time, and the changes lead to new patterns of trade and investments. Consequently, at any point in its lifetime, a product can be characterized by its relative cost, but the cost is a reflection of the location of the product along its life cycle, and changes in cost are reflections of the product transitioning from one point in time to another.

One of the assumptions of PCT is that the firms of advanced economies are fundamentally similar and have equal access to scientific knowledge. However, the firms do differ with regard to their ability to translate scientific knowledge into marketable new products, which is greatly impacted by the efficiency and effectiveness of the available information channels that connect firms and markets. Because of that difference between scientific knowledge and "how to make marketable products" knowledge within PCT, knowledge is an important factor in trade and investments, and it is not a free good. In other words, two firms, one in Finland, and another one in France, would have the same free access to scientific knowledge of chemistry and biology, but a Finnish company would have an advantage over a French company in developing a new ecologically friendly cleaning solution that could be profitably sold in Finland.

To present his argument, Vernon offers a reader to consider entrepreneurs in the United States, where capital is abundant as a resource but labor is considered to be scarce. Making his example very specific to those products that are "associated with high income and those which substitute capital for labor" (Vernon 1966, p. 193), Vernon proposes that American businessmen would be the first ones to identify and take advantage of market opportunities associated with such products. And while it seems obvious that proximity to local markets would allow producers in the United States to identify and design the new products ahead of international competitors, the cost factors alone could not explain why the new products would be manufactured locally. For Vernon, an explanation is inherent to the process of design and development of new products, where at the beginning, albeit temporarily, the design is not standardized yet—specifications have not been settled upon, the inputs have not been clearly identified, and the process of transformation of the inputs into the ready-to-use product is still very flexible. And it is this flexibility associated with the nonstandardized design that impacts what Vernon calls "locational choice," where local producers would have an advantage in choosing the set of inputs, differentiating, and pricing the new product that is afforded by the effectiveness and efficiency of local communication channels. During this *new product stage* [Vernon refers to "introduction of a new product" (p. 195) period in his work], no international trade takes place yet, for a product is manufactured locally for local consumption.

Things change, however, during the *maturing product* stage. Despite continuing product differentiation during this stage, a certain degree of standardization takes place and brings about an inevitable decline in the need for flexibility that was so vital during the new product stage. This increasing stability of the manufacturing process allows for settling on certain long-term commitments and partnerships and beginning developing economies of scale—again, having locational implications. For Vernon, it is during this stage that the new product becomes known beyond the local market—this is followed by an initially limited demand for the product on the international market. Inevitably, if the product is successful on the market elsewhere, local entrepreneurs would consider investing in the new product and setting up local manufacturing facilities. This is, of course, in addition to original American producers investing in production facilities in new locales. This process of establishing of new manufacturing facilities abroad by the original United States producers is one of the important characteristics of the maturing product phase.

The new period brings about abundance of a locally manufactured new product on the local market. This is followed by the exploration of servicing new third-country markets by American producers from new manufacturing locations, as well as possible export of the product back to the United States that is made possible by the inherent differences in cost of labor between the United States and the new locations. Furthermore, international investments for American producers become almost inevitable, for once the international trade involving the new product grows, the local governments and manufacturers start competing with the original producers—Vernon states that "[an] international investment by the exporter, therefore, becomes a prudent means of forestalling the loss of a market" (Vernon 1966, p. 200).

Vernon suggests that this turn of events, with the reversal of export–import activity, could explain the Leontief paradox. According to Vernon's hypothesis, indeed, a country like the United States would start exporting capital-intensive products (just as H-O models predict), but later in the process of international trade, in accordance with the Leontief finding, the United States would end up importing capital-intensive products.

The last phase of the product cycle, the *standardized product*, introduces less-developed countries as potential competitive players that possess a locational advantage to manufacture the new product. Vernon notes that this turn of events, where less-developed countries export labor-intensive products, is at odds with the H-O model. The resolution of this disagreement is offered by one of the fundamental assumptions of Vernon's model—that knowledge is not free, and the flow of relevant information is not free. The implication is that less-developed countries will not engage in manufacturing new products at the new and maturing product phases due to the costs incurred by acquiring knowledge about new distant markets and establishing reliable channels for the flow of marketing information. Things change, however, at the *standardized product* phase—the demands and the markets for the product are known, and the competition is primarily on the basis of price of the product. These greatly diminished costs of acquiring market knowledge and establishing information channels create "a necessary if not a sufficient condition for investment" (Vernon 1966, p. 203) in manufacturing locations in less-developed countries. In Vernon's opinion, such new locations are particularly suitable for standardized products, for at this phase, the products could be manufactured "on a vertically-integrated self-sustaining basis" (Vernon 1966, p. 203), requiring minimal interaction with a potentially unreliable local economy. What sorts of products are suitable for being manufactured in locales of less-developed economies? The suggested characteristics of a candidate product are as follows: labor intensive to manufacture, highly flexible in terms of pricing, and minimally reliant on external industries in its production process. Additionally, the fairly long life span of the candidate products would preclude obsolescence, and high value will address shipping costs. Vernon suggests that standardized textile products, as well as steel and fertilizers, among others, could serve as excellent examples of such products.

While Vernon offers multiple examples of investment patterns resulting in interregional movement of production facilities within developed countries (e.g., the United States, Italy, Britain, Ireland), he admits a scarcity of empirical evidence for such patterns in the context of less-developed economies. More evidence supporting PCT, however, could be found in the case of international trade patterns, where the examples of Japan and Taiwan developing overseas markets for standardized manufacturing products are consistent with Vernon's hypothesis.

Is the proposed model consistent with classical economic perspective? According to classical perspective, less-developed economies, where the capital is scarce and labor is abundant, should not serve as a target for investments in manufacturing sites. But according to the PCT model, the cost of capital may not present a deterrent for such investments in production facilities. Vernon offers two reasons that reconcile the difference. The first reason is the investment behavior of an international entrepreneur who considers not only the capital cost of the local manufacturing facility, but rather, an opportunity cost within "a very restricted range of alternatives" (p. 206). The second reason is that the assumption of scarcity of capital in less-developed economies, where capital markets "…typically consist of a series of water-tight, insulated, submarkets in which wholly different rates prevail..." (p. 206), may not apply in the case of international investors. Vernon supports this reasoning by citing the examples of India and Pakistan, where the largest firms were not constrained by the shortage of capital resources. Furthermore, if a project of developing a new facility in a less-developed country could be shown to be economically feasible, then this opens the door to "...public international lenders [who] tend to lend at near-uniform rates, irrespective of the identity of the borrower and the going interest rate in his country" (p. 206). As a result, the major problem of capital cost in less-developed economies is greatly reduced or resolved altogether.

While the results seem to be contrary to traditional economic theory, Vernon notes that it is not a new paradox suggesting the development of capital-intensive economies in capital-poor countries. Instead, the disagreement is due to the difference in contexts—while traditional economic theory applies to the general context, Vernon's model deals with an exception to the general context. In the words of the author, "All we are concerned with here is a modest fraction of the industry of such countries, which in turn is a minor fraction of their total economic activity" (p. 207). However, these seemingly anomalous exceptions, according to Vernon, are worth investigation further, for they could be systematic enough to represent a new characteristic of less-developed economies.

Because the economic conditions in the world have changed since PCT was first outlined in 1966, in 1979, Vernon offered a modification of PCT, which was primarily concerned with the original location of manufacturing of a new product (does not have to be in the United States) and a target group of customers (not only high-income group of US customers). Nevertheless, the fundamentals of the model, such as dynamic competitive advantage, investments abroad, and importance of economies of scale, remain intact.

Product Life Cycle Model: An Overview of the Stages and Common Criticisms

According to PCT, every product on the market could be placed in one of the three categories—new product, maturing product, and standardized product. The categories primarily have to do with the patterns of trade and investments associated with the product, and not the life of a product itself. The three categories of Vernon are commonly mapped to the four stages in the product life cycle (PLC) model—*introduction*, *growth*, *maturity* (sometimes partitioned into *maturity* and *saturation*—the later stage of *maturity*), and *decline*. This categories-to-stages mapping is important, for it provides a bridge between Vernon's categories of a product, which are primarily geared toward economics and finance, and the stages in the life of a product, which are marketing and business oriented. The importance of bridging these two perspectives is intuitive—the stages and categories, taken together, offer a more complete perspective on what is happening with the product over a period of time, for finance and economics go together with marketing and business.

For a critical review of the relevant literature, we refer our readers to Gardner (1987), who reported that the first conceptualization of the product life cycle was made by Kleppner (1931), and the actual term appeared for the first time in the work of Jones (1957). Unlike Vernon's conceptualization of a product as of a manufactured good, the concept of a product as it is used in the PLC model is broader—Gardner (1987) reports a virtual consensus on using the definition of Kotler (1984), according to which a product could refer to "physical objects, services, places, organizations and ideas" (Gardner 1987, p. 4). At this point, we offer an overview of each of the four stages of the PLC model. We must note that we are not covering the development stage of the new product. While this stage serves as a foundation for the new product, the product itself is not yet "visible" to the marketplace—there are no sales and, hence, no revenues associated with the new product.

Stage 1—Introduction

The first stage of the product life cycle is the most important one from the perspective of marketing. It is at this stage that the new product demonstrates its viability on the marketplace—the first responses from customers show that the design and development of the product are a success. By this point, it is good news for the company manufacturing the product, for sunk costs associated with the development, design, and first marketing of the product are usually significant. At this stage, the company manufacturing the product must make two decisions: the first one related to the pricing of the product and the second one related to the scale of production. In terms of the pricing, there are two options—the first one is to keep prices fairly high and rely on a somewhat limited group of first-time buyers of the product, and the second one is to keep the price relatively low in an attempt to reach a wider group of potential customers. The choice of an option is usually associated with the nature of the product—a rollout of a new discount mobile phone service may rely, originally, on lower-than-optimal pricing to gain traction, while an introduction of a premium concierge phone service may go an expensive, limited-target-group route. In any case, access to the market information is crucial—a local presence is a must, and marketing "boots on the ground" cannot be substituted by a more economical virtual presence. *Development* and *introduction* of the new product are phases of the PLC that correspond to the *new product* category of PCT. Given a positive reception of the new product by the marketplace, the company will face the challenge of expanding production capacity to a sufficient extent—this will mark arrival of the new stage.

Stage 2—Growth

Continuous increase in sales and rapidly rising revenues are a hallmark of the growth stage. This stage maps well to Vernon's *maturing product* category—the product becomes well known to the customers, and the manufacturer must allocate additional investments to ramp up production and benefit from economies of scale. At this point in time, the inputs required to manufacture the product are clearly identified, the partnership has been established, and the design of the product is settled upon—the value and supply chains have been set. The importance of this stage is hard to underestimate—if the product is well received and sales are increasing rapidly, then the producer must accommodate the current demand to sustain and build the stream of revenues. Any difficulty with increasing the production capacity would spell problems for the manufacturer—at this point, the product also becomes well known to the competitors, who start entering the market with their versions of the product. At this stage, the producer may increase promotional spending to increase market share, as well as invest in minor design changes of the product with the purpose of continuous differentiation of the product from competitors' offerings. It is also becomes important to not overcommit to production, for a continuous rapid increase in sales cannot be sustained indefinitely—when the growth of sales starts slowing down, the product enters the next stage—maturity.

Stage 3—Maturity

Regardless of the original popularity of the product, a time comes when the marketplace starts approaching the point of being saturated—the growth of sales is declining. The original manufacturer of the product fights with the competitors to maintain the share of the market that was hard earned during growth. Marketing expenses are very high at this stage, promotions are abundant, and the product is offered to the customers at lowered, sometimes significantly, prices. Nevertheless, the stream of revenues is at its height—especially for the company that reached its goal of securing the targeted share of the market. Any growth at this point is associated with taking a share of the market from competitors—often by means of creating new distribution channels. It is at this stage where we see the manufacturers branching out into similar products, updating the existing offerings, as well as innovating and evolving the original product. At this point, producers aim to extend the life of the product by almost any means—the direction of the manufacturers' pitch changes from advertising the product to get new customers, to promoting the product's quality and reliability. Regardless of all the efforts at this stage to keep the marketplace interested in the product, the inevitable comes—the market is saturated with the product, and the popularity of the product falls. This marks the arrival of the fourth phase—decline.

Stage 4—Decline

This stage of the product life cycle maps well to *standardized product* category of Vernon (1966). There are a few reasons for a decline in sales of the product—this could happen due to the introduction of a new substitute product, or be a result of a decreasing popularity among customers or increasing competition on the marketplace. The product could also become obsolete at this point—this would make the maintenance and service of the product more expensive as time goes by. Prices of the products at this stage are stable—no additional discounts are offered, and a relatively high price keeps generating profits for the producer while discouraging remaining loyal customers from continuing to purchase the product. The decline of sales also impacts the number and variety of distribution channels—the producer gradually trims down the distribution network until the minimum number of essential channels is left. Keeping efficiency of the remaining distribution channels high is very important, for profits are steadily declining. Eventually, the product is phased out.

Criticisms and Applications of Product Life Cycle Model

The common criticisms of the PLC model are associated with its limited predictive and explanatory power. Despite the presence of fairly well-delineated stages within the model, the real-life situation may not yield such clear-cut distinction of the stages, and the prediction of the duration of the stages themselves is not an easy task. The model, as presented, is time dependent, but the life of any of the real-world products is also impacted by a wide variety of factors—general business environment, trends of the marketplace, marketing and business decisions of the producers, as well as the nature of the product itself.

Nevertheless, the utility of the model is hard to dispute. For example, no product or service appears on the market from nowhere—it must be developed and introduced. The level of sales of a viable product does go up, and as time passes, the level goes down. In order for the level of sales to go up, the manufacturer must ramp up the available inventory and increase the scale of production. Thus, this general nonlinearity of the model could be applied to sales, revenues, manufacturing and distribution costs, as well as marketing expenses associated with the product, which makes it a useful planning tool for a business decision making. For example, if a new cloud-based storage service, originally developed and introduced in the United States, is to be introduced in Tanzania, then the local decision maker would have at his/her disposal the picture of a general pattern of expenses associated with such an undertaking.

Furthermore, the model allows for benchmarking—if the rollout of a new cell phone service in Tanzania was hampered by a slow growth of sales due to the problems with developing economies of scale, then it is likely that the cloud-based offering would encounter a similar obstacle. Additionally, by analyzing the levels of costs and revenues over the life of a product, the decision maker could obtain valuable insights regarding their efficiency of operations—it is only reasonable to assume that the levels of efficiency of marketing, manufacturing, and distribution expenditures (e.g., conversion of spent resources into revenues or some other sort of measurable outcome) vary from stage to stage. The PLC model could also serve as a strategic planning tool. The identified stages of the product allow decision makers to plan their course of action regarding competitors, pricing and promotion, as well as distribution of the product.

One of the attractive features of the model is its general applicability—as long as one takes a high-level perspective, the model applies equally well to products, services, and technologies. Let us consider some of the examples. Services of scriveners in the field of law were phased out with the advent of copy machines, as well as switchboard operators working for the phone companies being eventually replaced by automatic exchanges. The iPod Classic, audio and video cassette tapes, as well as dot printers all have traversed their product life cycles. Baseband communication technology, at one point the cutting edge, was replaced by the broadband approach, and punch cards are no longer used to operate computers. Some markets and industries (agriculture and manufacturing in the United States), too, go through stages from development to decline.

References

Gardner, D. M. (1987). The product life cycle: A critical look at the literature. In *Review of Marketing*, ed. M. Houston, American Marketing Association, Chicago, pp. 162–194.

Jones, C. (1957). Product development from the management point of view. In *Chicago Marketing's Role in Scientific Management*, ed. R. L. Clewelt, American Marketing Association.

Kleppner, O. (1931). *Advertising Procedure*. Prentice Hall Inc., New York.

Kotler, P. (1984). *Marketing Management: Analysis, Planning and Control*, 5th edition. Prentice-Hall, Inc., Englewood Cliffs, NJ.

Vernon, R. (1966). International investment and international trade in the product cycle. *The Quarterly Journal of Economics*, 80(2), 190–207.

Chapter 5

Overview on Decision Tree Induction

The chapter provides an overview of decision tree induction. Its main purpose is to introduce the reader to the major concepts underlying this data mining technique, particularly those that are relevant to the chapters that involve the use of this technique.

Introduction

Decision tree (DT) induction is a popular data mining modeling technique that is being increasingly used in information systems research (e.g., Wang & Chuang 2016; Müller et al. 2016; Mansingh et al. 2015a,b; Takieddine & Andoh-Baidoo 2014; Andoh-Baidoo et al. 2012; Osei-Bryson & Ngwenyama 2011; Lee 2010; Samoilenko 2008; Samoilenko & Osei-Bryson 2008; Zhou et al. 2004). A DT is an inverted tree structure representation of a given decision problem (e.g., Figure 5.1). There are two main types of DTs: classification trees (CTs) and regression trees (RTs). For a CT, the target variable takes its values from a discrete domain, while for an RT, the target variable takes its values from an interval or continuous domain (e.g., Osei-Bryson 2004; Ko & Osei-Bryson 2002; Torgo 1999; Kim & Koehler 1995; Breiman et al. 1984).

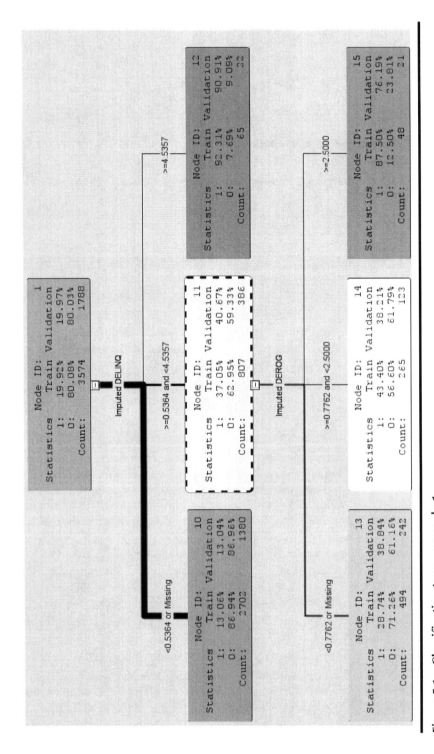

Figure 5.1 Classification tree—example 1.

Classification Tree

A DT can also be described as a model of a decision problem in the form of interpretable and actionable rules (see Figure 5.1). Associated with each leaf node of the DT is an *if–then* rule. For a given rule, the condition component of the rule [i.e., the potential predictor variable(s) and their values] is described by the set of relevant internal nodes and branches from the root node to the given leaf node; the action part of the rule is described by the relevant leaf node that provides the relative frequencies for each class of the target variable.

To generate a DT from a given data set, a single variable must be identified as the *target* (or dependent) variable, and the potential predictor variables must be identified as the *input* variables. We will illustrate with an example data set that is described in Table 5.1. The SAS Enterprise Miner data mining software was applied to this data set, resulting in the CT that is displayed in Figure 5.1. The reader may note that for several variables in this data set, there are rows with missing values (see "percent missing" column).

In some cases, a data set may have missing values for some variables in some records. Although the DT generation process can process records with missing values, many DT induction software packages (e.g., SAS Enterprise Miner) provide facilities for imputing missing values. In the case of SAS Enterprise Miner, when values are imputed for even one record for a given variable, the prefix "IMP_" is appended to the name of the variable. For this illustrative example, we will impute missing values for relevant potential predictor variables before generating a DT that is based on this data set (see Figure 5.1).

The reader may note that each branch of the DT is associated with a given variable (e.g., IMP_DELINQ) and a value or range of values for that variable (e.g., IMP_DELINQ < 0.5364). Let us consider the bottom leftmost leaf node (i.e., node ID = 13). The parent of this leaf node is node 11. The root node of the DT is node 1. From the root node to node 13, there are two connecting branches associated with the simple condition IMP_DELINQ $\in$ [0.5364,4.5347) and the simple condition IMP_DEROG < 0.7762. The conjunction of these two simple conditions is the condition component of the rule that is associated with node 13. For this rule, the target class 0 is the predicted class for this rule because it has the largest relative frequency of the two classes (i.e., 1: 28.74%; 0: 71.26%) in the relevant leaf node (i.e., node ID = 13).

Rules 1 and 2 are considered to be *sibling rules* because they are children of the same node (i.e., node ID = 1). Similarly, rules 3, 4, and 5 are considered to be sibling rules because they are children of the internal node, node ID = 12, that is associated with the condition IMP_DELINQ $\in$ [0.5364,4.5347) (Table 5.2).

Regression Tree

An RT is a DT in which the target variable takes its values from an interval or continuous domain, numeric. For each leaf, the RT associates the mean value and

Table 5.1 Example 1—Variables of the Illustrative HMEQ Data Set

Name	Role	Level	R...	...	Lower	Upper	Number of Levels	Percent Missing
BAD	Target	Binary	No	No	–	–	.	0
CLAGE	Input	Interval	No	No	–	–	.	5.167785
CLNO	Input	Interval	No	No	–	–	.	3.724832
DEBTINC	Input	Interval	No	No	–	–	.	21.25839
DELINQ	Input	Interval	No	No	–	–	.	9.731544
DEROG	Input	Interval	No	No	–	–	.	11.87919
JOB	Input	Nominal	No	No	–	–	6	4.681208
LOAN	Input	Interval	No	No	–	–	.	0
MORTDUE	Input	Interval	No	No	–	–	.	8.691275
NINQ	Input	Interval	No	No	–	–	.	8.557047
REASON	Input	Nominal	No	No	–	–	2	4.228188
VALUE	Input	Interval	No	No	–	–	.	1.879195
YOJ	Input	Interval	No	No	–	–	.	8.64094

Table 5.2 Ruleset of CT of Figure 5.1

			Target Variable BAD		
			Training Set Class Distribution		Predicted
Rule ID	Node ID	Condition Component of the Rule	1	0	
1	10	IMP_DELINQ < 0.5364	13.06%	86.94%	0
2	12	IMP_DELINQ ≥ 4.5347	92.31%	7.69%	1
3	13	IMP_DELINQ ∈ [0.5364, 4.5347) & IMP_DEROG < 0.7762	28.74%	71.26%	0
4	14	IMP_DELINQ ∈ [0.5364, 4.5347) & IMP_DEROG ∈ [0.7762, 2.5000)	43.40%	56.60%	0
5	15	IMP_DELINQ ∈ [0.5364, 4.5347) & IMP_DEROG ≥ 2.5000	87.50%	12.50%	1

the standard deviation of the target variable. Similar to CTs, each RT has a corresponding ruleset, an example of which is provided in Table 5.3.

- Rule 1 can be interpreted as follows: if LABEXP < 4.88069, then the target variable has a mean value of 5.32278 with a standard deviation of 0.78855.
- Rule 4 can be interpreted as follows: if NITCP2 < 6.82436 and LABEXP ∈ [.88069, 5.93812), then the target variable has a mean value of 6.1043 with a standard deviation of 0.31324.

Although RTs are similar to regression, the RT model is more like a step function, whereas the regression model involves a continuous linear function. Compared to regression models, RTs provide a model with better interpretability because the model involves interpretable English rules or logic statements. There have been instances where an RT has shown clues to data sets while a traditional linear regression analysis could not clearly indicate them (Breiman et al. 1984; Mansingh et al. 2015a,b).

DT Generation Process

The DT generation process involves a *growth phase* and, optionally, a *postpruning phase* (e.g., Kim & Koehler 1995). To avoid overfitting of the model to the training

Table 5.3 Ruleset of a Regression Tree

RuleID	Description	
1	IF LABEXP < 4.88069 COUNT = 44; **Target =** {AVE = 5.32278; SD= 0.78855}	THEN
2	IF 7.994195 ≤ LABEXP < 8.52146 COUNT = 79; **Target =** {AVE: 8.76191; SD: 0.31463}	THEN
3	IF LABEXP ≥ 8.52146 COUNT = 42; **Target =** {AVE: 9.64295; SD: 0.53401}	THEN
4	IF NITCP2 < 6.82436 & 4.88069 ≤ LABEXP < 5.93812 COUNT = 59; **Target =** {AVE: 6.1043; SD: 0.31324}	THEN
5	IF NITCP2 ≥ 6.82436 & 4.88069 ≤ LABEXP < 5.93812 COUNT = 63; **Target =** {6.52256; SD: 0.39495}	THEN
6	IF 5.93812 ≤ LABEXP < 6.48124 & NITCP2 < 7.496815 COUNT = 79; **Target =** {AVE: 6.63599; SD: 0.2681}	THEN
7	IF 6.48124 ≤ LABEXP < 6.906925 & NITCP2 < 7.496815 COUNT = 62; **Target =** {AVE: 7.0047; SD: 0.20182}	THEN
8	IF NITCP2 ≥ 9.462845 & 5.93812 ≤ LABEXP < 6.906925 N: 31; **Target =** {AVE: 7.63025; SD: 0.23262}	THEN
9	IF 6.906925 ≤ LABEXP < 7.42099 & IT3 < 5.37876 COUNT = 142; **Target =** {AVE: 7.6654; SD: 0.24763}	THEN
10	IF 7.42099 ≤ LABEXP < 7.994195 & IT3 < 5.378768 COUNT = 64; **Target =** {AVE: 8.0125; SD: 0.24741}	THEN
11	IF NITCP2 < 9.88379 & IT3 ≥ 5.378768 & 6.906925 ≤ LABEXP < 7.994195 N: 105; **Target =** {AVE: 8.19; SD: 0.31646}	THEN
12	IF 9.88379 <= NITCP2 & IT3 ≥ 5.378768 & 6.906925 <= LABEXP < 7.994195 COUNT = 30; **Target =** {AVE: 8.62255; SD: 0.31654}	THEN
13	IF 5.93812 ≤ LABEXP < 6.312265 & 7.496815 ≤ NITCP2 < 9.462845 COUNT = 32; **Target =** {AVE: 6.912; SD: 0.22931}	THEN
14	IF 6.312265 ≤ LABEXP < 6.906925 & 7.496815 ≤ NITCP2 < 9.462845 COUNT = 94; **Target =** {AVE: 7.25173; SD: 0.24957}	THEN

Note: AVE, mean; COUNT, number of training set records; SD, standard deviation.

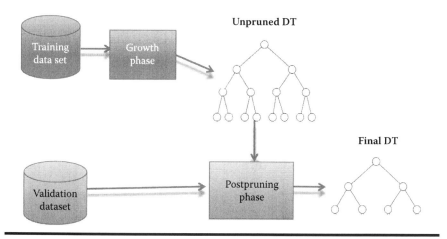

Figure 5.2 DT generation process.

data, for large data sets, the generation of a DT (illustrated in Figure 5.2) includes partitioning the model data set into either two parts (i.e., *training* and *validation*) or three parts (i.e., *training, validation*, and *test*). For small data sets, techniques such as *v*-fold (e.g., 10-fold) cross-validation allow for the entire data set to be used for both the growth and postpruning phases.

The growth phase involves generating a DT from the training data set such that either each leaf node is associated with a single value of the target variable or a further partitioning of the given leaf node would result in the number of records in one or both of the resulting child nodes being below some specified threshold. The DT that is generated in this phase is referred to as the *unpruned DT*. The postpruning phase aims to generalize the unpruned DT that was generated in the growth phase in order to avoid overfitting the final DT to the training data. The postpruning phase is described in more detail in sub-section "Postpruning."

Recursive Splitting

The growth phase involves recursive splitting of the training data set into progressively smaller subsets that are more homogenous with respect to the target variable. At each iteration, splitting decisions would be made automatically by a DT Induction algorithm, including

- What variable to split
- What is the best split
- When to *stop* splitting

Selection of the Splitting Method

A splitting method (e.g., Osei-Bryson & Giles 2004; Quinlan 1993; Taylor & Silverman 1993) is the component of the DT induction algorithm that determines both the attribute that is selected for a given node of the DT and also the partitioning of the values of the selected attribute into mutually exclusive subsets such that each subset uniquely applies to one of the branches that emanate from the given node. It is well known that there is no single splitting method that will give the best performance for all data sets. Table 5.4 displays the results of applying six different entropy-based splitting methods, including the popular gain-ratio (GR) method on 35 data sets. These results demonstrate that while some data sets are insensitive to the choice of splitting methods, others are very sensitive. Given that it is never known beforehand which splitting method will lead to the best DT for a given data set, it is advisable that the data miner explore the effects of different splitting methods. For CTs, these include Chi, Gini, and various entropy-based methods. For RTs, these include variance reduction and *F*-test.

Prepruning and Postpruning

Prepruning

Prepruning occurs during the growth phase. Its goal is to ensure that the resulting DT is not overfitted to the training data set. It attempts to stop growing the DT if the node is pure (i.e., all the training data set records associated with the given node have the same value for the target variable), the number of training data set records is below some threshold, or the split is not statistically significant at a specified level. It may also be used to limit the maximum number of branches from a node and the depth of the DT (i.e., maximum number of variables in a rule).

Postpruning

Postpruning (e.g., Osei-Bryson 2007; Fournier & Cremilleux 2002) occurs during the postpruning phase of DT induction. Given the unpruned DT that was generated in the growth phase, it creates a sequence of sub-trees of decreasing size; and applies an assessment criterion (e.g., best value of assessment measure, specified number of leaves, maximum number of leaves) on the validation data to determine the "best" subtree. For example if the specified assessment measure is the accuracy rate, then the selected "best" subtree is the one that has the highest accuracy rate against the validation data set. It follows that the selected "best" subtree is not independent of the training data set or the validation data set. For this reason, it is important that the distribution of the values of the target variable in the validation data set corresponds to the overall distribution of the values of the target variable.

Table 5.4 Example of Some Splitting Methods on Accuracy

Classification Accuracy Rate of CE and CAMI Families of Splitting Methods

Data Set			CE Family		CAMI Family				
ID	Name	GR	IG	CAMI	CAIR	EffCAMI_0	EffCAMI_1	Best-Worst
1	IRIS	**95.33**	94.67	94.67	94.67	94.67	94.67	0.66
2	Breast Cancer	72.49	72.49	72.49	72.49	**74.50**	**74.50**	2.01
3	Credit Approval	85.94	84.20	85.36	**86.96**	84.64	84.64	2.76
4	Car	92.48	**93.52**	**93.52**	**93.52**	92.48	92.48	1.04
5	Abalone	20.19	20.79	20.76	20.71	20.40	**21.88**	1.69
6	Wave	77.02	76.74	76.76	76.76	75.66	75.24	1.78
7	Glass	65.89	67.76	67.76	67.76	**71.96**	69.63	6.07
8	Soybean	92.09	87.56	89.02	89.02	92.09	**92.68**	5.12
9	Page Blocks	96.95	96.99	96.99	96.99	96.97	96.78	0.21
10	Mushroom	100.00	100.00	100.00	100.00	100.00	100.00	0.00
11	Wine	94.94	**95.51**	**95.51**	**95.51**	94.94	94.94	0.57
12	Yeast	54.78	53.03	53.03	53.03	**55.39**	54.72	2.36
13	Zoo	92.08	**94.06**	**94.06**	**94.06**	92.08	92.08	1.98

(Continued)

Table 5.4 (Continued) Example of Some Splitting Methods on Accuracy

Classification Accuracy Rate of CE and CAMI Families of Splitting Methods

ID	Data Set Name	CE Family GR	CE Family IG	CAMI Family CAMI	CAMI Family CAIR	CAMI Family EffCAMI_0	CAMI Family EffCAMI_1	Best–Worst
14	Pima	**74.09**	72.14	72.14	72.14	71.48	71.61	2.61
15	Nursery	97.11	97.10	97.10	**98.13**	97.11	97.11	1.03
16	Audiology	77.88	67.26	77.43	77.43	77.88	**79.65**	12.39
17	Heart	**77.78**	72.59	72.59	72.59	73.33	73.33	5.19
18	Hepatitis	79.35	78.06	**80.65**	**80.65**	80.00	80.00	2.59
19	Tumor	40.71	43.01	40.41	40.41	42.18	**44.25**	3.84
20	Chess	99.53	99.44	99.44	99.44	99.47	99.47	0.09
21	Letter	87.76	87.96	87.96	87.96	87.68	87.67	0.29
22	Segment	**97.14**	97.10	97.10	97.10	96.93	96.97	0.21
23	Sick	98.65	98.52	98.97	98.91	98.67	98.67	0.45
24	Sonar	74.04	73.08	73.08	73.08	**75.97**	**75.96**	2.89

(Continued)

Table 5.4 (Continued) Example of Some Splitting Methods on Accuracy

Classification Accuracy Rate of CE and CAMI Families of Splitting Methods

Data Set		CE Family		CAMI Family				
ID	Name	GR	IG	CAMI	CAIR	EffCAMI_0	EffCAMI_1	Best–Worst
25	Splice	93.98	93.67	93.67	93.51	93.67	93.67	0.47
26	Anneal	98.44	98.89	98.89	98.55	98.55	98.55	0.45
27	Autos	**82.44**	73.17	77.56	77.56	80.00	79.02	9.27
28	Colic	85.87	76.09	**85.60**	**85.87**	**85.87**	**85.87**	9.78
29	Hypothyroid	**99.58**	98.99	**99.50**	**99.50**	**99.55**	**99.55**	0.59
30	Ionosphere	**90.88**	88.60	88.60	88.60	90.03	89.46	2.28
31	Labor	75.44	71.93	80.70	**84.21**	78.95	78.95	12.28
32	Lymph	**77.03**	**77.03**	**77.03**	76.35	75.00	75.68	2.03
33	Vehicle	73.40	73.17	73.17	73.17	73.52	73.40	0.35
34	Vote	96.78	95.86	**96.55**	**96.55**	**96.55**	**96.55**	0.92
35	Vowel	78.38	83.84	83.84	**85.45**	79.49	78.59	7.07

Source: Osei-Bryson, K. M. & Giles, K., Splitting Methods for Decision Tree Induction: An Exploration of the Relative Performance of Two Entropy-Based Families, *Information Systems Frontiers* 8:3, 195–209, 2006.

Note: CE, Conditional Entropy; CAMI, Class/Attribute Mutual Information.

Recursive Splitting Example

We will use the data set of example 1 to illustrate how recursive partitioning would be done. Our example will involve two iterations.

Figure 5.3a applies to the first iteration.

- The "presplit: node data" row of this figure describes the statistics for the training and validation data sets. The target variable *BAD* has two classes, 1 and 0. The statistics indicate that the 19.94% of the records in entire training data set have the class of 1 for the target variable BAD, and 80.06% % of the records in entire training data set have the class of 0 for the target variable BAD. For the validation data set, 19.97% of its records have the class of 1 for the target variable BAD, and 80.03% % of its records have the class of 0 for the target variable BAD.
- The "splitting calculation results" row of this figure describes the value of the splitting metric (i.e., *worth*) that is associated with the split of each of the input variables that was determined by the splitting method as well as the number of branches from the root node that would result from each split. It can be seen that the IMP_DELINQ variable has the best value of worth and so would be selected as the variable to be used for the first split of the data. It should be noted that it is the entire training data set that is used to calculate the value of worth for each of the input variables.
- The "split: result" row of this figure displays the DT that results after splitting the entirety of the training and validation sets based on subranges of the IMP_DELINQ variable that were determined by the splitting method.

Figure 5.3b applies to the first iteration.

- The "presplit: node data" row of this figure describes the statistics for the training and validation data sets that are based on the DT that resulted from the first split. An examination of this DT indicates that for the training data set, the leftmost and rightmost child nodes (i.e., node IDs 13 and 15) are each highly dominated by one of the classes of the target variable BAD (i.e., class 0 for node 13 and class 1 for node 15), but for node 14, the corresponding domination is not as high. For illustrative purpose, iteration 2 will thus explore splitting the subset of the data that are associated with node 14.
- The "splitting calculation results" row of this figure is similar to that of the corresponding row of Figure 5.3a but is based on the subset of the training data set that is associated with the subrange of the IMP_DELINQ variable that is associated with node 14. It can be seen that the IMP_DEROG variable has the best value of worth and so would be selected as the variable to be used for the second split of the data.

Overview on Decision Tree Induction

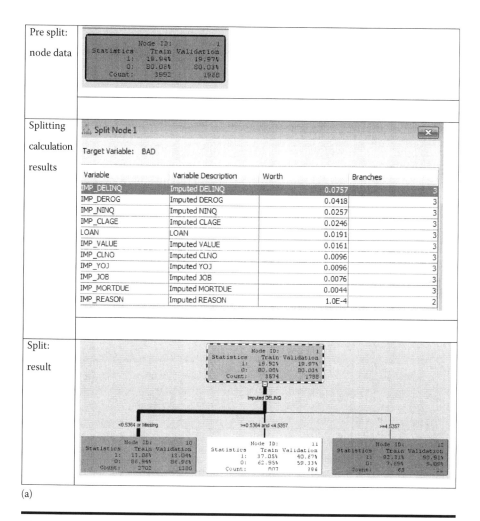

Figure 5.3 (a) Recursive partitioning—iteration 1. (*Continued*)

- The "split: result" row of this figure displays the DT that results after splitting the training and validation sets based on subranges of the IMP_DELINQ variable and the subranges of the IMP_DEROG variable that were determined by the splitting method.

Software Implementation of the DT Generation Process

Many DM software packages, (e.g., C5.0, SAS Enterprise Miner, IBM SPSS Miner, R, RapidMiner) provide facilities that make the generation of DTs a relatively easy

52 ■ *Theoretical Research Frameworks Using Multiple Methods*

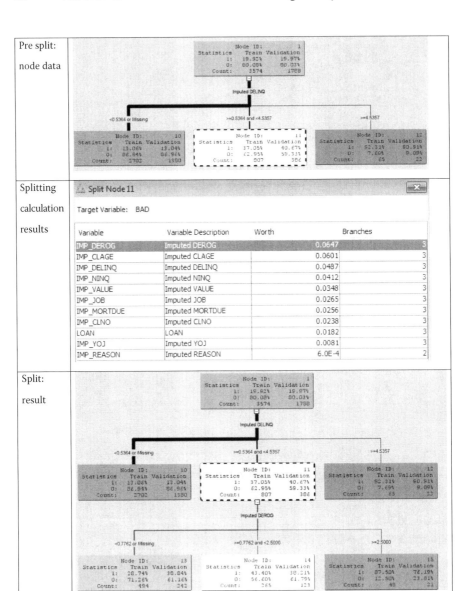

Figure 5.3 (Continued)   (b) Recursive partitioning—iteration 2.

Figure 5.4 Process flow diagram.

task. Figure 5.4 presents the SAS Enterprise Miner *process flow diagram* that is used to generate the DT from the given data set. Each node in this diagram performs a specific task: the *input* node (i.e., HMEQ) reads the given data set; the *impute* node is used to impute missing values; the *data partition* node partitions the input data set into training, validation, and test data sets; and the *tree* node generates the DT.

References

Andoh-Baidoo, F. K., Osei-Bryson, K. M., & Amoako-Gyampah, K. (2012). A hybrid decision tree based methodology for event studies and its application to e-commerce initiative announcements. *ACM SIGMIS Database*, 44:1, 78–101.

Breiman, L., Friedman, J., Olshen, R., & Stone, J. (1984). *Classification and Regression Trees*. Wadsworth Inc., Belmont, CA.

Fournier, D. & Cremilleux, B. (2002). A quality index for decision tree pruning. *Knowledge-Based Systems*, 15, 37–43.

Kim, H. & Koehler, G. (1995). Theory and practice of decision tree induction. *Omega*, 23:6, 637–652.

Ko, M. & Osei-Bryson, K.-M. (2002). A regression tree based exploration of the impact of information technology investments on firm level productivity. *European Conference of Information Systems*, 507–517.

Lee, S. (2010). Using data envelopment analysis and decision trees for efficiency analysis and recommendation of B2C controls. *Decision Support Systems*, 49:4, 486–497.

Mansingh, G., Rao, L., Osei-Bryson, K.-M., & Mills, A. (2015a). Profiling Internet banking users: A knowledge discovery in data mining process model based approach. *Information Systems Frontiers*, 17:1, 193–215.

Mansingh, G., Osei-Bryson, K. M. and Asnani, M. (2015b). Exploring the antecedents of the quality of life of patients with sickle cell disease: Using a knowledge discovery and data mining process model based framework. *Health Systems, Palgrave MacMillan*, 5:1, 52–65.

Müller, O., Junglas, I., vom Brocke, J., & Debortoli, S. (2016). Utilizing big data analytics for information systems research: Challenges, promises and guidelines. *European Journal of Information Systems*. doi:10.1057/ejis.2016.2

Osei-Bryson, K.-M. (2004). Evaluation of decision trees: A multi-criteria approach. *Computers and Operations Research*, 31:11, 1933–1945.

Osei-Bryson, K.-M. (2007). Post-pruning in decision tree induction using multiple performance measures. *Computers and Operations Research*, 34:11, 3331–3345.

Osei-Bryson, K.-M. and Kendall Giles, K. (2004). An exploration of a set entropy-based hybrid splitting methods for decision tree induction. *Journal of Database Management*, 15:3, 1–17.

Osei-Bryson, K.-M. & Giles, K. (2006). Splitting methods for decision tree induction: An exploration of the relative performance of two entropy-based families. *Information Systems Frontiers*, 8:3, 195–209.

Osei-Bryson, K.-M. & Ngwenyama, O. (2011). Using decision tree modelling to support Peircian abduction in IS research: A systematic approach for generating and evaluating hypotheses for systematic theory development. *Information Systems Journal*, 21:5, 407–440.

Quinlan, J. (1993). *C4.5 Programs for Machine Learning*. Morgan Kaufmann, San Mateo.

Samoilenko, S. (2008). Contributing factors to information technology investment utilization in transition economies: An empirical investigation. *Information Technology for Development*, 14:1, 52–75.

Samoilenko, S. & Osei-Bryson, K.-M. (2008). Increasing the discriminatory power of DEA in the presence of the sample heterogeneity with cluster analysis and decision trees. *Expert Systems with Applications*, 34:2, 1568–1581.

Takieddine, S. & Andoh–Baidoo, F. K. (2014). An exploratory analysis of Internet banking adoption using decision tree induction. *International Journal of Electronic Finance*, 8:1, 1–20.

Taylor, P. & Silverman, B. (1993). Block diagrams and splitting criteria for classification trees. *Statistics & Computing*, 3:4, 147–161.

Torgo, L. (1999). Predicting the density of algae communities using local regression trees. Proceedings of the European Congress on Intelligent Techniques and Soft Computing, EUFIT'99.

Wang, C. H., & Chuang, J. J. (2016). Integrating decision tree with back propagation network to conduct business diagnosis and performance simulation for solar companies. *Decision Support Systems*, 81, 12–19.

Zhou, L., Burgoon, J. K., Twitchell, D. P., Qin, T., & Nunamaker Jr, J. F. (2004). A comparison of classification methods for predicting deception in computer-mediated communication. *Journal of Management Information Systems*, 20(4), 139–166.

Chapter 6

Overview on Cluster Analysis

This chapter provides an overview of cluster analysis. Its main purpose is to introduce the reader to the major concepts underlying this data mining technique, particularly those that are relevant to Chapters 17, 18, and 20 which involve the use of this technique.

Introduction

Clustering is a popular data mining (DM) technique that attempts to automatically partition a data set into a meaningful set of mutually exclusive clusters (or segments). In recent years clustering has been increasingly used in information systems research (e.g., Trivedi et al. 2016; Swobodzinski & Jankowski 2015; van Dam & van de Velden 2015; Ayanso et al. 2014; Blooma et al. 2014; Chen et al. 2012; Samoilenko & Osei-Bryson 2010; Okazaki 2006; Osei-Bryson & Joseph 2006; Rai et al. 2006; Boley et al. 1999). While clustering can be done directly by humans without the use of DM software, DM-based clustering attempts to segment the data into natural clusters that are relatively homogenous with respect to some similarity metric (e.g., Euclidean distance). The characteristics of the resulting clusters are thus not subjectively determined. However, in some cases, the resulting clusters may have no natural meaning to the user.

56 ■ Theoretical Research Frameworks Using Multiple Methods

There are several reasons for doing clustering, two major categories of which are as follows:

1. Finding a set of natural clusters, and the corresponding description of each cluster. This is relevant if there is the belief that there are natural groupings in the data. In some cases, such as fraud detection, there is interest in finding segmentations that include outlier clusters (e.g., Aggarwal & Yu 2001). For some other cases, there may be a preference in finding a segmentation that includes a pair of clusters that provide the lowest mean and highest mean for each variable. And still for other cases, there may be a preference for finding segmentations that include certain specified variables as important discriminating variables between the clusters. For each of these cases, it is possible that multiple segmentations could apply. Thus, it might not be appropriate to use a single clustering algorithm and/or parameter setting to find the appropriate set of natural clusters.
2. Improving the performance of predictive modeling (see Figure 6.1) and other DM techniques when there are many competing patterns in the data. For this case, there is an interest in obtaining a segmentation that will result in an improvement in performance and also offer a convenient description of

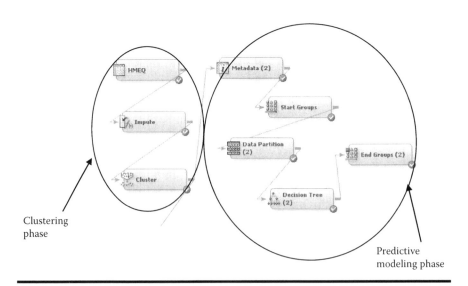

Figure 6.1 Process flow for using clustering to improve predictive modeling.

the clusters so that it will facilitate the assignment of new observations to the appropriate predictive model (Mansingh et al. 2015). However, it is possible that multiple segmentations could apply.

In addition to its use by practitioners, in recent years, clustering has also been applied as a research tool for exploring important research problems in information systems and other areas (Balijepally et al. 2011). For example, Rai et al. (2006) used cluster analysis to explore various questions including whether the assimilation of electronic procurement innovations increases procurement productivity; Okazaki (2006) used it to characterize mobile Internet adopters, and Wallace et al. (2004) used it to characterize software project risks. For these applications of clustering, the given researcher was interested in obtaining natural grouping and corresponding descriptions. Typically, these research projects did not involve the exploration of multiple clustering algorithms and parameter settings but involved the use of default parameter settings. As noted by Balijepally (2006): "A vast majority of IS studies have however neither reported the algorithm used in the study nor the distance measure used, though some improvement has been noticed over the two time periods. Non-reporting of such basic requirements of cluster analysis leads to suspicion that researchers could be blindly using the default settings in the computer packages without a clear understanding of the methodology or the implications of the decision choices involved therein." Yet it is known that for a given clustering algorithm, different parameter settings could result in different segmentations, several of which could be relevant to the given research question(s).

Understanding the Output of Clustering

Different approaches may be used to understand the output of clustering, including the following:

1. Building a decision tree (DT) with the cluster label as the target variable (see Figure 6.2) and using the associated rules to conveniently describe each cluster as well as explain how to assign new records to the correct cluster (e.g., Mathers & Choi 2004, Mansingh et al. 2015).
2. Examining the distribution of variable values from cluster to cluster. Typically, this involves the domain expert(s) doing comparison of cluster means (see Figure 6.3) for the relevant variables (e.g., Wallace et al. 2004).
3. Visual inspection by a domain expert of a graphical two-dimensional (2-D) representation of the clustering output in order to assess the validity of the results (e.g., Bittman & Gelbrand 2009; Kimani et al. 2004)
4. A hybrid of the aforementioned approaches.

58 ◼ *Theoretical Research Frameworks Using Multiple Methods*

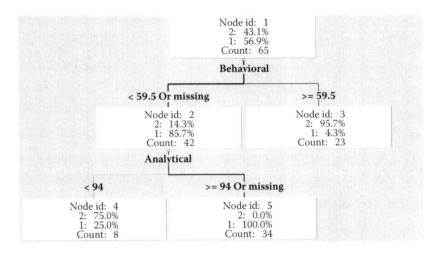

Node = 3	IF **Behavioral** >= 59.5 THEN
	Number of observations = 23
	Predicted: **_SEGMENT_ = 2 = 0.96**
	Predicted: _SEGMENT_ = 1 = 0.04

Node = 4	IF **Behavioral** < 59.5 AND **Analytic** < 94 THEN
	Number of observations = 8
	Predicted: **_SEGMENT_ = 2 = 0.75**
	Predicted: _SEGMENT_ = 1 = 0.25

Node = 5	IF **Behavioral** < 59.5 AND **Analytic** >= 94 THEN
	Number of observations = 34
	Predicted: _SEGMENT_ = 2 = 0.00
	Predicted: **_SEGMENT_ = 1 = 1.00**

Figure 6.2 Example of a DT that describes a segmentation with two clusters (i.e., "1," "2"). Input variables: *analytical, behavioral, conceptual, directive.*

Cluster	Analytical	Behavioral	Conceptual	Directive
1	105.46	43.81	73.03	75.49
2	83.00	69.82	75.04	72.14
Overall: mean	95.78	55.02	73.89	74.05

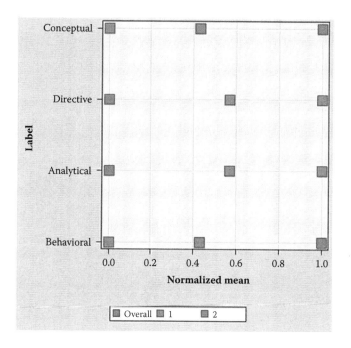

Relative importance of variables with regards to differentiating between the clusters

Variable	Analytical	Behavioral	Conceptual	Directive
Relative weight	0.946	1.000	0.000	0.711

Figure 6.3 Example of cluster means that describes a segmentation with two clusters. Input variables: *analytical, behavioral, conceptual, directive.*

Clustering Algorithms

There are numerous algorithms available for doing clustering, which may be categorized in various ways, including the following: hierarchical (e.g., Murtagh 1983; Ward 1963) or partitional (e.g., Chen et al. 2004; Boley et al. 1999; MacQueen 1967; Bock 1996), model based (e.g., Osei-Bryson & Joseph 2006; Osei-Bryson & Inniss 2007), deterministic or probabilistic (e.g., Bock 1996), hard or fuzzy (e.g., Bezdek 1981; Dave 1992), and two-phase. Hierarchical methods generate a hierarchy of clusters from the given data set using either a top-down or bottom-up iterative process. Partitional methods (e.g., k-means, k-median) divide the given data set into a user-specified number of clusters. An example of a two-phase method is a process that involves a partitional method in the first phase and a hierarchical method in the second phase where the output from the first phase is used as input for the second phase. It should also be noted that neural network (NN) based clustering algorithms (e.g., Kohonen 1995) have also been proposed.

k-Means: divides the data set into k clusters

Step 1: Pick k seed points as the initial clusters' *centroids*.
Step 2: Assign each object to the cluster whose centroid is closest to the given object.
Step 3: Let the new centroid of each cluster be the *mean* of objects in the cluster.
Step 4: If the new centroid of each cluster is the same as the old centroid or is sufficiently close, then *terminate*; otherwise, *repeat* steps 2 through 4.

The final set of clusters generated is sensitive to the choice of the initial cluster centroids.

Hierarchical methods may be categorized as being agglomerative or divisive:

Agglomerative	• A bottom-up approach that merges pairs of clusters. • Start with each data point in its own cluster. • At each step, merge the closest pair of clusters.
Divisive	• Top-down approach involving binary division of clusters. • Start with all data-points in an all-inclusive cluster. • At each step, split a cluster until each data point is in its own cluster.

Example of an agglomerative algorithm: average-link algorithm for generating g clusters from m data points

Step 1: Assign each data point to its own cluster, so that there are m clusters. A partitional method could be used to do this assignment.
Step 2: Merge the most similar pair of clusters into a single cluster. The result is that there is now one less cluster.
Step 3: Compute the distance between the new cluster and each of the old clusters.
Step 4: Repeat steps 2 and 3 until there are g clusters.

Table 6.1 Some Distance Metrics

Minkowski Distance Metrics	
$q = 1$: mean absolute deviation • Distance$(x,y) = \Sigma_i \|x_i - y_i\|$ • Resistant to outliers • Centroid value is the median	**$q = 2$: Euclidean distance** • Distance$(x,y) = (\Sigma_i(x_i - y_i)^2)^{1/2}$ • Sensitive to outliers. • Centroid value is the arithmetic mean
$q > 1$ and < 2: • Distance$(x,y) = (\Sigma_i(x_i - y_i)^q)^{1/q}$ • Resistant to outliers	**$q > 2$:** • Distance$(x,y) = (\Sigma_i(x_i - y_i)^q)^{1/q}$ • Sensitive to outliers
Percent Disagreement Metric • This measure is particularly useful if the data for the dimensions included in the analysis are categorical/nominal in nature. • This distance is computed as distance(x,y) = percentage of corresponding component variables that are in disagreement (i.e., $x_{ik} \neq y_{jk}$).	

Similarity Metrics

Distances are normally used to measure the similarity or dissimilarity between two data points. Commonly used metrics include the *Minkowski distance*, cosine, *percent disagreement*, and correlation. The general form of the Minkowski distance is

$$d(i, j) = \sqrt[q]{\left(\left|x_{i1} - x_{j1}\right|^q + \left|x_{i2} - x_{j2}\right|^q + \ldots + \left|x_{ip} - x_{jp}\right|^q\right)}$$

where $i = (x_{i1}, x_{i2}, \ldots, x_{ip})$ and $j = (x_{j1}, x_{j2}, \ldots, x_{jp})$ are two p-dimensional data objects, and q is a positive integer. Examples for various values of q are given in Table 6.1.

Evaluating the Output of Clustering Algorithms

Typically, for a given data set, different algorithms may give different sets of clusters, so it is never clear which algorithm and which parameter settings (e.g., number of clusters) are the most appropriate. For as noted by Jain et al. (1999), "There is no clustering technique that is universally applicable in uncovering the variety of structures present in multidimensional data sets." They thus raised the following questions: "How is the output of a clustering algorithm evaluated? What characterizes a 'good' clustering result and a 'poor' one?" Ankerst et al. (1999) also commented, "Most of the recent research related to the task of clustering has been directed towards efficiency. The more serious problem, however, is effectivity, i.e., the quality or usefulness of the result."

The Issue of Quality: Assessing Cluster Validity

Jain et al. (1988) describe cluster validity as the assessment of the set of clusters that is generated by the given clustering algorithm. They note that there are three approaches for assessing validity:

1. *External assessment*, which involves comparing the generated segmentation (i.e., set of clusters) with an a priori structure, typically provided by some domain experts.
2. *Internal assessment*, which attempts to determine if the generated set of clusters is "intrinsically appropriate" for the data. Several techniques have been proposed for internal assessment of cluster validity (e.g., Bezdek 1981; Dunn 1974; Gordon 1999; Kaufman & Rousseeuw 1990; Osei-Bryson 2005; Ramze Rezaee et al. 1998; Tibshirani & Walther 2005), with the central idea behind these approaches being that a valid, high-quality segmentation should consist of clusters that are both cohesive and well separated, although some techniques give greater emphasis to one of these properties.
3. *Relative assessment*, which involves comparing the relative performance of pairs of segmentations (i.e., two sets of clusters) based on some performance measures (e.g., Dubes 1983; Jain & Dubes 1988) and measuring their relative performance.

The Issue of Usefulness: Goals for Clustering

There are several possible goals for a clustering exercise, including the following:

1. *Segmentation should include (exclude) outliers*: This problem has several applications, including fraud detection (e.g., Aggarwal & Yu 2001).
2. *Segmentation should include a pair of clusters that provide the lowest and highest means for each variable*: An example of this problem is a theory-building study (Wallace et al. 2004) where the interest was in finding clusters that provide the characteristics that could be used to describe low-, medium-, and high-risk projects. In this case, a pair of clusters was found that provided the smallest and largest means for almost all the variables.
3. *Segmentation should include user-specified variables as important discriminating variables*: There are several reasons why this goal may be important to users. For example, as noted by Huang et al. (2005), "It is well-known that an interesting clustering structure usually occurs in a subspace defined by a subset of the initially selected variables. To find the clustering structure, it is important to identify the subset of variables." This could also be relevant for theory-building exercises. Balijepally (2006) suggested, "Studies where the clustering variables are tightly linked to theory are considered deductive. If the variables are generated based on expert opinion, the approach is

deemed cognitive … Legitimacy accorded to the pursuit of theories from reference disciplines in IS research could be one reason working in favor of adopting a deductive approach. … One suggestion for improvement would be to consider using a cognitive approach to variable selection over a pure inductive approach. This involves tapping expert opinions either from other IS researchers, IS practitioners, or both."
4. *Segmentation should include several clusters for which the differences of the means of such clusters with regard to a specified nondiscriminator variable and the overall corresponding mean for the entire data set are statistically significant, with some of these differences being positive and others negative.*

Overview of a Process Model for Clustering

A common presumption that underlies approaches for doing clustering on a given data set is that there is a single optimal *segmentation* (i.e., partitioning) that is independent of the objectives of the end user. For example, Halkidi et al. (2002) speaks of the "'optimal' clustering scheme as the outcome of running a clustering algorithm (i.e., a partitioning) that best fits the inherent partitions of the data set." So how would this "optimal" partitioning (i.e., segmentation) be identified if two or more segmentations appear to have approximately the "best" fit. Further, would this choice be based on the goals of the clustering exercise (e.g., fraud detection vs. marketing profiling vs. theory building)? It appears to us that when clustering is used for knowledge discovery, what is the best fit may not be independent of the goals of the clustering exercise. Thus, Kim (2007) suggests that for some situations, "comparison between clustering outputs of two methods should be interpreted only from managerial perspectives, rather than from numerical perspectives based on evaluation metrics such as intra-cluster compactness or inter-cluster separability." This position has been recognized in many knowledge discovery via data mining (KDDM) process model methodologies including cross industry standard procedure for data mining (CRISP-DM) (e.g., Sharma et al. 2012; Sharma & Osei-Bryson 2010; Kurgan & Musilek 2001; Shearer 2000), for as noted by Kurgan and Musilek (2006), with regard to DM, "The general research trends were concentrated on the development of new and improved DM algorithms rather than on the support for other KD activities … Before any attempt can be made to perform the extraction of this useful knowledge, an overall approach that describes how to extract knowledge needs to be established."

Interest in the development of KDDM process models arose after DM practitioners and researchers became aware of the need for formal DM process models that prescribe the journey from data to discovering knowledge. The KDDM process has been described in various ways (e.g., Sharma & Osei-Bryson 2010; Kurgan & Musilek 2006; Shearer 2000) but essentially consists of the following steps: application domain or business understanding (which includes definition of

DM goals), data understanding, data preparation, DM, evaluation (e.g., evaluation of results based on DM goals), and deployment.

Osei-Bryson (2010) presented a KDDM process model that was adapted for clustering exercises. Here we present an overview description of this process model in which the business understanding phase is repurposed as a research problem understanding phase (Table 6.2).

Table 6.2 Phases of the KDDM Process Model for Clustering

Phase	Description
Research problem understanding	a. Define *research problem* objectives b. Learn current solutions and domain terminology c. Translate research problem objectives into *clustering goals* d. Initialization activities including selection of specific clustering algorithm(s), *cluster validity criterion* (i.e., external, internal or relative assessment), plus their parameter settings for use in the *modeling* phase
Data understanding	Collecting the data; verification and exploration of the data with respect to completeness, redundancy, plausibility, missing values, sample statistics (e.g., mean, standard deviation, and range of each numeric variable), and usefulness of the data with respect to the research problem objectives and clustering goals
Data preparation	Preparation of the final *modeling data set*, which will be fed into DM tools and includes data and attribute selection, cleaning, construction of new attributes, imputation of missing values, and relevant data transformations
Modeling	a. Application of the clustering algorithm(s) to the modeling data set based on specifications in the research problem understanding phase b. Results of the clustering algorithms pruned using the selected cluster validity criterion
Evaluation	a. Results of modeling phase automatically pruned and ordered based on the user-specified clustering goals b. Researcher uses specified search procedure for exploring the pruned results set of part (a) of this phase
Deployment	Presentation of the discovered knowledge in a customer-oriented way

Software Implementation of the Clustering Process

Many DM software packages (e.g., SAS Enterprise Miner, IBM SPSS Miner, R, RapidMiner) provide facilities that make cluster analysis a relatively easy task. Figure 6.4 presents the SAS Enterprise Miner *process flow diagram* that is used to generate clusters from the given data set. Each node in this diagram performs a specific task: the *file import* node reads the given data set; the *impute* node is used to impute missing values; and the *cluster* node generates the clusters. For this data set (which was a Excel spreadsheet), the *file import* node was used to read the data; if the data set had been a SAS data set, then the *input* node would have been used to read the data set. The following narrative, as well as Figures 6.5 and 6.6, provide additional elaboration.

Figure 6.4 Process flow diagram.

Property	Value
General	
Node ID	Clus
Imported Data	
Exported Data	
Notes	
Train	
Variables	
Cluster Variable Role	Segment
Internal Standardization	Standardization
Number of Clusters	None
Specification Method	Standardization
Maximum Number of Clusters	Range
Selection Criterion	

Range: subtracts the minimum and divides by the range (i.e., maximum − minimum), so that the resulting values have a minimum of 0 and a maximum of +1. The concern is variables with larger ranges could have a larger impact just because of that fact; therefore the variables should be transformed so that equal relative distances are of equal practical importance.

Standard Deviation: subtracts the *mean* and divides by the *standard deviation*, so that the resulting values have a mean of zero (0) and a standard deviation of 1. The concern is that variables with large absolute variances could have a greater effect than those with small variances if all variables are not standardized.

Figure 6.5 Some internal standardization options.

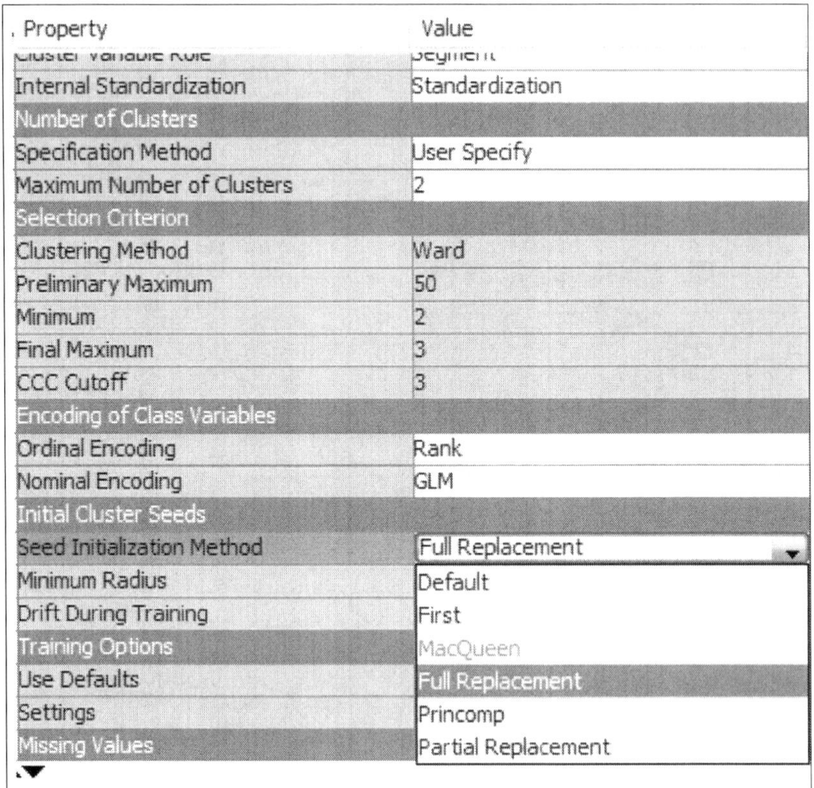

Property	Value
Cluster Variable Role	Segment
Internal Standardization	Standardization
Number of Clusters	
Specification Method	User Specify
Maximum Number of Clusters	2
Selection Criterion	
Clustering Method	Ward
Preliminary Maximum	50
Minimum	2
Final Maximum	3
CCC Cutoff	3
Encoding of Class Variables	
Ordinal Encoding	Rank
Nominal Encoding	GLM
Initial Cluster Seeds	
Seed Initialization Method	Full Replacement ▼
Minimum Radius	Default
Drift During Training	First
Training Options	MacQueen
Use Defaults	Full Replacement
Settings	Princomp
Missing Values	Partial Replacement

Number of clusters:

For the number if clusters, the user may specify either a specific number or a range (i.e., *minimum* and *final maximum*). If a range is specified and the clustering software uses an Agglomerative Approach then the agglomerative *clustering method* must be specified. For some datasets, the final output is sensitive to this choice. For example, at a given iteration the Ward method merges the pair of clusters whose merger gives the minimum increase in the variation while the centroid method merges the pair that has the most similar centroids.

Seed initialization method:

The final partitioning generated by k-Means algorithm is sensitive to the choice of initial *Cluster centroids*. It is therefore important that adequate experimentation be done with the various cluster centroid initialization options that are offered by the relevant software.

Figure 6.6 Number of clusters, agglomerative clustering method, and seed initialization method.

Table 6.3   Examples of Imputation Methods

Centroid based	• Mean—the arithmetic average. • Median—middle value. • Midrange—the maximum plus the minimum divided by 2. • The mean and median are useful for a symmetric distribution. • The median is less sensitive to extreme values than the mean or midrange. Therefore, the median is preferable when you want to impute missing values for variables that have skewed distributions. • The median is also useful for ordinal data.
Distribution based	• Replacement value is based on the probability distribution of the nonmissing observations. • This imputation method typically does not change the distribution of the data very much.
Tree imputation	• Other variables are used to predict the value of the given variable. • This imputation technique may be more accurate than simply using the variable mean or median to replace the missing values.

Role of the *impute* node: The clustering process ignores records in which any of the identified potential discriminator variables do not have a value. In such a situation, it may therefore be useful to apply an appropriate imputation method before applying the clustering algorithm(s). So if the data set has missing values for any of the potential discriminator variables, then an *impute* node should be used to generate replacement values (see Table 6.3).

Cluster Node: Some Important Parameter Settings

- **Internal standardization:** The primary concerns that motivate the application of internal standardization of the potential discriminator variables are that (1) equal distances should be of equal practical importance and (2) variables with large variances tend to have a greater effect than those with small variances if all variables are not standardized. Figure 6.5 presents some of the options.
- **Seed initialization:** Some clustering algorithms (such as k-means) are sensitive to the set of initial cluster centroids that is selected and so are sensitive to the method that is used to generate the set of initial cluster centroids. Figure 6.6 displays options for the SAS Enterprise Miner software.

References

Aggarwal, C. & Yu, P. (2001). Outlier detection for high dimensional data. *Proceedings of the 2001 ACM SIGMOD International Conference on Management of Data*, 37–46.

Ankerst, M., Breunig, M., Kriegel, H.-P., & Sander, J. (1999). OPTICS: Ordering points to identify the clustering structure. *Proceedings of ACM SIGMOD'99 International Conference on the Management of Data*, Philadelphia, PA, pp. 149–160.

Ayanso, A., Cho, D. I., & Lertwachara, K. (2014). Information and communications technology development and the digital divide: A global and regional assessment. *Information Technology for Development*, 20(1), 60–77.

Balijepally, V. (2006). *Application of Cluster Analysis in Information Systems Research: A Review*. Prairie View A&M University.

Balijepally, V., Mangalaraj, G., & Yengar, K. (2011). Are we wielding this hammer correctly? A reflective review of the application of cluster analysis in information systems research. *Journal of the Association for Information Systems*, 12:5, 375–413.

Basilevsky, A. (1983). *Applied Matrix Algebra in the Statistical Sciences*, Elsevier, New York.

Bezdek, J. (1981). *Pattern Recognition with Fuzzy Objective Function Algorithms*. Plenum Press, New York.

Bittman, A. & Gelbrand, R. (2009). Visualization of multi-algorithm clustering for better economic decisions—The case of car pricing. *Decision Support Systems*, 47:1, 42–50.

Blooma, M. J., Huy, T. D., & Wickramasinghe, N. (2014). Healthcare social question answering: Concept mapping and cluster analysis based on graph theory. *Proceedings of the 25th Australasian Conference on Information Systems, 8th–10th December 2014*, Auckland, New Zealand.

Bock, H. (1996). Probability models in partitional cluster analysis. *Computational Statistics and Data Analysis*, 23, 5–28.

Boley, D., Gini, M., Gross, R., Han, E., Hastings, K., Karypis, G., Kumar, V., Mobasher, B., & Moore, J. (1999). Partitioning-based clustering for web document categorization. *Decision Support Systems*, 27(3), 329–341.

Chen, L., Zou, L. J., & Tu, L. (2012). A clustering algorithm for multiple data streams based on spectral component similarity. *Information Sciences*, 183(1), 35–47.

Chen, S.-C., Ching, R., & Lin, Y.-S. (2004). An extended study of the k-means algorithm for data clustering and its applications. *Journal of the Operational Research Society*, 55, 976–987.

Dave, R. (1992). Generalized fuzzy C-shells clustering and detection of circular and elliptic boundaries. *Pattern Recognition*, 25, 713–722.

Dubes, R. (1983). Cluster analysis and related issues. In *Handbook of Pattern Recognition & Computer Vision*, eds. C. Chen, L. Pau, and P. Wang, World Scientific Publishing Co., Inc., River Edge, NJ, pp. 3–32.

Dunn, J. (1974). Well separated clusters and optimal fuzzy partitions. *Journal of Cybernetics*, 4, 95–104.

Gordon, A. (1999). *Classification*. Chapman and Hall, New York.

Halkidi, M., Batistakis, Y., & Vazirgiannis, M. (2002). Cluster validity methods: Part 1. *ACM SIGMOD Record*, Berlin, Germany, 31:2, 40–45.

Huang, J., Ng, M., Rong, H., & Li, Z. (2005). "Automated variable weighting in k-means type clustering. *IEEE Transactions on Pattern Analysis and Machine Intelligence*, 27:5, 657–668.

Jain, A. & Dubes, R. (1988). *Algorithms for Clustering Data*. Prentice-Hall Advanced Reference Series. Prentice-Hall, Inc., Upper Saddle River, NJ.

Jain, A., Murty, M., & Flynn, P. (1999). Data clustering: A review. *ACM Computing Surveys*, 31:3, 264–323.

Kaufman, L. & Rousseeuw, P. (1990). *Finding Groups in Data*, Wiley, New York.

Kim, Y.-S. (2007). Weighted order-dependent clustering and visualization of web navigation patterns. *Decision Support Systems*, 43:4, 1630–1645.

Kimani, S., Lodi, S., Catarci, T., Santucci, G., & Sartori Vidamine, C. (2004). A visual data mining environment. *Journal of Visual Languages and Computing*, 15, 37–67.

Kohonen, T. (1995). *Self-Organizing Maps*. Springer.

Kurgan, L. & Musilek, P. (2006). A survey of knowledge discovery and data mining process models. *The Knowledge Engineering Review*, 21:1, 1–24.

MacQueen, J. (1967). Some methods for classification and analysis of multivariate observations. In *Proceedings of the 5th Berkeley Symposium on Mathematical Statistics and Probability*, eds. L. M. Lecam & J. Neyman, University of California Press, Berkeley, California, USA, pp. 281–297.

Mansingh, G., Rao, L., Osei-Bryson, K. M., and Mills, A. (2015). Profiling Internet banking users: A knowledge discovery in data mining process model based approach. *Information Systems Frontiers*, 17:1, 193–215.

Mathers, W. & Choi, D. (2004). Cluster analysis of patients with ocular surface disease, blepharitis, and dry eye. *Archives of Ophthalmology*, 122, 1700–1704.

Mettler, T. (2013). Explorative clustering of clinical user profiles: A first step towards user-centered health information systems. *Proceedings of the 21st European Conference on Information Systems*.

Murtagh, F. (1983). A survey of recent advances in hierarchical clustering algorithms which use cluster centers. *Computer Journal*, 26, 354–359.

Okazaki, S. (2006). What do we know about mobile internet adopters? A cluster analysis. *Information & Management*, 43:2, 127–141.

Osei-Bryson, K.-M. (2005). Assessing cluster quality using multiple measures. *The Next Wave in Computing, Optimization, and Decision Technologies*, 371–384.

Osei-Bryson, K. M. (2010). Towards supporting expert evaluation of clustering results using a data mining process model. *Information Sciences*, 180(3), 414–431.

Osei-Bryson, K. M. & Inniss, T. R. (2007). A hybrid clustering algorithm. *Computers & Operations Research*, 34(11), 3255–3269.

Osei-Bryson, K. M. & Joseph, A. (2006). Applications of sequential set partitioning: A set of technical information systems problems. *Omega*, 34(5), 492–500.

Rai, A., Tang, X., Brown, P., & Keil, M. (2006). Assimilation patterns in the use of electronic procurement innovations: A cluster analysis. *Information & Management*, 43:3, 336–349.

Ramze Rezaee, M., Lelieveldt, B., & Reiber, J. (1998). A new cluster validity index for the fuzzy c-mean. *Pattern Recognition Letters*, 19, 237–246.

Samoilenko, S. & Osei-Bryson, K.-M. (2010). Determining sources of relative inefficiency in heterogeneous samples: Methodology using cluster analysis, DEA and neural networks. *European Journal of Operational Research*, 206:2, 479–487.

Sharma, S., Osei-Bryson, K. M., & Kasper, G. M. (2012). Evaluation of an integrated knowledge discovery and data mining process model. *Expert Systems with Applications*, 39(13), 11335–11348.

Sharma, S. & Osei-Bryson, K. M. (2010). Toward an integrated knowledge discovery and data mining process model. *The Knowledge Engineering Review*, 25, 49–67.

Shearer, C. (2000). The CRISP-DM methodology: The new blueprint for data mining. *Journal of Data Warehousing*, 5(4), 13–22.

Swobodzinski, M. & Jankowski, P. (2015). Evaluating user interaction with a web-based group decision support system: A comparison between two clustering methods. *Decision Support Systems*, 77, 148–157.

Tibshirani, R. & Walther, G. (2005). Cluster validation by prediction strength. *Journal of Computational and Graphical Statistics*, 14:5, 11–28.

Trivedi, N., Asamoah, D. A., & Doran, D. (2016). Keep the conversations going: Engagement-based customer segmentation on online social service platforms. *Information Systems Frontiers*, 1–19.

van Dam, J. W. & van de Velden, M. (2015). Online profiling and clustering of Facebook users. *Decision Support Systems*, 70, 60–72.

Wallace, L., Keil, M., & Rai, A. (2004). Understanding software project risk: A cluster analysis. *Information & Management*, 42, 115–155.

Ward, J. (1963). Hierarchical grouping to optimize an objective function. *Journal of the American Statistical Association*, 58, 236–244.

Chapter 7

Overview on Data Envelopment Analysis

The purpose of this chapter is to offer a general nontechnical overview of data envelopment analysis (DEA), while the comprehensive coverage of the subject of this chapter can be found in Charnes et al. (1994) and Cooper et al. (2004).

Introduction

Data envelopment analysis (DEA) is a *nonparametric* method of analysis—it does not require any assumptions regarding the data distribution, and it does not assume that the data distribution could be defined in terms of a finite number of parameters. Thus, unlike a parametric method, DEA is *context specific* in terms of the interpretation of the results of the analysis, which are restricted to the sample and should not be generalized beyond the sample.

The subject of the analysis of DEA is a set of decision-making units (DMUs), a group of entities that may be very different *physically* but *logically* are alike. Any collection of entities that operates with the same set of the inputs and produces the same set of outputs, be it a person, a firm, or a country, could be designated as a set of DMUs. As long as an investigator could identify a group of entities that are conceptually alike—as long as the entities could be meaningfully labeled the same way and described using the same set of attributes, inputs and outputs—an investigator could conduct DEA on that group. The concept is not entirely unlike that of a database table, which provides a common set of attributes for each record in that table.

The common set of inputs and outputs ensures a commonality of all DMUs in the sample and is referred to as a *DEA model*. The general guideline to creating a DEA model is that it has to make sense—the relationships between inputs and outputs could be complex, or unknown, but it should be possible to conceptualize them as a "resource–product" construct.

This method, which is nonparametric in the sense that DEA is entirely based on the observed input–output data, was originated as a collection of techniques for measuring the relative efficiency of a set of DMUs with unknown or unavailable price values for data inputs and outputs (Sengupta 1996). It is important to note that an input–output model of DEA is a *black-box model*—the method does not require an investigator to assume any functional relationships between inputs and outputs. For example, if we consider a group of students to be a set of DMUs, then we can describe them in terms of a simple input–output model where number of hours studied is an input and *test grade* is an output. Regardless of gender, age, classification, etc., as long as we can agree that the students are alike in terms of having the same types of inputs and outputs, we can subject the sample to DEA. All we need to know is the inputs and outputs of the DEA model—we do not have to know how the hours of studying actually translate into the test grades or whether, indeed, there is a proven relationship between the hours and the grades. In this sense, DEA answers the question of *what* (what is the efficiency of the conversion of DMUs inputs into outputs) and not the question of *how* (how the inputs are converted into outputs). Thus, the explanatory power of DEA is clearly limited.

DEA does not allow accounting for a uniqueness of DMUs—an inclusion of an entity with a unique input or output would provide an unfair advantage to that entity and penalize the other entities in the set. The general rule regarding the choice of inputs and outputs is intuitive and is based on the assumption that a DMU has a goal of minimization of the levels of inputs and maximization of the levels of outputs. For example, it makes sense to expect that a student would like to minimize the number of hours spent studying and would aim to maximize the grade for the test.

DEA Model: Common Guidelines and Assumptions

In choosing a DEA model, an investigator should consider a set of criteria that refer to the type, availability, number, and relative importance of the inputs and outputs comprising the model.

One of the common assumptions is that all DMUs in the sample come from a similar context and operate under the same conditions of availability of resources. This implies that no chosen input should greatly favor some of the DMUs (e.g., in terms of ease of access) over others, and no DMU should have a clear technological superiority over others (e.g., a construction crew using shovels should not be compared with a crew using excavators).

It would appear that a DEA model would benefit from having a large number of inputs and outputs—after all, it seems reasonable to expect that the greater number of variables describing DMUs would allow for a better representation. However, as the number of variables increases, the discriminatory power of DEA decreases (Dyson et al. 2001). Consequently, the choice of the inputs and outputs should be made judiciously. The general guideline regarding the size of the sample is to have at least as many DMUs as twice the product of inputs and outputs of the DEA model. Thus, an investigator should have at least 40 DMUs in the sample if the DEA model is composed of 4 inputs and 5 outputs.

Some variables comprising a DEA model could be more important than other variables. An investigator could address this by using a *weighting scheme*—assigning different weights to different inputs or outputs. For example, given a DEA model based on the framework of neoclassical growth accounting, a researcher may create a model consisting of *labor force* and *capital investment* as inputs and gross domestic product (*GDP*) as an output. Given the assumption of the product cycle theory that in the context of developed countries, capital is abundant and labor is scarce, DEA would allow for operationalizing this assumption via assignment of a greater weight to labor force than to capital investment. However, in the absence of a clear reason for a preferential treatment of one variable over another, an equal weighting scheme is commonly used as a default.

DEA: General Approach and Types of Efficiency

In 1978, Charnes, Cooper, and Rhodes introduced the original DEA model, now commonly referred to as the *CCR model* (for Charnes, Cooper and Rhodes). CCR allowed for representing multiple inputs and outputs of a DMU model as a single abstract *virtual* input and a single abstract *virtual* output. This allowed for expressing any DEA model in the form of an input–output ratio, which, in turn, allowed for using the value of the ratio to express the efficiency of each DMU in the sample. By comparing the value of the ratio of a DMU across the set, it is possible to gauge how efficient a DMU is relative to other DMUs in the set (hence the term *relative efficiency*).

The comparison of the values of relative efficiencies of DMUs in the sample yields a ranking of DMUs in the set, and the highest-ranking DMUs are assigned a score of 1 and considered to be 100% efficient. The process of comparison relies on the optimization method of *linear programming* (LP), and the type of LP used in the comparison impacts the interpretation of the scores that DMUs receive. In the case of *nonrelaxed LP*, the values of inputs and outputs are restricted to integers, while in the case of *relaxed LP*, the integer constraint is relaxed by allowing the inputs and outputs to take interval values.

If a DMU received a score of less than 1 in the case of nonrelaxed LP, then this automatically indicates that there are other DMUs that could generate the given level of outputs using lower levels of the inputs, or, conversely, could use the given level of the inputs to generate a higher level of the outputs. However, in the case of relaxed LP, a DMU with a score of less than 1 is not automatically considered to be inefficient—that DMU could still be considered *weakly efficient* if there is no other DMU that is better given the levels of every input or output of the model. The *weak efficiency* is signified by the presence of *nonzero slacks*—the presence of some unutilized amount of inputs or outputs.

Regardless of the type of LP used in comparison, the score of the relative efficiency does not have to be unique to a DMU; thus, a given set could have as many relatively efficient DMUs as there are DMUs with the perfect score. If we imagine all DMUs plotted on a two-dimensional graph, then the relatively efficient DMUs would be located at the boundary of the plotted data, and by connecting the boundary points representing the efficient DMUs, we can get an *efficiency frontier* that *envelops* the rest of the data (hence the name data *envelopment* analysis). At this point, one can easily contrast regression analysis, which aims to accommodate most of the data points, with DEA, which aims to identify the outlier points of the sample.

For all intents and purposes of DEA, a DMU is completely represented by its "size"—the set of values of the inputs and outputs comprising the DEA model. For example, if we consider basketball players to be DMUs with the input–output model *minutes played* and *points scored*, then the size of DMU1 (PlayerA) could be (7, 2), and the size of DMU2 (PlayerB) could be (35, 18). The size of a DMU has an important implication on its *scale efficiency*. If a DMU is efficient to the point where any changes in the values of the inputs and outputs will result in a decrease in efficiency, then that DMU is *scale efficient*.

The problem of comparing DMUs of different scales (it does not seem to be appropriate to compare an NBA rookie who plays 7 minutes per game with an NBA star who plays 35 minutes) is countered via a concept of *technical efficiency*. This concept is based on the assumption that no DMU in the sample is operating beyond its optimal capacity—basically, that all DMUs in the sample are of the same scale.

In those cases where inputs and outputs could be monetized, a concept of an overall, or *allocative*, efficiency could be applied. A measure of allocative efficiency, which is also referred to as *price efficiency*, might not correspond to the estimates of technical efficiency—a DMU that is technically efficient may not end up being price efficient. Illustrations are plentiful and intuitive—a baker A may be the most efficient relative to the other bakers in the set in baking M loaves over N hours, but the picture may change very quickly dependent on the salary of the baker and the price of the loaf.

DEA Models: Common Orientations

An orientation of a DEA model is determined by the scope of control—there could be a situation where inputs of a DMU are controllable, or where the outputs are controllable, or where control could be exercised over inputs and outputs simultaneously. To fit the three scenarios, DEA offers three options to an investigator—input orientation, output orientation, and base orientation.

An *input-oriented* DEA model is based on the assumption that the inputs are controllable and deals with the minimization of the levels of inputs required to obtain a given level of outputs. In the case of an input-oriented DEA model, a DMU is relatively efficient if two conditions hold: (1) it is not possible to lower the level of any of its inputs without impacting other inputs, and (2) it is not possible to lower the level of any of its inputs without decreasing the level of the outputs.

Let us consider the example of basketball players whose goal is to score 20 points in shortest period of time using the smallest number of attempts. The output is predefined—the number of required points is not under the control of the players, but the inputs are under control. This scenario fits an input-oriented DEA model. Clearly, the most relatively efficient player would be one who has a combination of minimum attempts *and* minutes, and not the one who, let us say, minimizes number of attempts at the expense of minutes played.

An *output-oriented* DEA model, on the other hand, deals with the efficiency of the production of outputs under the condition of controllable outputs and is concerned with maximization of the level of the outputs per given level of the inputs. In the case of an output-oriented DEA model, a DMU is considered efficient if it is not possible to increase the level of any of its outputs without affecting other outputs or increasing the level of any of its inputs. Let us consider the example of investigating relative efficiencies of basketball teams that are given a game to play. Under this scenario, the inputs might be considered fixed—a team at any time is limited to the same number of players on the court, as well as to the number of minutes that the team could play. This example lends itself well to being investigated via an output-oriented model, for only the outputs of the model—the number of points scored—could be controlled by each team, while the values of inputs remain unchanged.

A *base-oriented* model has a dual orientation—it assumes the presence of control over inputs *and* outputs within the model and is concerned with the optimization of the balance of the inputs and outputs. Consequently, a base-oriented DEA model deals with the efficiency of the input utilization and the output production. Let us consider the example of comparing coaches of National Basketball Association (NBA) teams in terms of their relative base-oriented efficiency. Based on this scenario, every coach could control the inputs (the number of team members playing and the number of minutes the players spend on the court) as well as the outputs (average points scored in the game by each player are known). This sort of situation will be best analyzed via a base-oriented DEA model, for inputs and outputs are not fixed. It is easy to see that the most efficient coach is not the one who won most of the games or the one that used all of his stars on the court every game, but the one who won most of the games using his best players the least.

DEA: Types of Models

There are four commonly encountered DEA models: CCR, BCC, additive, and multiplicative.

As was mentioned earlier, CCR was the original DEA model that was introduced in 1978 by Charnes, Cooper, and Rhodes. The most appropriate context for using CCR is when there is a reason to believe that DMUs in the sample operate under the condition of *constant return to scale*—when a change in the levels of inputs results in a proportional change in the outputs, such as doubling a number of machines resulting in doubling the amount of widgets produced. The condition of constant returns to scale has an implication on the shape of the efficiency frontier, which would be in the form of a single straight line. While the CCR model works well under the assumption of constant returns to scale, it does not allow for dealing with the contexts of increasing or decreasing returns to scale—where the increase in the inputs results, correspondingly, in greater or lesser proportional increase in the outputs.

The *BCC* (stands for *B*anker, *C*harnes, and *C*ooper) model distinguishes between technical and scale inefficiencies by estimating pure technical efficiency at the given scale of operation and identifying whether the possibilities of increasing, decreasing, or constant returns to scale are present. The main difference between the original model and the BCC model is that the latter is more flexible because it allows for considering the variable returns to scale. This, however, plays an important role in determining what DMUs are considered to be relatively efficient within both models.

The BCC model is less strict than the CCR—a DMU is relatively efficient if it is scale efficient and technically efficient in the case of CCR, while the same DMU would be considered relatively efficient if it is only technically efficient in the case of BCC. Consequently, if a DMU is relatively efficient within the CCR model, then that DMU is also efficient within the BCC model, but not vice versa. Additionally, unlike a straight-line efficient frontier of CCR, the enveloping surface produced by the BCC model would be formed by the segments connecting the outlying relatively efficient DMUs.

Similarly to the CCR model, the relative efficiency of a DMU within the BCC model is dependent on the zero slacks (e.g., no unused inputs and all possible outputs produced) and the value of the output–input ratio being equal to 1. This means two things: first, all inefficiencies are due to the presence of the slacks and the value of output–input ratio being other than 1, and second, for both models, a relatively efficient DMU is efficient regardless of the orientation of the model.

While the purpose of CCR and BCC models is concerned with the problems of minimization of the inputs *or* maximization of the outputs, the purpose of the *additive* DEA model is to estimate the relative efficiency of DMUs by means of *simultaneous* minimization of the inputs and maximization of the outputs. Also, unlike CCR and BCC models that rely on proportional minimization of the inputs or maximization of the outputs, additive DEA models allow individual inputs and outputs to change at different rates.

Unlike the DEA models that rely on addition to aggregate the inputs and the outputs into the ratio to calculate the scores of the relative efficiency, *multiplicative* DEA models, rarely used in practice, rely on multiplication. This approach results in a log-linear envelopment of the efficiency frontier, where the shape of the enveloped surface is not uniform (e.g., straight line of CCR or convex surface of BCC) but may consist of concave and convex regions. One of the benefits of using multiplicative models is that, unlike the distinct options of constant, decreasing, or increasing returns to scale relied on by CCR and BCC models, multiplicative models allow for more precise estimation of the return to scale.

DEA: Malmquist Index

While the investigator could use DEA to evaluate the relative efficiency of a DMU, he/she also may be interested in *changes* in the relative efficiency of that DMU over time. The score of a DMU could remain unchanged, decrease, or increase over time—and this would indicate no changes, negative changes, or positive changes in *productivity* relative to other DMUs in the sample. Under the assumption of constant return to scale, the link is fairly simple and straightforward—positive changes in efficiency indicate improvements in productivity, and negative changes indicate decline. A Malmquist index (MI) constructed via DEA offers an investigator a valuable tool for identifying changes in relative efficiency of the DMUs over time.

Within the framework of neoclassical growth accounting, for example, economic growth is explained by two factors. The first factor is an increase in capital and labor, which is subject to the law of diminishing return. The second factor is a growth in productivity, which is not subject to the law of diminishing returns and is thus preferable because it allows for attaining the sustained economic growth. The growth in productivity could be assessed via the use of MI. Essentially, the approach relies on performing DEA at two points in time and analyzing the difference that took place.

For any DMU, the period between two points in time is represented by the distance that the DMU traveled between these points, let us say, Year1 and Year2. That traveled distance is represented by the MI and is reflective of changes in the DMUs' productivity. There are two important changes that take place over a period of time and comprise the overall change. The first change is a movement of the efficiency frontier itself—the change in the position of the enveloping surface formed by the relatively efficient DMUs. The second change is a movement of a DMU relative to the efficiency frontier.

Let us consider the example of a train departing from the station and a passenger sitting at the back of the train. As time passes and the train reaches its destination, the distance travelled by the train is representative of the movement of the efficiency frontier. However, during this time, the passenger could also decide to move within the train—let us say to move from the last car to the middle of the train. The movement of the passenger is reflective of the changing position of the passenger relative to the efficiency frontier.

Thus, a change in the position of each DMU could be perceived as consisting of two components that, taken together, comprise the value of MI. The first component, the change in distance of each DMU relative to the efficient frontier, reflects the *changes in efficiency* (EC) of a DMU. The second component, the change in position of the efficient frontier itself, reflects the *changes in technology* (TC) that took place over a period of time.

Conceptually, therefore, the mechanism of estimating changes in productivity of a DMU using DEA is intuitive—the position of a DMU changes over time and is measured by means of MI. The change in the position of a DMU, and the corresponding value of MI, is comprised of two components—EC and TC. There are three possible scenarios regarding the changes in productivity—the value of MI could be greater than, equal to, or less than 1. A value of MI equal to 1 indicates no change in productivity, while a value of greater than 1 or less than 1 reflects, respectively, a growth or a decline in productivity.

Estimation of the *sources* of changes in productivity, however, is more complex precisely due to MI being a composite index comprised of EC and TC components. Let us consider some of the options based on the comparison of a group of DMUs—lumberjacks using chainsaws to cut trees. Let us consider a scenario of MI > 1 for lumberjack Bob, which is indicative of Bob's growth in productivity:

1. EC > 1 and TC > 1. Bob's growth in productivity is balanced, an indication that his efficiency as well as his chainsaw are superior.
2. EC > 1 and TC = 1. Bob's growth in productivity is due to him becoming more efficient, while using a chainsaw adequate to keep up with the work.
3. EC > 1 and TC < 1. Bob is becoming more productive due to improvements in his efficiency, while his chainsaw is keeping him back.
4. EC = 1 and TC > 1. Bob did not change his own level of efficiency, but his chainsaw has improved significantly.
5. EC < 1 and TC > 1. Bob's efficiency has decreased, but overall, the productivity growth was obtained due to superior technology—his chainsaw.

Similar scenarios could be developed for cases of MI = 1, and MI < 1. The importance of interpreting sources of growth in productivity via MI and its two components has a practical significance, for it allows for a precise allocation of the resources where they are needed the most. For example, in scenario 5, it is not wise to give Bob an even better chainsaw than the one he has now; instead, it would make sense to spend some money on improving his skills.

References

Charnes, A., Cooper, W. W., Lewin, A. Y., & Seiford, L. M. (1994). *Data Envelopment Analysis: Theory, Methodology and Applications*. Kluwer Academic Publishers, Norwell, MA.

Cooper, W. W., Seiford, L. M., & Zhu, J. (2004). Data envelopment analysis: History, models and interpretations. In *Handbook on Data Envelopment Analysis*, eds. W. W. Cooper, L. M. Seiford, & J. Zhu, Chapter 1, Kluwer Academic Publishers, Boston, pp. 1–39.

Dyson, R. G., Allen, R., Camanho, A. S., Podinovski, V. V., Sarrico, C. S., & Shale, E. A. (2001). Pitfalls and protocols in DEA. *European Journal of Operational Research*, 132(2), 245–259.

Sengupta, J. K. (1996). Data Envelopment analysis: A new tool for improving managerial efficiency. *International Journal of Systems Science*, 27(12), 1205–1210.

Chapter 8

Overview on Structural Equation Modeling

The purpose of this chapter is to offer a general, necessarily brief and nontechnical, overview of structural equation modeling (SEM)—a methodology representing the second generation of multivariate analysis. Unlike the statistical tools of the first generation, exemplified by such techniques as cluster analysis, multivariate regression, principal component analysis (PCA), and others, SEM allows for answering multiple interrelated research questions within a single analysis. Use of SEM allows researchers to posit the presence of the relationships between the multiple unobserved, or *latent*, variables, where every latent variable is associated with multiple observed variables, often called *indicators* or *measures*. The corresponding basic structure of SEM is illustrated by Figure 8.1.

Let us consider a very simplistic example—suppose we are interested in investigating whether the people who like their jobs have a healthier lifestyle than those who do not like their jobs. We can start by positing the existence of two concepts—*job satisfaction* and *healthy lifestyle*. These two concepts are latent variables in the realm of SEM—they are constructs that make intuitive sense but could not be directly observed. However, this is not a problem, because we can represent our two unobserved constructs via some observed variables that we can actually measure. We can chose to represent job satisfaction via three measures—*length of workweek, duration of commute to work*, and *pay amount*, and we can represent healthy lifestyle via measures *hours of exercise per week, amount spent on healthy food*, and *amount spent on fitness equipment*. This gives us an SEM model with two latent variables and six measures—the hypothesized relationship between the unobserved constructs is referred to as a *structural model*, and the representation of each of the constructs by the indicators is referred to as a *measurement model*.

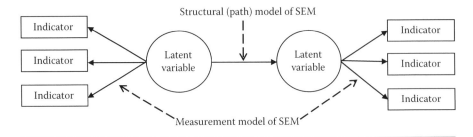

Figure 8.1 Basic structure of SEM.

Within our structural model, job satisfaction could also be referred to as an *exogenous* variable—within our model, this construct is not influenced by any other constructs. However, because we hypothesized that healthy lifestyle is influenced by job satisfaction, healthy lifestyle would be referred to as an *endogenous* variable. Simply put, exogenous variables are independent variables, endogenous variables are dependent variables in the SEM model, and the direction of an arrow connecting the two indicates a *path*—the direction of causality in the model. Paths depicted by double-headed arrows indicate correlations without a causal interpretation.

In the case of having three or more latent variables, models could contain examples of *mediation* and *moderation*. To illustrate these concepts, we introduced two latent constructs—*work social environment* and *family environment*. We do not claim any validity of the model depicted in Figure 8.2—it simply allows us to depict both concepts within the same model.

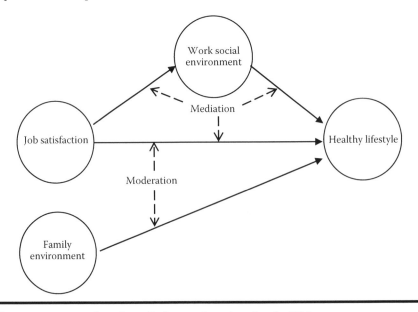

Figure 8.2 Examples of mediation and moderation in SEM.

Moderation is depicted by the presence of a variable (or variables) that impacts the relationship between the other two constructs. For example, despite family environment not being causally associated with job satisfaction, its presence will impact the strength of the relationship between job satisfaction and healthy lifestyle. Consequently, if the strength of the relationship between job satisfaction and healthy lifestyle is impacted by the level of family environment, we can state the presence of a moderating effect.

Mediation refers to a model reflecting the presence of a direct effect between exogenous and endogenous variables, as well as indirect effects between an exogenous and a *mediator* variable, and between a mediator and endogenous variables. In the previous model, latent variable work social environment serves a role of a mediator variable, and the impact of the presence of the mediator variable on the strength of the direct relationship between job satisfaction and healthy lifestyle is referred to as a *mediational effect*.

SEM: Reflective and Formative Measurement Models

We start our overview of the measurement model of SEM by discussing the differences between *reflective* and *formative* measurement models (see Figure 8.3 for an illustration). The most obvious visual difference is that of diagrammatical representations. The reflective measurement model has arrows directed from a latent construct to the measures, while the formative measurement model would have the directions of the arrows reversed—pointing from the measures to the latent constructs.

In the case of reflective models, the assumption is that the changes in the latent variable are *reflected* by the changes in indicators that represent the variable. In the case of our previous example, we represented job satisfaction via three measures length of workweek, duration of commute to work, and pay amount. If the model is reflective, then the changes in a person's job satisfaction will be reflected by the changes in the values of the correlated observable measures—the changes in the

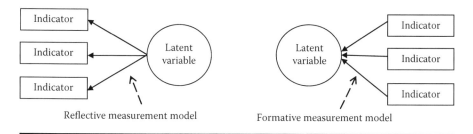

Figure 8.3 Examples of representation of formative and reflective measurement models.

latent construct would cause changes in the indicators. The important point to note is that in the case of a reflective measurement model, the indicators representing the latent construct are supposed to be correlated. However, in the case of our example, length of workweek, duration of commute to work, and pay amount may or may not be intercorrelated. We can drop the assumption of correlation between the measures and consider that instead of length of workweek, duration of commute to work, and pay amount *reflecting* job satisfaction, the indicators are *forming* the latent construct. This would allow us to construct a formative measurement model without any assumptions regarding the patterns of intercorrelation between the measures. In the case of the formative model, changes in the values of indicators would cause changes in the latent construct.

There are some conceptual considerations for choosing the appropriate type of measurement model. For example, in the case of a reflective model, we can chose a number of indicators that reflect the construct, and we can replace those indicators with appropriate substitutes without impacting the conceptual nature of the latent construct. In the case of a formative model, however, adding, removing, replacing a measure would impact the conceptual representation of the latent variable. Simply put, latent variables of reflective models are stable—we posit their independent existence regardless of the measures used, whether latent variables of formative models are dynamic and existence-dependent on the measures that form them.

SEM: Model Specification

Given a large data set comprising many variables (and SEM is a large-sample technique), it is often not easy to understand the meaning that the data set carries. SEM analysis is one of the ways of assigning a meaning to a data set that could be perceived as placing a predefined template of constructs and their relationships over the actual data set. That template is an SEM model, and one of the goals of SEM is to establish a correspondence between two models—the true model contained in the data and the specified a priori SEM model.

Model specification of SEM is the process associated with identifying all the relevant constructs and their relationships. If the model is specified on the basis of an existing theory or a framework, then SEM analysis is *confirmatory* in nature. If the actual, true model of the variables and their relationships in the data set conforms to the hypothesized model, then the SEM model is properly specified. However, if the relationships between the variables in the data set are not in accordance with the proposed model, then the SEM model is *misspecified*. The difference between a properly specified and a misspecified SEM model is caused by a *specification error*—due to leaving out important variables or due to including unimportant variables, or both.

If the original SEM model is misspecified, then, given the presence of the theoretical support for other options, the alternative model or models could be used.

Finally, the last resort in model specification is via *model discovery*—the process of altering the original SEM model to fit the data. It must be noted, however, that the statistical correspondence of the model to the data is not a sufficient condition—the model has to make theoretical sense. Thus, a misspecified model that is theoretically sound is not acceptable, and neither is a properly specified model that has no theoretical support—both conditions must be present.

After the creation of the research model by an investigator, the process of SEM could be perceived as consisting of two parts. The first part involves testing of the measurement model and primarily deals with the validation of the latent constructs included in the model. Once validity of the measurement model has been established, testing of the structural model, the second part of SEM, takes place. Assessment of the structural model involves testing of the hypothesized relationships between the latent constructs of the research model. The results of the assessment are based on the significance of the structural paths, which could be estimated by using different methods, such as general least squares (GLS), ordinary least squares (OLS), maximum simulated likelihood estimation (MSL), partial least squares (PLS), and others. Taken together, the structural and the measurement models comprise the complete structural equation model.

SEM: Two Common Approaches

There are two common approaches to SEM, covariance based and variance based. It is considered that these two approaches differ with regard to two criteria. The first criterion is the strength of the assumption of the normality of data distribution, where a covariance-based approach relies on a more rigid assumption of normality than a variance-based approach does. There also seems to be a consensus that a covariance-based approach requires a larger data set than a variance-based method. The second criterion refers to the strength of the theory underlying a path model of SEM. A covariance-based approach is preferred in the case of a strong existing theory, where researchers aim for further development and refinement of a framework. A variance-based approach, on the other hand, is considered to be more suitable for the cases of theory building.

Simply put, the two approaches differ with regard to what is being fit to what—covariance-based SEM aims to fit the data to the fairly strict hypothesized model, while variance-based SEM attempts to fit a malleable structural model to the true model represented by the sample. We now take a look at the two methods in more detail.

The covariance-based approach is based on the objective of minimization of the difference between the covariance matrix of the sample and the covariance matrix of the model. Thus, this approach is also commonly called *factor based*, for the goal is to maximize the fit of the model by means of minimizing the *unique* variance. It is because of this goal of optimization of the fit that the covariance-based approach is considered to be more suitable for situations where the investigator operates under the guidance of a strong and well-developed theory.

In contrast to the covariance-based approach that attempts to optimize fit, the variance-based approach attempts to optimize predictive capability of the research model relative to the sample. The optimization of the prediction is achieved by estimating, as close as possible, the parameters of the model by means of the minimization of the residual variances of the variables in the model. Thus, this method is commonly referred to as *component based*, for the approach assumes that all the measured variance is useful for explanatory purposes.

SEM: Preliminary Data Analysis and Factor Analysis

The first phase of SEM is referred to as testing of the measurement model and is concerned with validating the latent constructs. The process of validation involves assessing how well the latent constructs are reflected by their indicators. The second phase of SEM deals with testing of the structural model and entails assessing the statistical significance of the paths between the latent constructs, where each path is represented by a null hypothesis.

Factor analysis (FA) is a tool that is commonly used in testing of the measurement model. However, before an investigator can conduct an FA in order to determine if the constructs and their measures demonstrate a specific pattern of loadings and align in the same direction and the measures associated with a given latent construct load together on the same factor, some preliminary analysis must be performed.

The purpose of the preliminary data analysis is to determine whether the data set could be factor-analyzed—whether the number of variables in the set could be reduced to a smaller number of factors that would still account for a large amount of variance in the data set. There are two commonly used tests to determine whether the data set is suitable to be factor-analyzed—the *Kaiser–Meyer–Olkin (KMO) test of sampling adequacy* and the *Bartlett test of sphericity*. The results of the tests are commonly reported together, for the both tests analyze the data set from different but complementary perspectives—the Bartlett test looks at whether the number of variables could be reduced to a smaller number of factors, and KMO looks at whether the extracted factors account for a significant amount of variance in the data set.

The Bartlett test is used to assess whether all the variables in the data set are *orthogonal* to each other—this would indicate that we have as many factors as there are variables in the set. The assessment is based on the test of hypothesis that compares the actual correlation matrix of the data set with the *identity* matrix—a theoretical matrix in which all of the diagonal elements are 1 and all off-diagonal elements are close to 0. If the result of the Bartlett test allows an investigator to reject the null hypothesis at the 5% level or better, then the data set is considered to be suitable for FA.

The KMO test assesses the degree to which the variables in the model measure common factors—the test estimates the proportion of a common variance shared by the variables. The lower the proportion, the better suited the data to be factor-analyzed. The scale of the KMO test is from 0 to 1, where values above 0.6 are commonly considered to indicate the sampling adequacy of the set. Simply put, the higher the score of KMO, the greater the amount of variance in the data set that is accounted for by extracted factors.

Once the preliminary analysis of the data set has been concluded, with the results of the KMO and Bartlett test indicating that the data set is suitable to be factor-analyzed, the investigator may begin conducting FA, which is instrumental for the purposes of the assessment of the measurement model.

Fundamentally, FA is a dimension-reduction technique. Given a set of variables in the data set, FA allows an investigator to determine if the number of the variables (or *items*) could be reduced to a smaller number of factors, such that multiple variables *load* on the same factors. *Factor loading* is a correlation between the extracted factor and a variable. For example, given a data set consisting of 15 variables, FA may allow for reducing the number of variables to, let us say, 5 factors—common dimensions that are reflected by the variables. Factors of FA correspond to latent variables of SEM, and the variables/items of FA are the measures/indicators of SEM. There are multiple methods allowing for extracting factors from the data set, with PCA, the maximum likelihood (ML) method, and principal (also called common) factor analysis (PFA) being the most common.

It is expected that the variables representing the factor would have high loadings on that factor and low loadings on all other factors in the model. In order to maximize high and minimize low item loadings, the *rotated component matrix* is commonly used—this allows for simpler solutions that are easier to interpret. There are two common rotation techniques—*orthogonal* and *oblique*. Orthogonal rotation results in uncorrelated factors, while oblique rotation yields correlated factors.

There are two types of FA—exploratory factor analysis (EFA) and confirmatory factor analysis (CFA). The basic difference between the two approaches is their purpose. EFA aims to explore whether the number of items (variables) in the data set could be reduced to a smaller number of meaningful factors—latent constructs. In this sense, EFA is a data analytic tool allowing for discovering of common themes in the data set. While EFA is not performed with the preconceived structure to be discovered in mind, an investigator may interpret the extracted factors in the light of a theory or a framework. CFA, on the other hand, is driven by the goal of testing a hypothesis of the existence of relationships between the items and the factors that is in accordance with the established framework or a theory.

Thus, given a well-structured and a theoretically sound questionnaire that aims to assess a number of factors, where each factor is assessed by multiple items, an investigator would apply CFA to confirm that the structure of the questionnaire is indeed in accordance with the theoretical model that was used to create it. But, if given a large number of survey questions, items that do not necessarily correspond to specific factors, an investigator would chose EFA to attempt to identify if the items of the questionnaire asses some common themes—factors. Extraction of the factors from the data set allows an investigator to proceed with the assessment of the measurement model of SEM.

SEM: Assessment of the Measurement Model

The process of evaluating the adequacy of the measurement model is a three-step process that involves assessing the unidimensionality, validity, and reliability of the model.

The purpose of the assessment of *unidimensionality* of the model is to make sure that items have acceptably high loadings on their extracted factors—constructs. Loadings above 0.6 for the established items and above 0.5 for the new items are considered to be acceptable. The items with loadings below the baseline are eliminated, and the data set is factor-analyzed again. This process is repeated until the unidimensionality of the model is achieved—all the items exhibit acceptably high positive loadings on their constructs.

The general purpose of the validity assessment is to test whether the items do indeed measure what they are supposed to measure for their factors and represent what they are supposed to represent for their constructs. The assessment of validity of the measurement model involves assessing convergent, discriminant, and construct validity.

Construct validity refers to the fitness of the measures of the constructs to accurately measure their target. Construct validity could be assessed through the measures of convergent and discriminant validity, which examine the degree to which measures of a latent construct converge (share a proportion of their variance) and are distinct (how the measures of a construct are different from others).

Convergent validity assessment tests whether the items in the measurement model are statistically significant. The evaluation of the measure of internal consistency is commonly used for assessing convergent validity of the measures. The evaluation process involves assessment of the magnitude of the loadings of each of the individual items as well as assessment of the loadings of the measures on their own constructs. It is expected that the measures representing their constructs exhibit high loadings on that construct and low loadings on the other constructs in the model. Additionally, the average variance extracted (AVE) and composite reliability (CR) of the construct could be used to assess the convergent validity—the value of AVE should be above 0.5 and the value of CR above 0.7.

Discriminant validity assessment tests how well individual items represent their own construct and how distinct they are from other items in the model.

Clearly, any presence of *cross-loadings* (high loadings of an item on multiple constructs) will automatically indicate a discriminant validity–related problem. The common assessment approach involves comparison of the squared correlation between two constructs with their individual AVE estimates—if the values of AVE are greater than the values of squared correlations, the models has passed the test of discriminant validity.

Validity assessment allows us to test the accuracy of the measures in the model; however, the measures could be accurate (valid) but not consistent (not reliable). Reliability assessment tests the extent of consistency with which the items of the measurement model measure their latent constructs. Internal reliability, CR, and AVE are criteria that could be used in assessing the reliability of measurement model.

Test of *internal reliability* (internal consistency) assesses the consistency of the items in measuring their construct—we would expect that the measures representing their latent construct would have similar values. Cronbach alpha is commonly used to assess internal consistency of a model, with values above 0.7 being generally acceptable.

CR assessment tests the reliability, as well as internal consistency, of a construct. Values of CR of 0.6 or higher are considered to be good indicators of CR of a construct.

Finally, AVE indicates the average percentage of the variance of a construct that is explained by its individual measures. A value of AVE above 0.5 is required for every construct in the model to pass the test of reliability of the measurement model.

The successful evaluation of the adequacy of the measurement model allows an investigator to proceed further with the assessment of the structural model.

SEM: Assessment of the Structural Model

The test of the significance of the hypothesized relationships between the constructs of the research model requires assessing the paths of the structural model. However, the process of the assessment of the structural model differs between covariance-based (or, factor-based) and variance-based (or, component-based) approaches.

The factor-based approach commonly relies on ML estimation (e.g., LISREL- LInear Structural RELations), which aims to determine the extent to which the hypothesized path model is consistent with the true model of the data set. Fundamentally, the assessment is based on the comparison of the covariance matrix of the structural model with the covariance matrix of the actual data set. Because all of the model's parameter estimates are calculated simultaneously, the ML method is referred to as a *full information method*.

Thus, because the ML estimation of the measurement and the structural model in a factor-based approach is done simultaneously, there is no real separation of the assessment into two parts. The fit of the structural model is usually evaluated by a combination of the results of the chi-square test (comparison of the actual distribution of the data with the hypothesized one) and one or more goodness-of-fit statistics offered by any SEM software.

The component-based approach employs the least-squares (LS) method (e.g., PLS) to test the validity of the structural model. Unlike the *full model variable* estimation of the ML approach, LS is based on a *block variable* estimation approach, where each block corresponds to a group of measures representing their latent variable. During the first step of estimation, test of the measurement model, the relationship between measures and a construct is assessed, and in the second step, test of the structural model, the relationships between the latent variables in the structural model are assessed.

Basically, this approach aims to assess how well the causal structure embedded in the path model works when applied to the data set. Running the analysis of the structural model yields the path coefficients between the constructs in the model. The significance of the path coefficients is then evaluated by running a bootstrap or jackknife procedure to estimate standard errors. Once t values for each path have been obtained, the significance level of each path is established using a two-tailed t-distribution table.

As we can see, these two approaches differ in their subject of the assessment of estimation accuracy—while ML evaluates the accuracy of the parameter estimates within the path model, LS assesses the accuracy of the prediction of the dependent (endogenous) variables by the independent (exogenous) variables in the path model.

Regardless of the approach used, one of the common points of critique of SEM is associated with the causal structure depicted by the path model. An investigator should always keep in mind that the path model represents correlational data, and any interpretation of the presence of correlation in terms of causality should be made with care, even if supported by the data.

Chapter 9

Overview on Artificial Neural Networks

The chapter provides an overview of artificial neural networks, with a focus on directed (or unsupervised) learning problems such as classification or value estimation. Its main purpose is to introduce the reader to the major concepts underlying this data mining technique, particularly those that are relevant to Chapters 17, 20, and 21.

Introduction

Artificial neural network (NN) induction is a popular data mining modeling group of techniques that are being increasingly used in research in information systems and other business disciplines (e.g., Kwon et al. 2016; Wang & Chuang 2016; Colomo-Palacios et al. 2014; Samoilenko & Osei-Bryson 2010, 2013; Chan & Chong 2012; Hájek 2011; Khansa & Liginlal 2011; Ragothaman & Lavin 2008). An NN is primarily used in the modeling of unknown complex relationships in data.

In this overview, we will focus on NNs for directed-learning problems, though it should be noted that NNs can also be used for nondirected (i.e., unsupervised) learning activities such as clustering (e.g., Kohonen 1995). For a more comprehensive exposition on NNs, the reader may consult texts such as Bishop (1995), Hand et al. (2001), or Giudici (2005).

An NN consists of a set of nodes ("neurons") and links such that nodes are grouped into layers (i.e., input, output, and optionally, hidden):

- *Input layer*: This layer receives values (say x_i); it thus represents the input variables of the given data set, where a numeric or binary variable would be represented by a single input node, while a categorical variable would be represented by multiple nodes each corresponding to a possible value of the given variable.
- *Output layer*: This layer communicates the values of the output variables (say y_k) of the system to the environment. It should be noted that multiple output variables of the given data set can be represented in this layer.
- *Hidden layers*: There may also be one or more hidden layers. Each hidden layer node receives input values (say x_j) from input or hidden nodes and communicates its output (say y_j) to either hidden or output nodes.
- *Links*: The link (s,t) from node i to node j has a weight w_{st}.

An example of a common architecture of an NN is depicted in Figure 9.1, which could be represented as consisting of the three layers: input, hidden, and output layers, where each layer consists of one or more nodes. This type of architecture is referred to as a *multilayer perceptron* (MLP).

For directed-learning problems, training the NN involves adjusting the weights so that the difference between the target output and the output predicted by the NN is minimized. Within this directed-learning context, an NN can be considered

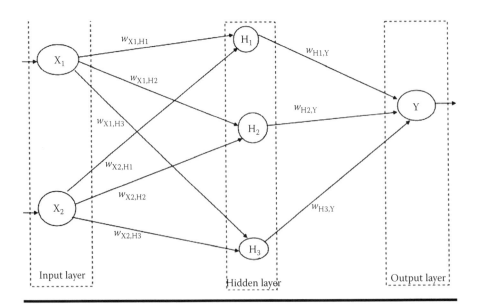

Figure 9.1 Example of the architecture of an artificial neural network.

to be similar to least-squares regression and can be viewed as an alternative statistical approach to solving the least-squares problem by minimizing the sum of squared errors where the set of *input neurons* is equal to the set of independent variables, while the *output neuron*(s) represents the dependent (i.e., target) variable(s). A linear regression model may be viewed as a feed-forward NN with no hidden layers and one output neuron with a linear transfer function where the weights connecting the input neurons to the single output neuron are analogous to the *coefficients* in a linear least-squares regression. NNs with one hidden layer resemble nonlinear regression models where the weights represent regression curve parameters. Within this context, training of an NN can be considered to have some similarity to the simultaneous solution of multiple interrelated regression models in order to generate the weights (or coefficients) that will minimize the relevant error function.

NNs are particularly useful for prediction problems where

- No mathematical formula is known that relates inputs to outputs
- Providing high-quality (e.g., accuracy) classification, estimation, prediction, or clustering is more important than providing an explanation
- There is lots of training data

NN Induction Process for Directed Learning

The NN induction process for directed learning (i.e., supervised learning) is an iterative process that involves the following steps:

1. Selecting an initial set of weights for the set of links between the nodes. This may be done using some random generation method.
2. Based on the current values of the input layer nodes and current set of weights, use a *forward-propagation* approach to generate input and output values of each node of the hidden and outer layers. For an NN with a single hidden layer, this begins with computation of the input value for each hidden layer node, followed by computation of the output for each hidden layer node, followed by computation of the input value for each outer layer node, followed by computation of the output for each outer layer node. The current error of each node of the output layer is calculated based on the difference between its actual and computed output values.
3. Determine if the conditions of the *termination test* are satisfied, and if they are, then terminate. Typically, this step includes calculating the current *total error* of all of the nodes of the output layer and comparing the differences between the weight vectors of the current and previous iteration.
4. Use a *backpropagation* approach to adjust the current set of weights in a manner that aims to further reduce the total error. A popular weight adjustment method involves a gradient descent approach. For an NN with a single hidden

layer, this begins with adjustment of the weights of relevant links between nodes of the hidden and output layers, followed by adjustment of the weights of relevant links between nodes of the input and hidden layers.
5. Repeat steps 2–5.

Typical termination conditions are as follows:

- For all weights w_{st}, the difference between the old and new values $\left|\left(w_{st}^{New} - w_{st}^{Old}\right)\right|$ is less than some specified threshold
- Error (e.g., misclassification rate, average squared error) is less than some specified threshold
- The prespecified number of iterations have been completed.

The NN Induction process for directed learning may be done in an *incremental mode* or a *batch mode*. In the incremental mode, steps 2–5 are done for each record (i.e., instance or observation) of the *training* data set. In the batch mode, step 2 is done for each input record, and steps 3–5 are done based on the entire batch of records of the training data set.

To avoid overfitting of the model to the training data, for large data sets, the generation of an NN includes partitioning the model data set into either two parts (i.e., *training* and *validation*) or three parts (i.e., *training*, *validation*, and *test*).

Computing the Inputs to the Nodes of the Hidden and Output Layers

Computation of the input value for each node of the hidden and output layers involves combining the output values from predecessor nodes into a single value using a specified *combination function*.

- Linear combination functions: Compute a linear combination of the weights of the arcs and the output values of the nodes feeding into the node, then add the bias value (the bias b_j acts like the intercept of a regression equation). Given s, a node of the output layer or a hidden layer, let In(s) be its set of predecessor nodes [i.e., there exists a link (r,s)], and then using a linear combination function, the input value of node k is $x_s = b_s + \Sigma_{r \in In(s)} w_{rs} y_r$, where y_r is the output value of node r.
- Radial combination functions: Compute the squared Euclidean distance between the vector of link weights and the vector of values feeding into the node and then multiply by the squared bias value (the bias acts as a scale factor or inverse width).

Table 9.1 Examples of Activation (or Transfer) Functions

Step	$y_r = 1$ if $x_r > T$; $y_r = 0$ if $x_r \leq T$, where T is the threshold value
Signum	$y_r = 1$ if $x_r > 0$; $y_r = 0$ if $x_r = 0$; $y_r = -1$ if $x_r < 0$
Logistic or sigmoid	$y_r = 1/(1+e^{-\lambda x_r})$, where the value of λ is specified by the NN designer
Hyperbolic tangent	$y_r = (e^{\lambda x_r} - e^{-\lambda x_r})/(e^{\lambda x_r} + e^{-\lambda x_r})$
Linear function	$y_r = \lambda x_r$
Threshold linear function	$y_r = (x_r - T)$ if $x_r > T$; $y_r = 0$ if $x_r \leq T$

Computing Outputs of the Nodes of the Hidden and Output Layers

For each node of the hidden layer or an output layer, the output value of the node is as follows:

- $y_s = g()$, where $g()$ is the transfer or activation function (Table 9.1)
- For a linear combination function, we could have

$$y_s = g(x_s) = g\left(b_s + \sum_{r \in In(s)} w_{rs} y_r\right)$$

Network Architectures

There are different architecture types for NNs, some of which are described in Table 9.2.

Weight Adjustment for the Backpropagation Algorithm for Multilayer Perceptron

The backpropagation algorithm makes changes to the weights by working backward from the output layer to the input layer. This is because the sensitivity of a node in layer $(l - 1)$ depends on the sensitivities of all the nodes in layer l. Several rules have been proposed for weight adjustments such as the Hebb, delta, and generalized delta rules for directed learning (see Table 9.3a), and the Kohonen rule for undirected learning. For the MLP, the activation function is a sigmoid function, so the Generalized Delta rule is applicable. Table 9.3b provides the corresponding weight adjustment formulas for the MLP.

Table 9.2 Some Types of Network Architectures

Multilayer perceptron (MLP)	• *Input layer*: any number of nodes • *Hidden layers*: one or more hidden layers with any number of nodes • *Combination functions*: uses linear combination functions in the hidden and output layers • *Transfer (or activation) function*: uses sigmoid activation functions in the hidden and output layers • *Link*: has connections between the nodes of the input layer and the nodes of the first hidden layer, between the nodes of successive hidden layers, and between the nodes of the last hidden layer and the nodes of the output layer

Radial basis function model (RBF)	• *Input layer*: has any number of nodes
	• *Hidden layer*: typically has only one hidden layer with any number of nodes
	• *Combination functions*:
	  ■ Hidden layer: uses *radial combination functions* in the hidden layer, based on the squared Euclidean distance between the *input vector* and the *weight vector*
	  ■ Output layer: uses linear combination functions in the output layer
	• *Transfer (or activation) function*: typically uses exponential or softmax activation functions in the hidden layer, in which case the network is a Gaussian RBF network
	• *Link*: has connections between the input layer and the hidden layer, and between the hidden layer and the output layer

Table 9.3a Formulas for Some Popular Weight Adjustment Rules for Directed Learning

Rule	Weight Adjustment Formula
Hebb	$w_{st} = \alpha x_s y_t$ where $\alpha \in [0,1]$ is called the learning parameter such that the higher the value of α, the faster the system learns
Delta	$\Delta w_{st} = \alpha y_s (y_t - \text{desired}_t)$ where $(y_t - \text{desired}_t)$ is the *error* for node t at the last iteration
Generalized Delta	$\Delta w_{st} = \alpha y_s (y_t - \text{desired}_t) g'(x_t)$ where $g'(x_t)$ is the derivative of the *transfer function* at x_t and $g()$ is the sigmoid, hyperbolic, or some other nonlinear function

Table 9.3b MLP Weight Adjustment Formulas Based on the Generalized Delta Rule

Link	Weight Change
Hidden layer node to output layer	$\Delta w_{jk} = \lambda(t_k - y_k)y_k(1 - y_k)y_j$
Input layer node to hidden layer node	$w_{ij} = \lambda y_j(1 - y_j)y_i \Sigma_{k \in O(j)} (t_k - y_k)y_k(1 - y_k)y_j$ where $O_{(j)}$ is the set of nodes into which node j feeds

Error Functions

The reader may recall that an NN is trained by minimizing the selected error function. As can be seen in Table 9.4, there are multiple possible error functions, with the relevance of each determined by the data type and, in some cases, also the data distribution of the target variable. It is therefore important that adequate data exploration of the modeling data set be conducted before training of an NN is attempted.

Software Implementation of the NN Induction Process

Many DM software packages (e.g., SAS Enterprise Miner, IBM SPSS Miner, R, Rapid Miner) provide facilities that make the generation of NNs a relatively easy task. Figure 9.2 presents the SAS Enterprise Miner *process flow diagram* that is used to generate the NN from the given data set. Each node in this diagram performs a specific task: the input node reads the given data set; the data partition node partitions the input data set into *training*, *validation*, and *test* data sets; and the neural network node generates the NN. Figure 9.3 presents an example of some of the output from the *impute* node, Figures 9.4–9.6 provide an elaboration of issues relevant to the *cluster* node, while Figures 9.7 and 9.8 present examples of outputs from this node.

Similar to the clustering process, the NN induction process ignores records (i.e., instance or observations) for which any of the identified potential predictor variables does not have a value. In such a situation, it may therefore be useful to apply an appropriate imputation method before applying the NN induction method(s). So if the data set has missing values for any of the potential predictor variables, then an *impute* node should be used to generate replacement values.

Table 9.4 Examples of Error Functions

Gamma distribution	Suitable for *skewed, positive interval targets* where the conditional standard deviation is proportional to the conditional mean
Poisson distribution	Suitable for *skewed, nonnegative interval targets*, especially counts of rare events, where the conditional variance is proportional to the conditional mean
Bernoulli distribution	Suitable for *binary targets*
Multiple Bernoulli	Default for *categorical (nominal or ordinal) targets*
Entropy	Cross or relative entropy for independent *interval targets* with values in the [0, 1] interval
Multiple entropy	Cross or relative entropy for *interval* targets that sum to 1 and have values in the [0, 1] interval; Also called Kullback–Leibler divergence
Normal distribution	Suitable for *unbounded interval targets* with constant conditional variance, no outliers, and a symmetric distribution; and for *categorical targets* with outliers
Huber M-estimator	Suitable for *unbounded interval targets* with outliers or with a moderate degree of inequality of the conditional variance but a symmetric distribution; and for *categorical targets* where the objective is to predict the mode rather than the posterior probability
Redescending M-estimators	Suitable for *unbounded interval targets* with severe outliers

Figure 9.2 Process flow diagram.

Overview on Artificial Neural Networks ■ 99

Variable Name	Role	Measurement Level	Order	Label	Use
BAD	Target	Binary			Y
IMP_CLAGE	Input	Interval		Imputed CL...	D
IMP_CLNO	Input	Interval		Imputed CL...	D
IMP_DEBTI...	Input	Interval		Imputed DE...	D
IMP_DELINQ	Input	Interval		Imputed DE...	D
IMP_DEROG	Input	Interval		Imputed DE...	D
IMP_JOB	Input	Nominal		Imputed JOB	D
IMP_MORT...	Input	Interval		Imputed M...	D
IMP_NINQ	Input	Interval		Imputed NI...	D
IMP_REAS...	Input	Nominal		Imputed RE...	D
IMP_VALUE	Input	Interval		Imputed VA...	D
IMP_YOJ	Input	Interval		Imputed YOJ	D
LOAN	Input	Interval			D

Figure 9.3 Example of the output of the *impute* node.

Property	Value
Architecture	Multilayer Perceptron
Direct Connection	No
Number of Hidden Units	3
Randomization Distribution	Normal
Randomization Center	0.0
Randomization Scale	0.1
Input Standardization	Standard Deviation
Hidden Layer Combination Function	Standard Deviation
Hidden Layer Activation Function	None
Hidden Bias	Range
Target Layer Combination Function	MidRange
Target Layer Activation Function	Default
Target Layer Error Function	Default
Target Bias	Yes
Weight Decay	0.0

Internal standardization of interval variables reduces the risk of the training process getting trapped in bad **Local Minima**, and thereby speeds up training.

○ **Range:** subtracts the minimum and divides by the range (i.e., maximum−minimum), so that the resulting values have a minimum of 0 and a maximum of +1.

○ **MidRange:** subtracts the midrange and divides by half the range, so that the resulting values have a minimum of −1 and a maximum of +1.

➢ **Standard deviation:** subtracts the *Mean* and divides by the *Standard Deviation*, so that the resulting values have a *Mean* of zero (0) and a *Standard Deviation* of one (1).

Figure 9.4 Selection of the internal standardization method.

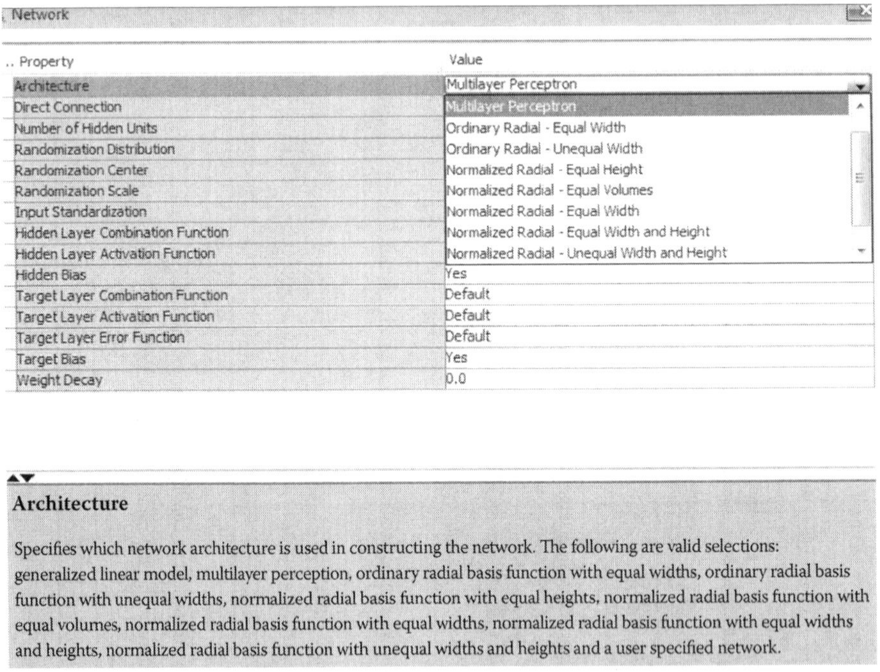

100 ■ Theoretical Research Frameworks Using Multiple Methods

Property	Value
Architecture	Multilayer Perceptron
Direct Connection	Multilayer Perceptron
Number of Hidden Units	Ordinary Radial - Equal Width
Randomization Distribution	Ordinary Radial - Unequal Width
Randomization Center	Normalized Radial - Equal Height
Randomization Scale	Normalized Radial - Equal Volumes
Input Standardization	Normalized Radial - Equal Width
Hidden Layer Combination Function	Normalized Radial - Equal Width and Height
Hidden Layer Activation Function	Normalized Radial - Unequal Width and Height
Hidden Bias	Yes
Target Layer Combination Function	Default
Target Layer Activation Function	Default
Target Layer Error Function	Default
Target Bias	Yes
Weight Decay	0.0

Architecture

Specifies which network architecture is used in constructing the network. The following are valid selections: generalized linear model, multilayer perception, ordinary radial basis function with equal widths, ordinary radial basis function with unequal widths, normalized radial basis function with equal heights, normalized radial basis function with equal volumes, normalized radial basis function with equal widths, normalized radial basis function with equal widths and heights, normalized radial basis function with unequal widths and heights and a user specified network.

Figure 9.5   Selection of the NN architecture.

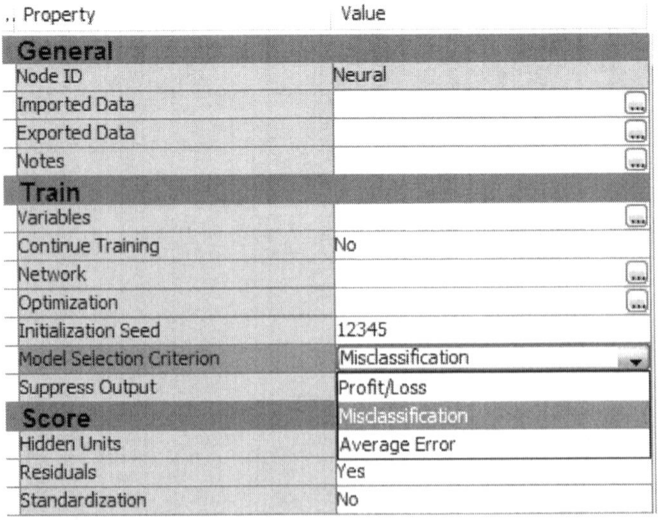

Property	Value
General	
Node ID	Neural
Imported Data	
Exported Data	
Notes	
Train	
Variables	
Continue Training	No
Network	
Optimization	
Initialization Seed	12345
Model Selection Criterion	Misclassification
Suppress Output	Profit/Loss
Score	Misclassification
Hidden Units	Average Error
Residuals	Yes
Standardization	No

Figure 9.6   Selection of the model assessment criterion.

Similar to how a regression model is described by its intercept and the coefficients of its statistically significant variables, an NN model is described by its final weights for each link in the model.

Label	From	Into	Weight
BIAS -> BAD1	BIAS	BAD1	1.160909
BIAS -> H11	BIAS	H11	1.910912
BIAS -> H12	BIAS	H12	0.150231
BIAS -> H13	BIAS	H13	-1.01868
H11 -> BAD1	H11	BAD1	-2.98893
H12 -> BAD1	H12	BAD1	-3.43671
H13 -> BAD1	H13	BAD1	1.951871
IMP_CLAGE -> H11	IMP_CLAGE	H11	0.773967
IMP_CLAGE -> H12	IMP_CLAGE	H12	-0.02724
IMP_CLAGE -> H13	IMP_CLAGE	H13	-0.53745
IMP_CLNO -> H11	IMP_CLNO	H11	-0.56901
IMP_CLNO -> H12	IMP_CLNO	H12	0.341885
IMP_CLNO -> H13	IMP_CLNO	H13	0.521662
IMP_DEBTINC -> H11	IMP_DEBTINC	H11	-0.91801
IMP_DEBTINC -> H12	IMP_DEBTINC	H12	0.001528
IMP_DEBTINC -> H13	IMP_DEBTINC	H13	0.509944
IMP_DELINQ -> H11	IMP_DELINQ	H11	-0.44102
IMP_DELINQ -> H12	IMP_DELINQ	H12	-0.35469
IMP_DELINQ -> H13	IMP_DELINQ	H13	-0.33153
IMP_DEROG -> H11	IMP_DEROG	H11	-0.0855
IMP_DEROG -> H12	IMP_DEROG	H12	-0.06201
IMP_DEROG -> H13	IMP_DEROG	H13	0.261679
IMP_JOBMgr -> H11	IMP_JOBMgr	H11	0.849417
IMP_JOBMgr -> H12	IMP_JOBMgr	H12	-0.3966
IMP_JOBMgr -> H13	IMP_JOBMgr	H13	-0.27825
IMP_JOBOffice -> H11	IMP_JOBOffice	H11	0.979337
IMP_JOBOffice -> H12	IMP_JOBOffice	H12	-0.18781
IMP_JOBOffice -> H13	IMP_JOBOffice	H13	-0.2118
IMP_JOBOther -> H11	IMP_JOBOther	H11	0.820051
IMP_JOBOther -> H12	IMP_JOBOther	H12	-0.3788
IMP_JOBOther -> H13	IMP_JOBOther	H13	-0.30545
IMP_JOBProfExe -> H11	IMP_JOBProfExe	H11	0.845857
IMP_JOBProfExe -> H12	IMP_JOBProfExe	H12	-0.37078
IMP_JOBProfExe -> H13	IMP_JOBProfExe	H13	-0.00485
IMP_JOBSales -> H11	IMP_JOBSales	H11	-2.9195
IMP_JOBSales -> H12	IMP_JOBSales	H12	0.627736

Figure 9.7   Example of final link weights of NN model.

Fit Statistics

Target	Target Label	Fit Statistics	Statistics Label	Train	Validation
BAD		_DFT_	Total Degrees o...	3992	
BAD		_DFE_	Degrees of Free...	3937	
BAD		_DFM_	Model Degrees...	55	
BAD		_NW_	Number of Esti...	55	
BAD		_AIC_	Akaike's Inform...	2927.659	
BAD		_SBC_	Schwarz's Baye...	3273.722	
BAD		_ASE_	Average Square...	0.107112	0.121695
BAD		_MAX_	Maximum Absol...	0.995955	0.995711
BAD		_DIV_	Divisor for ASE	7984	3936
BAD		_NOBS_	Sum of Frequen...	3992	1968
BAD		_RASE_	Root Average S...	0.32728	0.348848
BAD		_SSE_	Sum of Squared...	855.1841	478.9909
BAD		_SUMW_	Sum of Case W...	7984	3936
BAD		_FPE_	Final Prediction...	0.110105	
BAD		_MSE_	Mean Squared...	0.108609	0.121695
BAD		_RFPE_	Root Final Predi...	0.331821	
BAD		_RMSE_	Root Mean Squ...	0.329558	0.348848
BAD		_AVERR_	Average Error F...	0.352913	0.403813
BAD		_ERR_	Error Function	2817.659	1589.408
BAD		_MISC_	Misclassificatio...	0.136523	0.155996
BAD		_WRONG_	Number of Wro...	545	307

Figure 9.8   Example of fit statistics of NN model.

In addition to the issue of missing values in the data set, other issues, including the size of the data set, the characteristics of the data distributions of the variables, and the presence of outliers, may affect the performance of the NN induction process. These issues may be relevant not only to NN induction but also to other data analysis methods such as regression analysis, and so it would important to engage in data understanding and preparation activities before applying such data analysis techniques. Table 9.5 describes some of the approaches for dealing with these issues.

Table 9.5 Some Data Preprocessing Options

Issue	Action
Size	Sampling the input data source: Sampling is recommended for extremely large data sets because it can decrease training time tremendously. If the sample is sufficiently representative, then relationships found in the sample can be expected to generalize to the complete data set.
Data distribution characteristics	The use of appropriate input transformations can improve generalization and speed up training. Some common data transformations are as follows: • *First differencing* (i.e., using changes in a variable) can be used to remove a linear trend from the data. • *Logarithmic transformation* is useful for data that can take on both small and large values and is characterized by an extended right-hand tail distribution. Logarithmic transformations also convert multiplicative or ratio relationships to additive, which is believed to simplify and improve NN training. • *Ratios of input variables*. Ratios may highlight important relationships (e.g., financial statement ratios) while at the same time conserving degrees of freedom because fewer input neurons are required to code the independent variables.
Presence of outliers	Outliers in the input variables may be of concern because they can have undue influence on predictions. Thus, it is important to check for the presence of outliers and, if they are present, to remove them before attempting to do NN induction.

References

Amani, F. A. & Fadlalla, A. M. (2017). Data mining applications in accounting: A review of the literature and organizing framework. *International Journal of Accounting Information Systems*, 24, 32–58.

Bishop, C. M. (1995). *Neural Networks for Pattern Recognition*. Oxford University Press.

Chan, F. T. & Chong, A. Y. (2012). A SEM–neural network approach for understanding determinants of interorganizational system standard adoption and performances. *Decision Support Systems*, 54(1), 621–630.

Colomo-Palacios, R., González-Carrasco, I., López-Cuadrado, J. L., Trigo, A., & Varajao, J. E. (2014). I-Competere: Using applied intelligence in search of competency gaps in software project managers. *Information Systems Frontiers*, 16(4), 607–625.

Dag, A., Oztekin, A., Yucel, A., Bulur, S., & Megahed, F. M. (2017). Predicting heart transplantation outcomes through data analytics. *Decision Support Systems*, 94, 42–52.

Dunham, M. H. (2006). *Data Mining: Introductory and Advanced Topics*. Pearson Education, India.

Giudici, P. (2005). *Applied Data Mining: Statistical Methods for Business and Industry*. John Wiley & Sons, New Jersey, USA.

Hand, D., Manila, H., & Smyth, P. (2001). *Principles of Data Mining*. MIT Press, Cambridge, MA, USA.

Hájek, P. (2011). Municipal credit rating modelling by neural networks. *Decision Support Systems*, 51(1), 108–118.

Khansa, L. & Liginlal, D. (2011). Predicting stock market returns from malicious attacks: A comparative analysis of vector autoregression and time-delayed neural networks. *Decision Support Systems*, 51(4), 745–759.

Kohonen, T. (1995). *Self-Organizing Maps*. Springer, Berlin.

Kwon, H. B., Roh, J. J., & Miceli, N. (2016). Better practice prediction using neural networks: An application to the smartphone industry. *Benchmarking: An International Journal*, 23(3), 519–539.

Ragothaman, S. & Lavin, A. (2008). Restatements due to improper revenue recognition: A neural networks perspective. *Journal of Emerging Technologies in Accounting*, 5(1), 129–142.

Samoilenko, S. & Osei-Bryson, K. M. (2010). Determining sources of relative inefficiency in heterogeneous samples: Methodology using cluster analysis, DEA and neural networks. *European Journal of Operational Research*, 206(2), 479–487.

Samoilenko, S. & Osei-Bryson, K.-M. (2013). Using data envelopment analysis (DEA) for monitoring efficiency-based performance of productivity-driven organizations: Design and implementation of a decision support system. *Omega*, 41(1), 131–142.

Shawver, T. J. (2005). Merger premium predictions using a neural network approach. *Journal of Emerging Technologies in Accounting*, 2(1), 61–72.

Vellido, A., Lisboa, P. J., & Vaughan, J. (1999). Neural networks in business: A survey of applications (1992–1998). *Expert Systems with Applications*, 17(1), 51–70.

Wang, C. H. & Chuang, J. J. (2016). Integrating decision tree with back propagation network to conduct business diagnosis and performance simulation for solar companies. *Decision Support Systems*, 81, 12–19.

Zhang, S., Tjortjis, C., Zeng, X., Qiao, H., Buchan, I., & Keane, J. (2009). Comparing data mining methods with logistic regression in childhood obesity prediction. *Information Systems Frontiers*, 11(4), 449–460.

Zhou, L., Burgoon, J. K., Twitchell, D. P., Qin, T., & Nunamaker Jr, J. F. (2004). A comparison of classification methods for predicting deception in computer-mediated communication. *Journal of Management Information Systems*, 20(4), 139–166.

Chapter 10

Information Systems Fitness and Risk in IS Development: Insights and Implications from Chaos and Complex Systems Theories

Introduction

The failure of information systems (IS) development is a widely known and extensively researched phenomenon (Akkermans & van Helden 2002; Lyytinen & Robey 1999; Milis & Mercken 2002), but a consistently high number of failures demonstrate that it is still not clear how to succeed (KPMG 2003). Researchers agree that inadequate assessment and management of project risks increases the chances of failure (Keil et al. 1998; Ropponen & Lyytinen 2000; Wallace et al. 2004). It was also noted that identification of risk factors is one of the prerequisites for risk management (Keil et al. 1998), but it is not a trivial task to accomplish (Wallace et al. 2004), in part due to the inherent diversity of IS development (ISD) projects (Glass 2003). We suggest the existence of two types of risk factors associated with ISD projects. The first type refers to *contextual* risk factors that are specific to a type or a setting of a project. These risk factors, such as those associated with short cycle time systems development (Baskerville & Pries-Heje 2004), could be generalized and foreseen to only a limited extent. The second type refers to *inherent* risk factors that are common to all ISD efforts. These risk factors are associated with the fundamental attitudes and approaches toward systems development; not entirely unlike those associated with "persistent characteristics of ISD" (Kautz et al. 2007); these risk factors could be and should be foreseen and anticipated during ISD projects. We suggest that one of the inherent risk factors during ISD projects is associated with functionalism-based engineering approach to ISD. Practitioners of the field rarely sacrifice orthodox positivist approach to ISD, and traditional functionalist methodologies are often selected because of their maturity, popularity, and widespread acceptance. The reasons why the process and product of ISD could fail are many, but we concentrate on a single factor, namely a complexity of the structure and behavior of an IS. Consequently, the unit of analysis in our study is an IS, or its subset (hereto forth "IS"), that is involved in the process of ISD, and the focus of the study is a complexity of the behavior of an IS during the process of ISD. We define an IS as a "system that comprises people, machines, and/or methods organized to collect, process, transmit, and disseminate data that represent user information" (ATIS 2000).

Generally, the functionalist perspective deals with complexity by relying on reductionism, which assumes a linearity of the relationships between the parts constituting the whole. This approach does the job done by "abstracting away" nonlinearity and substituting a complex system (CS) with a "generalized, ideal model" (Shaw 1981). Consequently, the justification of our inquiry is that *mainstream functionalist methodologies that rely on reductionism cannot adequately deal with the complexity of nonlinearity that characterizes an IS*. We outline the research problem addressed in this paper as follows: *What are some of the means that could augment traditional approaches to ISD that can manage the complexity of the behavior of an IS?* To answer our research question, we adapt a two-phase approach. First, we argue that chaos theory (CT) can provide a solid theoretical foundation for researching the process and the product of ISD. The premise of the argument is that an IS can be perceived as a complex nonlinear dynamic system with a pattern of behavior that falls under the purview of CT. Previously, applicability of CT to the study of social systems has been successfully argued for by Dhillon and Ward (2002), Tsoukas (1998), and McDaniel and Walls (1997). Second, we argue that CS theory (CST) can provide important insights regarding the management of a complexity of the process of ISD. The crux of the argument is that CS exhibits self-organizing behavior that can be indirectly manipulated through the management of interdependencies between the system's components. Axley and McMahon (2006), Campbell-Hunt (2007), and Tsoukas (1998) have successfully argued the importance of the issues of complexity and applicability of complexity sciences to the study of organizations.

While an overall focus of our inquiry is in line with the established stream of research relating CT and CST to organizational contexts (Harvey & Reed 1996; Ketterer 2006; Morcol 1996), the incorporation of CT and CST into a single perspective is novel. We formulate the research question of this study as follows: *What insights can CT offer regarding the behavior of an IS during the process of ISD, and what insights can CST provide regarding the management of that behavior?* The justification of our approach is intuitive, for a theoretical foundation can offer a generalizable context-independent perspective on a context-dependent subject. The reason for using two theoretical perspectives on CS is quite simple also: while CT provides us with insights regarding the behavior of a CS, it does not deal with the *management* of that behavior. Consequently, we use CST to get insights regarding the ways by which the behavior of a CS can be managed. To substantiate the arguments, the paper proceeds as a sequence of five parts. Section "Conceptual Foundations" presents an overview of the characteristics of nonlinear dynamic systems and of the major tenets of CT; it also presents an argument that an organizational IS is a nonlinear dynamic system. Section "CT and ISD" inquires into the areas of ISD that may benefit from the insights provided by CT. Section "Insights and Implications for ISD" develops some of the implications of the insights that are relevant to the process of ISD. Section "IS and ISD from the Perspective of CST" offers a justification of CST perspective to inquire into the area of ISD. Section "Fitness Landscapes and ISD" offers some hypothetical examples illustrating how insights from CT and CST can be utilized during the process of ISD. A brief conclusion follows.

Conceptual Foundations

The conceptual foundation of this research is informed by CST and CT. The major premise of this study posits that IS can be viewed as a complex nonlinear dynamic system behavior which can (1) become chaotic and (2) be managed by manipulation of interdependencies between the system's components. To support our inquiry, we use the framework of one of the better tools of the positivist science, namely, hypothetico-deductive logic. In Table 10.1, we provide propositions (expressed as standard-form syllogisms) that we employ to substantiate our arguments.

In CST, the term *system* refers to any entity that consists of multiple components and can be viewed from a dual perspective, as a single thing or a collection of constituting parts. Absence of a universally accepted definition of complexity makes a CS hard to define, but easy to recognize, for such systems exhibit a common set of basic characteristics (Morel & Ramanujam 1999). The first characteristic refers to the number of elements comprising it. Simon (1962) defines a CS as a system made up of a large number of interacting components. The interactions between the elements of a CS are associated with the presence of a feedback mechanism; this causes the behavior of the system to be hard to predict due to nonlinearity of the feedback-controlled type of a behavior (Casti 1994). A large number of components interacting in a nonlinear fashion give rise to the second characteristic of CS, that of *emergent properties*, which refers to the appearance of independently observable and empirically verifiable patterns of the collective behavior of the system (Morel & Ramanujam 1999). A system is *dynamic* if its state or behavior changes with time in

Table 10.1 Propositions Supporting the Inquiry

Section	Major Premise	Minor Premise	Conclusion
2.0 Conceptual foundations	Complex nonlinear dynamic system can exhibit chaotic behavior that is studied by means of CT.	IS is a complex nonlinear dynamic system.	IS can exhibit chaotic behavior that can be studied by means of CT.
3.0 Chaos theory and IS development	CT can describe the general pattern of behavior of complex nonlinear dynamic system.	IS is a complex nonlinear dynamic system.	CT can describe the general pattern of behavior of an IS.
4.0 Insights and implications for ISD	There are limited methods that can affect the behavior of complex nonlinear dynamic system.	IS is a complex nonlinear dynamic system.	There are limited methods that can affect the behavior of an IS.
5.0 IS and ISD from the perspective of CST	Complex nonlinear dynamic systems are capable of self-organizing behavior that can be indirectly manipulated by means of changes in internal parameters.	IS is a complex nonlinear dynamic system.	Behavior of an IS can be indirectly manipulated by means of changes in internal parameters.
Overall argument	Complex nonlinear dynamic system can exhibit chaotic and self-organizing behavior, complexity of which can be indirectly manipulated by the changes in internal parameters of the system.	IS is a complex nonlinear dynamic system.	IS can exhibit chaotic and self-organizing behavior, complexity of which can be indirectly manipulated by the changes in internal parameters of the system.

a *linear* or *nonlinear* fashion. The systems that we are interested in are *deterministic* in terms of the cause and effect; if we define a complex dynamic system through a set of equations, then the variables would relate to each other in a nonprobabilistic way, i.e., only in terms of "0" or "1." There is an important distinction between *determinism* of a system and its *predictability*; while earlier states in a deterministic system do *determine* later ones, our knowledge of earlier states does not necessarily allow us to *predict* later states of the system. Because provided the above characterizations are quite general, we would expect a great variability among the nonlinear dynamic systems, along with the variability of the patterns of behavior that they exhibit. The subject of study of CT is a large class of complex nonlinear dynamic systems capable of exhibiting chaotic pattern of behavior.

We define CT as a *qualitative study of unstable aperiodic behavior in deterministic nonlinear dynamical systems* (Kellert 1993). It is a *qualitative* study because "in the case of non-linear equations expressing rates of change, no general formula exists for arriving at solutions for successive points in time" (Kamminga 1990). CT is the study of *unstable behavior* because a nonlinear dynamic system responds to even small perturbations in significant ways; for example, chaotic systems are *hypersensitive* to initial conditions (Dooley & Van de Ven 1999). Finally, the behavior is *aperiodic* because of the nature of nonlinear interactions of the system's variables. Viewing a complex nonlinear dynamic system through the lens of CT allows us to perceive such a system as consisting of two components. The first component is a unique set of *initial conditions*. The second component is a set of formulae representing the *dynamic* part of the system; it is a non-unique *rule* defining the *intended trajectory* of the system's states. The difference in initial conditions forces a unique evolutionary path of a system. It happens because inevitable small errors made in specifying the initial state of a system grow and eventually dominate the future behavior of the system, causing the actual path to deviate farther and farther from that defined by the *rule* trajectory. This is a point where a relative complexity, or a *dimension* designating a number of variables required to describe the system at the given state, plays an important role. The feedback mechanisms of the CS with the dimension of three or more prevent the errors from growing to infinity. Instead, a CS can go through extremely complicated behavior where it "stretches and folds over onto itself, like a baker kneading dough" (Radzicki 1990). When mapped on the *phase space*, an abstract geometrical space representing states of the system in time, the behavior of a CS could exhibit three different patterns. Namely, it may tend to the *state of equilibrium*, it may tend to *repetitive periodic behavior*, or it may behave *chaotically*.

Behavior of a system can be represented in terms of an *attractor*, which is a "set of points in the phase space of a dynamical feedback system that defines its steady state motion" (Radzicki 1990). Every attractor has a *basin of attraction*, or a set of points in the space of system variables that evolve to a particular attractor. A periodically fluctuating system is said to have a *strange* attractor. Strange attractors are *chaotic* when the trajectories in the phase space, from two points very close on the attractor, diverge exponentially due to the system's sensitive dependence on initial conditions and small perturbations in control parameters. The second important insight provided by CT is that while the evolutionary path of chaotic system is not predictable, the *pattern of the path* could be predicted (Dooley & Van de Ven 1999). Another insight provided by CT is that a system might have multiple attractors associated with multiple patterns of behavior. A CS can transition from a semistable to a chaotic state. This happens when a value of a key parameter of a system increases and exceeds a threshold value, forcing a single outcome basin to expand into two distinct causal fields. This process of doubling of a number of basins, or *bifurcation*, can safely continue up to the point of a system having eight basins of attraction. However, the next bifurcation marks the *onset of chaos* (Feigenbaum 1978). Young and Kiel (1994) comment that "the remarkable thing is that this precipitous rush into an infinite number of outcome basins is universal over all natural systems so far studied."

To argue the point that an IS could be investigated through the lens of CT, we follow a systematic approach of eligibility assessment. First, we assess the criterion of complexity. Because it takes more than three variables to describe an IS at a given state, it qualifies as CS in terms of the number of components. The next criterion refers to the nonlinearity of a CS. Being a social system, an IS consists of nonlinear relationships (Radzicki 1990). To qualify as a CS, nonlinearity must give rise to emergent properties. There are a variety of ways by which emergent properties can manifest themselves in any IS: as a mosaic e-mail correspondence (Lee 1994), as a network bottleneck, or as a social power relationship (Markus 1983). These emergent properties change with time, suggesting the dynamic nature of IS. An IS is a deterministic system; we "know" that the increased bandwidth must contribute to the throughput, and we "know" that improved virus protection must contribute to the security of the system. New help desk operator positions "deterministically" relate to the increase in customer support, and new software engineers are hired not because such step "probably" would result in the increased software production. All examples may exhibit a different degree of cause and effect, but that the cause and effect relationships exist between them is certain and not "probable." Let us keep in mind that determinism refers to the defined relationship between variables in the system. This means we may choose to relate deterministically a position of software engineer and a level of software production. However, determinism does not imply predictability; consequently, we cannot predict whether hiring of a software engineer shall indeed result in the intended outcome. Overall, IS qualifies as a complex nonlinear deterministic dynamic system. Thus, the behavior of an IS can be inquired into through the lens of CT. This intent is hardly surprising, for Dooley and Van de Ven (1999) state that an increasing interest of organizational scholars focused "on the notion that at times an organization may be viewed as behaving chaotically," as well as on the "implications of chaos—that the system is deterministic and hypersensitive to small perturbations." We direct the interested reader to the works of Jayanthi and Sinha (1998), Dooley et al. (1995), Stacey (1992), and Thietart and Forgues (1995).

CT and ISD

From the perspective of CT, ISD can be viewed as a process of matching the actual state of an IS with the intended one by means of manipulating the behavior of the existing IS. In terms of CT, the process of ISD concerned with a qualitative transformation of the current system's state to a new dynamical state (Young & Kiel 1994). We know that while a system goes through the process of consecutive bifurcations prior to the region where deterministic chaos sets in, the number of natural outcomes in which a system may be found changes (for illustration, see Figure 10.1). However, internal parameters (i.e., structure) of the system stay the same. This means in a prechaotic region, the same internal configuration of the system might produce up to eight different outcomes. However, as the system enters a state of chaos, new windows of order appear; this represents the emergence of entirely new organizational forms (Young & Kiel 1994).

The implications are obvious for ISD; if we desire an entirely different form of IS, then the process of development must go through the chaotic stage. This leads

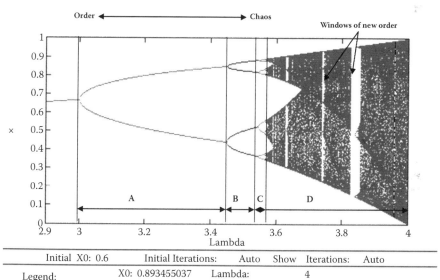

Legend:
Prior to region A (Feigenbaum number < 3) system could be thought of as having torus attractor, indicative of a single pattern of system's behavior.
System in region A could be thought of as having a butterfly attractor, which is indicative of two outcome basins of attraction.
In regions B and C the system has 4 and, correspondingly, 8, outcome basins of attraction.
Region D marks the onset of the deterministic chaos with the emergence of new windows of order.

Figure 10.1 Bifurcations resulting from the increase in the value of the Feigenbaum number. (Adapted from Young, T. R. and Kiel, L. D. *Control, Prediction, and Nonlinear Dynamics*. No. 28, 1994. Distributed as part of the Red Feather Institute Series on Non-Linear Social Dynamics. The Red Feather Institute, 8085 Essex, Weidman, Michigan, 48893.)

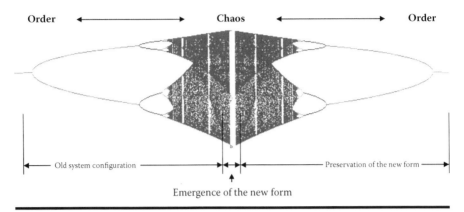

Figure 10.2 Order–chaos–order transformation.

us to Proposition 1: *The process of ISD will have a chaotic component if the goal of the process is to obtain a qualitatively new form of IS.* Consequently, such process of ISD can be perceived as taking an IS through the "order–chaos–order" transformation as depicted in Figure 10.2.

Conversely, if the goal of ISD is to change the behavior of the IS while preserving the existing structure, then the process of development does not have to have a chaotic part. This leads us to Proposition 2: *The process of ISD does not have to have a chaotic component if the goal of the process is to change the system's behavior while preserving the existing structure.* This means that even if in the context of the real world the process of ISD described in Proposition 2 could become chaotic, according to CT, such process *does not have to become chaotic*, while the process described in Proposition 1 does. Irrespective of whether the goal of ISD is to produce a new system or to elicit a new pattern of behavior from the existing system, the implementation process involves a change in the existing system's behavior. According to CT, the behavior of a system, as well as the change in its behavior, is unique to each system. Therefore, we put forward Proposition 3: *The process of ISD is unique to each organization and is driven by a unique set of factors for each organization that influence changes in the behavior of IS in a unique way.*

Insights and Implications for ISD

Three requirements are essential for managing the behavior of a chaotic system: (1) understanding of nonlinearity, (2) appreciation of the sensitivity of the system to its *initial conditions*, and (3) understanding of a non-average behavior as a source of change (Kiel 1997). Concerning the first requirement, Senge (1990) and Holland (1995) argue that it is of great importance to find the *leverages* (Senge 1990) or *lever points* (Holland 1995) of the system that could be subject to the *butterfly effect*. Once these leverages are found, even a "small targeted change may produce larger scale results compared to comprehensive change efforts that may squelch an organization's or social system's capacity for adaptive response" (Kiel 1997). And while Holland (1995) acknowledges that it is not a trivial task to find the system's leverage points, Kiel (1997) responds that the best approach is to use multiple possible leverages and hope that at least some of them will work. The second requirement, a system's sensitivity to initial conditions, raises concerns regarding the effectiveness of the commonly utilized practices of benchmarking, transferring of best practices, and searching for prepackaged solutions. According to CT, we cannot expect the same intervention produce the same effect in two different systems, and the suggestion of Kiel (1997) that "managers identify the elements unique to their environments prior to the implementation of another jurisdiction's best practice" seems to be well warranted. The last requirement refers to the appreciation for a *non-average* (Prigogine & Stengers 1984), unusual event that "pushes the boundaries of existing structures and processes and leads the way for new forms of organizational response and evolution after bifurcating events" (Kiel 1997). Such an event will most likely manifest itself in the form of an outlier, produced by a CS in which behavior on the brink of change is "neither normally distributed nor regular" (Parker & Stacey 1994). The implication of these requirements for the managing of the system's behavior is clear, for they serve as indicators that the behavior and structure of the system are about to change. Consequently, we propose the following Implication 1: *The process of ISD that aims to result in a qualitatively new form of IS should start from identifying the system's unique characteristics, then proceed by means of small targeted changes, while being guided by non-average events.*

What might it mean in the context of an IS? First, that the process of ISD must start from identifying unique characteristics of the organization and organizational IS. The importance of the identification of the setting-specific starting conditions is obvious, for they represent unique qualities that must be preserved during the process of transformation. Conversely, managers may decide to concentrate on negative characteristics (starting conditions) and direct their effort on their elimination during the transformative changes. Second, once the behavior of an IS became chaotic, managers must avoid any full-scale actions, for the system at that point is too vulnerable to sustain a major stress. An analogy of an *induced coma* comes to mind, for patients in intensive care units whose condition is too unstable to treat are often placed in the state of induced coma. Any medical intervention at that point is minimal, and the goal is to stabilize the condition of the patient so the treatment might start or resume. Similarly, in an IS that entered the state of the chaotic behavior, all changes and activities must become incremental and as widely dispersed through the scope of the project as possible. This also would assist in the search for lever points. In the context of the firm, "incremental and dispersed activities" may mean changes not on the organizational or project level but on the level of departments, functional units, or even single individuals. Third, a "guidance by non-average events" suggests that prior to the process of ISD, an organization and its IS must be monitored in order to record the average behavior of the system. Once a non-average event took place, it must be investigated and closely monitored. Such a non-ordinary event could manifest itself in the form of unusual network bottlenecks, or an uncharacteristic number of customer complaints, or a sudden increase in e-mail communication across the organization. Overall, it appears that no augmentation to formal ISD methodology can be used to control the behavior of an IS once it became chaotic, for all the possible means of control are setting-specific and cannot be prescribed in advance.

The next implication that we develop is pertinent to the management of the system that approaches the edge of chaos. The managing of the behavior of such a system is a process of controlled increase in gradual chaos induced into the system. There are three ways to control chaos: (1) by altering the system's parameters in order to reduce uncertainty and increase predictability, (2) by the application of small perturbations to the chaotic system to try to cause it to organize, and (3) by changing the relationship between the system and its environment (Kiel 1995). At this point, we offer Implication 2: *The process of ISD that aims to result in the change of the behavior of existing IS could prevent entry of the system into chaotic region by either altering the internal parameters of the system, or altering the system's relationships with the environment, or applying small perturbations to the system.* Let us consider the importance of the insights and implications provided above to the process of ISD. According to Implication 1, no formal ISD methodology can adequately control the process of ISD once the process became chaotic. However, can we control the behavior of a system that approaches the edge of chaos or emerges from the chaotic region? Let us consider a way of controlling chaos through the application of small perturbations (i.e., reassigning workers from one team to another, changing the structure of the workday, etc.). The important point to make here is that by applying small perturbations to the system, we *hope* that small change in some of the system's parameters could trigger the increase in the system's organization. Therefore, it is rather an indirect way of controlling chaos. Incorporation of this insight into ISD methodology is problematic (1) because it would require a methodology to include the capability of searching for multiple lever points, (2) because it would require a capability to affect the lever points, and (3) because it would require an upfront knowledge of the results that the impact on lever points may produce. Let us consider the way of controlling chaos by means of changing the relationship between the system and the environment. This requires continuous tracking of the relationship between critical conditions in the environment and key organizational parameters. As changes occur, the system's parameters are adjusted in a continuous feedback process. This presents a problem for ISD, first because most of the projects have a defined scope, and second because it will require the boundaries of an ISD methodology to expand and include critical conditions in the environment. We argue, therefore, that *the complexity of the process of ISD can be controlled by means of altering the internal parameters of the system*. To clarify, a change in internal parameters refers to a structural change of the system, achieved by means of altering the number of components that comprise the system and/or the density of the relationships between the components. We summarize implications of the insight provided by CT and CST for the process of ISD in Table 10.2.

Information Systems Fitness and Risk in IS Development ■ 117

Table 10.2 Implications of CT and CST for ISD

Goal	Means/Requirements	Implications for ISD
Management of the behavior of a chaotic system	1. Understanding of nonlinearity 2. Appreciation of the sensitivity of the system to its *initial conditions* 3. Understanding of a non-average behavior as a source of change	1. The process of ISD that aims to result in a qualitatively new form of IS should start from identifying the system's unique characteristics, then proceed by means of small targeted changes, while being guided by non-average events. 2. The process of ISD that aims to result in the change of the behavior of existing IS could prevent entry of the system into chaotic region by altering the internal parameters of the system, or altering the system's relationships with the environment, or applying small perturbations to the system.
Control of chaos	1. Alteration of the system's parameters in order to reduce uncertainty and increase predictability 2. Application of small perturbations to the chaotic system to try to cause it to organize 3. Change of the relationship between the system and its environment	1. Controlling the complexity of the process of ISD by means of application of small perturbations or by changing the relationship between the system and its environment is problematic. 2. The complexity of the process of ISD can be controlled by means of altering the internal parameters of the system.

IS and ISD from the Perspective of CST

The concept of an organization as of a CS is deeply rooted in organizational theory (Katz & Kahn 1966; Ashby 1968) and by now is well established (Carley 1995; Stacey 1996; Thietart & Forgues 1995). Research in the area of CS produced multiple important insights (Anderson 1999), such as that CS tends to exhibit a pattern of self-organization (Fontana & Ballati 1999) causing CS to evolve within its environment toward order (Kaufmann 1993).

Self-Organization of the CSs

Morel and Ramanujam (1999) describe self-organization as a *process of spontaneous creation of a complex structure that emerges due to the dynamics of the CS*. Our choice of self-organizing aspect of the behavior of CSs is not incidental, for it is one of the dominant research paradigms in the area of CS (Morel & Ramanujam 1999). Self-organization is a natural response of a CS to the constraints of its environment, which takes place naturally and automatically, and bears the purpose of increasing the level of *fitness* of the system, i.e., its efficiency and effectiveness. Such self-organizing behavior is endemic to any CS, be it a disparate group of people, or an organization (Thompson 1967), or central nervous system of the animals; as long as the system is complex enough, the pattern of self-organization will emerge. Modern organizations and IS exist within hypercompetitive environments (D'Aveni 1994; Illinitch et al. 1998), where the speed of self-organization becomes a strategic issue of being able to advance faster than competitors do (Brown & Eisenhardt 1998). It is certain that the level of fitness of organizational ISD effort will play a decisive role in a business world's version of survival of the fittest.

Fitness Landscapes

Wright (1931) has originally introduced the concept of fitness landscapes in biology. Simplistically, this is an idea of mapping a structure of the system to its level of fitness. Kaufmann (1993) has extended works of Wright and introduced the concept of *ruggedness* as a characteristic of a fitness landscape. According to Kaufmann, two variables, N, the number of system components, and K, the number of interconnections between N components, can characterize the system's fitness landscape. One of the points of Kaufmann's theory is that the increase in K gives rise to the *rugged landscapes* containing multiple fitness peaks, proliferation of which prevents the system from ever adapting to its optimal level. Kaufmann calls such detrimental increase in the number of system's interconnections a *complexity catastrophe*. Because of a complexity catastrophe, a system becomes caught in multiple suboptimal fitness peaks. According to Kaufmann, the optimal evolution of a CS takes place between the *edge of catastrophe* and the *edge of chaos*. In other words, the system must be complex enough to evolve, but not too complex to be unwieldy.

Levinthal and Warglien (1999) suggested that the "concept of a fitness landscape has rather natural analogues in the domain of social and economic phenomena" and applied Kaufmann's (1993) ideas of fitness landscapes to the domain of organizations. They proposed that because the process of self-organization is context-dependent, the dynamics of the process could be influenced through the manipulation of the context. This approach, similar in spirit to that of Simon (1962) and Thompson (1967), is based on the varying of the density of the interdependencies of the system that affect the ruggedness of the system's fitness landscape. Consequently, as the manipulation of interdependencies takes place, it gives rise to a variety of different patterns of a system's behavior. This manipulation allows a "landscape designer" to understand the implications of a given landscape for the behavior of the system and to find out the desired dynamics of the behavior (Levinthal & Warglien 1999). It turned out that the settings with low interdependencies produce structure-to-fitness mappings with the main features of single-peak landscapes, which are indicative of highly predictable dynamics of each individual component of the system. Such low-interdependence systems generate settings where local actions promote global improvement of the system, and even "a step in wrong direction… will involve only a minor degradation of global performance" (Levinthal & Warglien 1999). The authors call such landscape design *robust* and suggest its importance for organizations where "continuous improvement policies are sought." On the other hand, an increase in interdependencies between the system's components eventually results in the rugged, multipeaked landscapes. While this automatically brings along the problem of coordination, it also "encourages non-incremental search and exploration" that are "necessary when 'breakthrough' and 'platform' innovations" are desired during the new product development (Levinthal & Warglien 1999). The relevance of the study described above to the area of ISD becomes clear when the authors state that the process of "[manipulating] distance, social and geographic, among actors can serve to tune interdependencies" in such a manner that by "reducing distances, via technology or other means, one increases the apparent ruggedness of the landscape." In modern organizations, IS provides the mean by which multiple organizational components communicate with each other. Moreover, it is possible for an IS to recursively tune its own interdependencies. The implication of this perspective is that an IS becomes the medium that controls the density of interrelationships K between the N components of an ISD team. Naturally, an increase in the number of interdependencies would contribute to the increase in complexity of ISD process, while a decrease will contribute to the reduction of complexity.

We argue that communication channels of IS can be used to control the level of fitness of the system and the complexity of the process of ISD. This is how, from a theoretical standpoint, a methodology employed to conduct ISD could be augmented to manage the complexity of the process of ISD. From the perspective of the paradigm of self-organization of CST, *information system fitness* can be defined as the *structural capability of IS to increase or maintain its level of performance as a response to the constraints of the environment*. While determination of what constitutes a level of performance of IS is context-dependent, IS fitness is clearly associated with the capability to control the level of the complexity of an IS. At this point, we offer the following solution to the research question of this study: *Based on insights offered by CT and CST, a complexity of the behavior of an IS during the process of ISD can range from order to chaos and can be managed by manipulation of interdependencies between the system's components.*

Fitness Landscapes and ISD

The intent of this part of the paper is to illustrate, by using three hypothetical examples, how insights provided by the landscape theory (Kaufmann 1993) can be used to reduce the complexity of the process and product of ISD. First, let us discuss what "order" and "chaos" may mean in a real-life context. While it is possible to describe an IS by some sort of equation, or to identify formally when its behavior becomes chaotic, we doubt that practitioners engaged in the process of ISD require such a degree of precision. Instead, we suggest a less formal, based on a *perception of control*, approach in evaluating the behavior of an IS. Consequently, instead of using and being guided by such formal terms as *order*, *edge of chaos*, and *chaos*, practitioners of the field may find it more beneficial to operate in terms of, respectively, *perception of control*, *perception of the loss of control*, and *perception of the lost control* over the behavior of an IS. Nevertheless, all insights and implications developed in this paper still apply, for our goal was to find a way of managing complexity of the behavior of an IS, rather than to identify precisely at what point the behavior of an IS becomes chaotic.

According to the suggested guidelines, communication channels used during the process of ISD may proliferate and lead to an increase in the ruggedness of the system's landscape, as long as *perception of control* exists. Once the *perception of the loss of control* appears, the number of communication channels can increase further; this will lead the system into the chaotic region characterized by the *perception of the lost control*. Conversely, a number of communication channels can be reduced to prevent the system from entering a chaotic region; this will bring the behavior of the system back in the state of order, as determined by the *perception of control*. Based on the insights that we obtained in our study, we can offer the following response to the broad research question stated in Introduction: *Traditional approaches to ISD could manage the complexity of the behavior of an IS better if augmented by the capability of managing the communication channels of an IS.*

Figure 10.3 offers a summarized version of our inquiry and its results.

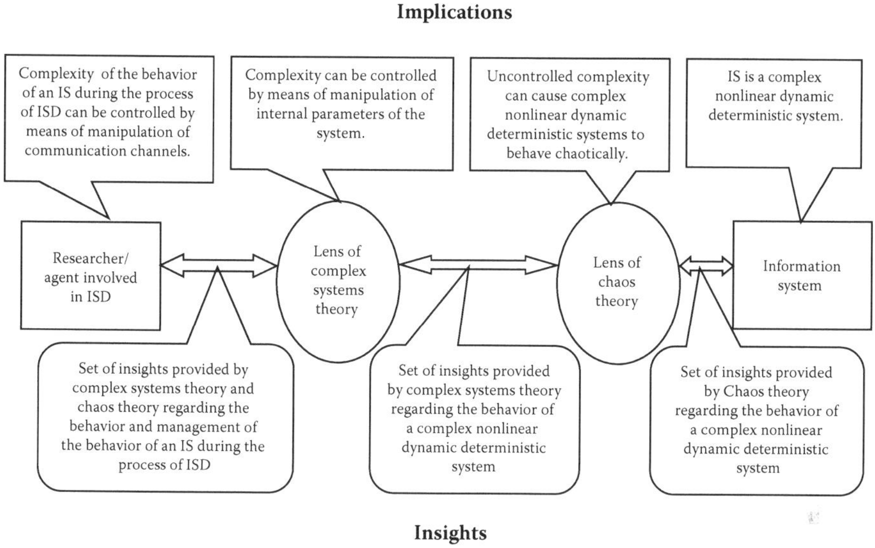

Figure 10.3 Insights and implications of the inquiry.

Scenario One: Fitness Landscapes and ISD Process

The first example refers to the Analysis or Design phase of ISD process. The commonality between these two phases is that in both cases the local agents of the ISD team are engaged in the search for the global solution. At the time when the cross-functional teams create different versions of the conceptual or physical design of the future product, the existing IS can contribute to the increase in the ruggedness of the landscape by opening all available communication channels. Consistent with the Levinthal and Warglien (1999) proposition, an increase in the level of communication will serve the purpose of decreasing the social distance, with the resulting increase of the apparent ruggedness of the landscape. Once the design teams start converging on a particular idea for the new product, the period of "non-incremental search and exploration" is over; at this point an IS can start contributing to the "smoothing" of the landscape. This may be achieved by restricting the lines of communication and flows of information, so the single-peaked type of landscape can emerge and a global optimum pursued by the ISD team. While, arguably, some of these methods are already in place, we suggest that the capabilities of any IS in this regard can be explored and utilized to a higher degree. The manipulation of the communication streams and information routes by an IS must take place consciously and deliberately.

Scenario Two: Fitness Landscapes and ISD Product

There are two useful insights that CST can provide to manage the complexity of the ISD product. First, landscape theory (Kauffman 1993) suggests against proliferation of the communication channels within CS. Such perspective seems to be contrary to the usual intent of IS developers, who, in their endless pursuit of provision of "information on demand," are bent on making the new IS as accessible to the users as possible. To prevent an IS from becoming trapped in rugged landscapes, it has to have some sort of regulatory capabilities to ration the flow of information. Regarding this issue, Levinthal and Warglien (1999) state that "the use of e-mail within an organization is increasingly productive until an overcrowding level is reached, beyond which 'infoglut' makes electronic communication a source of painful loss of time and productivity." Consequently, the complexity of the product of ISD can also be managed by controlling the number of communication channels between the system's components. The second insight is that an IS would benefit from loose coupling of its components, that is, from having multiple components (Kauffman's N) that could function independently (i.e., in the pursuit of a global optima), but could be interconnected when needed (i.e., when the presence of the rugged landscapes is desired). One of the problems associated with IS projects is brought by the issues of the technical complexity of the product arising from the "designed-in" interdependencies of the components. It is possible to conceive that the structure characterized by a loose coupling, based on the minimization of the interdependencies, would allow for minimizing the complexity of the system. A tight coupling of ERP-type systems, on the other hand, gives an example of an IS that by its design increases the ruggedness of the organizational landscape to the maximum.

Scenario Three: Fitness Landscapes and the Process and Product of ISD

There are two general insights regarding how manipulation of the communication channels can be used to manage the complexity of the behavior *and* the structure of an IS. The first insight refers to the *communication structure* within an IS. When the behavior of IS is stable, a communication structure within the system can be allowed to be represented by an "open door" policy, where subordinates are free to meet with their supervisors at a time of mutual convenience and discuss their problems, ideas, etc. Such structure promotes an increase in density of communication channels within an IS. Once the behavior of an IS becomes unstable, a more stringent policy of communication becomes advisable. Such policy could be based on the rule that only problems that are currently relevant to the IS can be brought up and discussed during the scheduled meetings.

The second insight refers to the *reporting structure* within an IS. During the period of stable behavior, an IS involved in the process of ISD can be represented by cross-functional teams, having a matrix-type internal organization. Such structure increases the number of communication channels and leads to an increase in the ruggedness of the fitness landscape. However, once an IS displays the pattern of behavior characterized by the perception of the loss of control, the reporting structure must be able to regress to bureaucratic hierarchy. The more rigid reporting structure will decrease the complexity of the structure of the system, consequently affecting the behavior of the system as well.

Conclusion

In this paper, we used CT to obtain insights and implications regarding the general pattern of behavior of the process of ISD. According to CT, the process of ISD can exhibit a wide variety of behaviors, ranging from stable to chaotic. The purpose of this study was to inquire into ways by which the complex behavior of the ISD process can be managed. We have argued that the traditional methodologies are not adequately equipped for dealing with the complexity associated with the process of ISD. To overcome these limitations, we employed a paradigm of self-organization of CST in order to obtain insights regarding the management of the process of ISD. Results of the investigation suggest that by varying the number of open communication channels, we can control the complexity of a system's structure and, consequently, manage its behavior. We would like to summarize some of the contributions of our study. From a theoretical standpoint, our inquiry demonstrated that such context-dependent undertaking as a process of ISD could be approached in an objective, context-independent way if perceived from the vantage points of CT and CST. Thus, regardless of the setting, we are able to foresee a general pattern of the behavior of the process and offer a theoretically sound way of managing it. Consequently, the main theoretical contribution of the paper is that *based on the insights provided by CT and CST, we outlined a general framework that allows predicting and managing general patterns of the behavior of an IS during the process of ISD*.

We suggest that our research carries some value for practitioners as well. Conclusions of our study not only allow managers to prepare in advance for the expected increase in the complexity of the behavior of the process of ISD, but also provide some practical suggestions regarding the management of the complex behavior. Moreover, because the results of the study are generic, practitioners of the field can incorporate our findings into any methodology employed in their setting. We are not aware of an ISD methodology that explicitly considers a range of possible behaviors of the ISD process or incorporates any tools designed for dealing with the complexity of the behavior of the process of ISD. We suggest that the results of our study fill in the void by offering managers and decision makers a theoretically sound, yet practical, tool allowing for managing the complexity of an IS during the process of ISD. Consequently, the main practical contribution of the paper is that *based on the insights provided by CT and CST, manipulation of communication channels allows for managing of the complexity of the behavior of an IS during the process of ISD.*

How do this study and its findings fit into the stream of empirical research dealing with risks and problems in ISD? Keil et al. (2002) defined a risk factor as a "condition that forms a serious threat to the successful completion of an IT project." A management of risk factors is of critical concern to organizations involved in ISD (Wallace et al. 2004). In order to be able to manage risk factors, they have to be identified first (Keil et al. 1998); however, this is not a trivial task in the absence of a theory predicting relevant risks (Stahl et al. 2003). The results of our inquiry demonstrate that the CST and CT allow for identifying complexity as a common risk factor endemic to all ISD projects and contexts. More research is required to identify how complexity can be operationalized, quantified, and explicitly taken into consideration during the process of ISD. According to Lyytinen and Rose (2006), organizations that develop and maintain IS innovate in a chaotic manner and organize in contradictory ways. The authors note that ISD organizations "need to garner specific capabilities associated with exploration and exploitation and how to balance them." We believe that the results of our study offer insights regarding the nature of those capabilities and ways of transitioning smoothly between the periods of exploration and exploitation that contributes to the agility of an ISD organization (Lyytinen & Rose 2006). However, additional research required to propose the actual design of the dynamic network of communication channels contributing to "an ISD organization's ability to sense and respond swiftly to technical changes and new business opportunities."

Mathiassen et al. (2005) note that when an organization involved in the software process improvement "moves up the maturity ladder, new organizing structures are required to ensure management and control" and that this process requires developing "new incentive structures to promote change." Our findings suggest, first, that creation of such new organizing structures should benefit from the robust design, and, second, that new incentive structures should be based on the promotion of the global improvement rather than the local goals. The determination of what elements of organizational structure contribute to its robustness as well as what types of incentives reward global performance are among some of the avenues for future studies.

There is a clear relationship between the diversity and the complexity of the environment. Kautz et al. (2007) identified the diversity as a "fundamental challenge in ISD generally," which, nevertheless, can be "handled in different ways at different contextual levels by either absorbing or reducing diversity." The authors identify communication, coordination, and division of labor as some of the coping mechanisms allowing for dealing with the diversity, and ask, "What kind of theoretical and empirical research contributions might describe and prescribe this absorption and reduction of diversity in practice?" We believe that this investigation offers a sought-after theoretical basis allowing one to utilize communication and coordination as sound mechanisms that allow for coping with diversity and complexity of the process and product of ISD. More research is needed, however, to determine the diversity-causing factors, for it would often appear that the eye of the beholder determines the difference between homogeneity and heterogeneity of the ISD environment.

In their study of short cycle time systems development, Baskerville and Pries-Heje (2004) note that the anticipation of the chaotic conditions during the project contributes to the effectiveness of the software development process. Results of our study demonstrate that CT allows us to forecast the presence of chaotic conditions, albeit only in those projects where such conditions are unavoidable.

Our research is not without its limitations. First, despite establishing a common theoretical foundation for ISD projects, we cannot account for setting-specific factors that often are the main reason for failure. Second, the insights and implications obtained in this study are yet to be tested in the real-world ISD project. Nevertheless, we hope that the contributions of our study outweigh its shortcomings.

Acknowledgment

Material in this chapter previously appeared in "Information Systems Fitness and Risk in IS Development: Insights and Implications from Chaos and Complex Systems Theories," *Information Systems Frontiers* 10:3, 281–292.

References

Akkermans, H. & van Helden, K. (2002). Vicious and virtuous cycles in ERP implementation: A case study of interrelations between critical success factors. *European Journal of Information Systems* 11, 35–46.

Anderson, P. (1999). Complexity theory and organization science, *Organization Science*, Special Issue: *Application of Complexity Theory to Organization Science* 10(3), 216–232.

Ashby, R. (1968). Some consequences of Bremermann's limit for information processing systems. In *Cybernetic Problems in Bionics*, eds. H. Oestreicher et al., pp. 69–76, Gordon and Breach, New York.

ATIS (2000). ATIS Telecom Glossary 2000. Available at http://www.atis.org/tg2k/t1g2k.html

Axley, S. & McMahon, T. (2006). Complexity: A frontier for management education. *Journal of Management Education* 30(2), 295–315.

Baskerville, R. & Pries-Heje, J. (2004). Short cycle time systems development. *Information Systems Journal*, 14, 237–264.

Brown, S. & Eisenhardt, K. (1998). *Competing on the Edge: Strategy as Structured Chaos*. Harvard Business School Press, Boston.

Campbell-Hunt, C. (2007). Complexity in practice. *Human Relations* 60(5), 793–823.

Carley, K. (1995). Computational and mathematical organization theory: Perspective and directions. *Computational and Mathematical Organization Theory* 1(1), 39–56.

Casti, J. (1994). *Complexification: Explaining a Paradoxical World through the Science of Surprise*. HarperCollins, New York.

D'Aveni, R. (1994). *Hypercompetition: Managing the Dynamics of Strategic Maneuvering.* Free Press, New York.

Dhillon, G. & Ward, J. (2002). Chaos theory as a framework for studying information systems. *Information Resources Management Journal* 15(2), 320–337.

Dooley, K, Johnson, T. & Bush, D. (1995). TQM, chaos, and complexity. *Human Systems Management* 14, 1–16.

Dooley, K. & Van de Ven, A. (1999). Explaining complex organizational dynamics. *Organization Science* 10(3), 358–375.

Feigenbaum, M. (1978). Quantitative universality for class of non-linear transformations. *Journal of Statistical Physics* 19, 25–52.

Fontana, W. & Ballati, S. (1999). Complexity, *Complexity*, 4(3), 14–16.

Glass, R. (2003). A mugwump's-eye view of web work: Choosing a point of entry into contemporary software development debate. *Communications of the ACM* 46, 21–23.

Harvey, D. & Reed, M. (1996). Social science as the study of complex systems. In *Chaos Theory in the Social Sciences*, eds. L. D. Kiel & E. Elliott, University of Michigan Press, Ann Arbor.

Holland, J. (1995). *Hidden Order: How Adaptation Builds Complexity.* Addison-Wesley, Reading, MA.

Illinitch, A., Lewin A. & D'Aveni, R. (1998). New organizational forms and strategies for managing in hypercompetitive environments. In *Managing in Times of Disorder: Hypercompetitive Organizational Responses.* Sage Publications, Thousand Oaks, CA. Introduction xxi–xxxiv.

Jayanthi, S. & Sinha, K. (1998). Innovation implementation in high technology manufacturing: Chaos-theoretic empirical analysis. *Journal of Operations Management*, 16, 471–494.

Kamminga, H. (1990). What is this thing called chaos? *New Left Review*, 181, 49–59.

Katz, D. & Kahn, D. (1966). *The Social Psychology of Organizations.* John Wiley, New York.

Kauffman, S. (1993). The origins of order. In *Self-Organization and Selection in Evolution*, Oxford University Press, New York.

Kautz, K., Madsen, S. & Nørbjerg, J. (2007). Persistent problems and practices in information systems development. *Information Systems Journal*, 17, 217–239.

Keil, M., Cule, P., Lyytinen, K., & Schmidt, R. (1998). A framework for identifying software project risks. *Communications of the ACM*, Chicago. 41(11), 76–83.

Kellert, S. H. (1993). *In the Wake of Chaos.* University of Chicago Press.

Ketterer, J. (2006). Chaos and complexity: The uses of the foremost metaphor of the new millennium. *International Management Review*, 2(3), 34–52.

Kiel, D. (1995). Chaos Theory and Disaster Response Management: Lessons for Managing Periods of Extreme Instability. What Disaster Response Management Can Learn From Chaos Theory? Conference Proceedings, May 18–19, 1995, ed. Gus A. Koehler.

Kiel, D. (1997). Embedding chaotic logic into public administration thought: Requisites for the new paradigm. *Public Administration and Management: An Interactive Journal*, 2(4), 26–39. Available online at http://www.pamij.com/kiel.html

KPMG (2003). KPMG's International 2002-2003 Programme Management Survey. Available at http://www.kpmg.com.au/Portals/0/irmprm_pm-survey2003.pdf

Lee, A. S. (1994). Electronic mail as a medium for rich communication: An empirical investigation using hermeneutic interpretation. *MIS Quarterly*, 18(2), 143–157.

Levinthal, D. & Warglien, M. (1999). Landscape design: Designing for local action in complex worlds, *Organization Science, Special Issue: Application of Complexity Theory to Organization Science*, 10(3), 342–357.

Lyytinen, K. & Robey, D. (1999). Learning failure in information systems development. *Information Systems Journal*, 9(2), 85–101.

Lyytinen, K. & Rose, G. (2006). Information system development agility as organizational learning. *European Journal of Information Systems*, 15, 183–199.

Markus, M. L. (1983). Power, Politics and MIS Implementation, *Communications of the ACM*, 26, 430–444.

Mathiassen, L., Ngwenyama, O. K., & Aaen, I. (2005). Managing change in software process improvement. *IEEE Software*, 22(6), 84–91.

McDaniel, R. & Walls, M. (1997). Diversity as a management strategy for organizations: A view through the lenses of chaos and quantum theories. *Journal of Management Inquiry*, 6(4), 363–375.

Milis, K. & Mercken, R. (2002). Success factors regarding the implementation of ICT investment projects. *International Journal of Production Economics*, 80(1), 105–117.

Morcol, G. (1996). Fuzz and chaos: Implications for public administration theory and research. *Journal of Public Administration Research and Theory*, 6, 315–325.

Morel, B. & Ramanujam, R. (1999). Through the looking glass of complexity: The dynamics of organizations as adaptive and evolving systems. *Organization Science*, 10(3), 278–293.

Parker, M. & Stacey, R. (1994). *Chaos, Management and Economics: The Implications of Non-linear Thinking*. Institute of Economic Affairs, London.

Prigogine, I. & Stengers, I. (1984). *Order out of Chaos*. Bantam, New York.

Radzicki, M. (1990). Institutional dynamics, deterministic chaos, and self-organizing systems. *Journal of Economic Issues*, 24(1), 57–102.

Ropponen, J. & Lyytinen, K. (2000). Components of software development risk: How to address them? A project manager survey, *IEEE Transactions on Software Engineering*, 26(2), 98–113.

Senge, P. (1990). *The Fifth Discipline: The Art and Practice of the Learning Organization*. Doubleday, New York.

Shaw, M. (1981). *ALPHARD: Form and Content*. Springer-Verlag, New York.

Simon, H. (1962). *The Sciences of the Artificial*. MIT Press, Cambridge, MA.

Staccy, R. (1992). *Managing the Unknowable*. Jossey-Bass, San Francisco.

Stacey, R. (1996). *Complexity and Creativity in Organizations*. Berrett Koehler Publishers, San Francisco.

Stahl, B. C., Lichtenstein, Y., & Mangan, A. (2003). The limits of risk management: A social construction approach. *Communications of the International Information Management Association*, 3(3), 15–22.

Thietart, R. & B. Forgues. (1995). Chaos theory and organization. *Organization Science*, 6(1), 19–31.

Thompson, D. (1967). *Organizations in Action*. McGraw-Hill, New York.

Tsoukas, H. (1998). Introduction: Chaos, complexity and organization theory. *Organization*, 5(3), 291–313.

Wright, S. (1931). Evolution in Mendelian populations. *Genetics*, 16, 97–159.

Young, T. R. & Kiel, L. D. (1994). *Control, Prediction, and Non-linear Dynamics*. No. 28. Distributed as part of the Red Feather Institute Series on Non-Linear Social Dynamics. The Red Feather Institute, 8085 Essex, Weidman, Michigan, 48893.

Chapter 11

Design of the Research Workbench for Investigations Relying on Multitheoretical Support

In the previous chapters, we have offered to the reader an overview of the framework of neoclassical growth accounting, product cycle theory (PCT) and product life cycle model (PLC), as well as complex systems and chaos theories (CST and CT). It is our position that an investigation of complex entities would be more informed if a researcher, explicitly or implicitly, uses a multitheoretical foundation to support his/her study.

Let us consider an economic unit, an organization, to be a target of an inquiry. The scale of an organization is not important—it could be a firm, or an industry, or an economy, as long as we describe the subject of the study using the same model. In our case, we use the Cobb–Douglas production function to depict the high-level structure and behavior of the organization—this allows us to explain the economic growth and to identify the sources of such growth. At the same time, considering an organization from the perspective of CST and CT, we know that its internal structure and behavior would change over time under the competitive pressures that would impact the relevant control parameters. This means that the paths according to which capital and labor impact the economic outcomes would change over time—what worked in past may not work tomorrow. Thus, we should expect the development of new emergent properties—capital would impact the economic bottom line via new avenues.

As an economic unit, an organization would interface with the business environment via products or services that require investments—and the patterns of the investments, according to PCT, would change over time. Furthermore, over the period of the lifetime of a product or a service, according to PLC, the expenses and revenues, as well as the efficiency of the production of revenues or services, would also change. The benefit of using multiple theoretical perspectives, explicitly or implicitly, is that they offer to an investigator a variety of vantage points to assess the issue of interest. But, while the multitheoretical perspective allows us to mitigate a limitation associated with the representation of the target of the study, another problem remains.

A common limitation of a scientific inquiry is associated with the availability of the tools—data analytic methods, techniques, and methodologies. In order to conduct a broad inquiry, an investigator must have access not only to various models, frameworks, and theories describing the target of an investigation but also to a variety of the tools and techniques that could be used in the analysis.

We suggest that a toolbox of a researcher could contain tools and techniques that could also be available to the target of the study. This means that those tools, techniques, and methodologies that could be used by an economic entity to manage its performance could also be used by the investigators to analyze the structure, behavior, and performance of an organization. In this chapter, we demonstrate that an organization could rely on principles of cybernetics to design an analytic support system that would allow for monitoring its performance. Furthermore, we demonstrate some of the options allowing for implementing functionality of such a system—it is those tools and techniques that could be utilized by investigators in their inquiries, and it is those tools and techniques that we utilized in our investigations.

As a result of using a multitheoretical perspective supported by multitechnique, multitool methodologies, research inquiries could be designed to be more comprehensive than would otherwise be possible. This, simply put, would result in investigations that are theoretically rigorous and practically relevant.

Performance Analysis for Complex Systems

A perspective on organizations as complex systems (CSs) sensitive to the disturbances of the environment and characterized by periods of unstable behavior is by now well established. Recently, this perspective has also been extended to the context of organizational information systems (Samoilenko & Osei-Bryson 2007a). But while previous inquiries offered a set of insights and implications regarding the functionality of the control system capable of managing the unstable behavior of organizations, past studies shed no light regarding the possible design of such a system.

In this chapter, we aim to obtain a set of insights regarding the possible structural design of an analytic toolbox capable of analyzing the behavior of an economic unit, where *a behavior* is *a pattern of activities associated with the maintenance of an organizational goal.* We rely on the assumption of relativity of an organizational goal and focus on economic units that consider the states of their internal and external organizational environment in formulation of their strategies. Especially, we concentrate on the context where the achievement of an organizational goal is dependent on the level of performance of the organization, commonly measured in terms of the levels of the efficiency of utilization of inputs, effectiveness of the production of outputs, and efficiency of conversion of inputs into outputs. Resultantly, we limit the scope of our inquiry to *productivity-driven organizations.*

Due to the relativity of the concepts of efficiency and effectiveness of performance, productivity-driven organizations must take into consideration performance of competitors. However, the dynamic nature of the business environment will cause the levels of performance of competing organizations to change over time, which will require reassessment of the values of the levels of effectiveness and efficiency of an organization relative to its competitors. There is an apparent link between significant changes in productivity of the competitors of an organization and changes in the business environment—if productivity of the competitors has improved, then a productivity-driven organization must respond with its own improvements in productivity.

Calls for improvements in the levels of effectiveness and efficiency are endemic to productivity-driven organizations. Significant changes in the levels of effectiveness and efficiency often require structural reorganizations (e.g., adoption of Enterprise Systems, business process reengineering, etc.) that bring about the periods of unstable behavior, which, if not managed, escalates and becomes chaotic (Samoilenko 2008). Granted, some improvements in productivity do not require any structural transformation, but simply call for a gradual type of improvement in the level of performance (e.g., TQM, BPI, etc.). However, in the absence of perfect scalability, the appropriate course of action leading to improvements will change over time, primarily due to the law of diminishing returns. Resultantly, in a dynamic business environment, any static model that is used to describe the relationship between inputs and outputs will have a limited life span. In the absence of an adaptive mechanism that allows for discovering the new pathways for improving overall organizational performance, a productivity-driven organization will engage in the process of search and exploration, during which the number of the possible states or behaviors of an organization will proliferate. While periods of search and exploration are common to dynamic CSs, these periods also bring about the danger of a system not converging on the global maximum, and settling, instead, on multiple suboptimal local maxima. This outcome of the search and exploration process will result in instability of organizational behavior and overall suboptimal performance of an organization.

Keeping this in mind, we suggest that a *performance analysis system* (PAS) can fulfill the role of a workbench capable of analyzing an organizational behavior. However, in order to do so, the design of PAS must take into consideration two questions, namely,

1. Relative to what context is the performance of an organization to be measured?
2. What are the determinants of the given level of the relative performance?

Keeping these questions in mind, in this chapter, we present the design *of a robust PAS that is capable of analyzing the behavior of an economic unit*. The required major functionalities of the system are as follows:

1. The PAS must be able to assess the business environment to identify the relevant group of competitors.
2. It must be able to assess the efficiency of the organization relative to the performance of the competitors.
3. It must be able to provide insights regarding the nature of performance inefficiencies.
4. It must offer insights regarding the ways of improving performance vis-à-vis competitors.

For the purposes of this chapter, we provide the following definitions: First, we define *robust design* of a PAS as *a design allowing for analyzing the unstable behavior of an organization*. Second, we define an *unstable behavior* of an organization as *a behavior that is characterized by the perception of the loss of control* (Samoilenko 2008) *over the process of the maintenance of the organizational goal caused by the precipitous increase in the number of the possible states or behaviors of an organization* (Heylighen & Joslyn 2001). The *analysis of a behavior* is defined as *a capability to assess the number of the possible states or behaviors of an organization.*

A state or a behavior of an economic unit, in turn, is determined by the set of *constraints*, and constraints serve the purpose of reducing the uncertainty about the system's state or behavior (Heylighen & Joslyn 2001). We define a constraint as *an attribute or a set of attributes that accurately represents a particular dimension of the business environment in the model that an organization uses in its decision-making process*. In line with this definition, we propose that an unstable behavior is unconstrained (e.g., the model is inaccurate), whether the stable behavior is constrained (e.g., the model is correct). We note that a constrained model does not have to be complete.

Finally, taking the aforementioned into consideration, we define PAS as *a medium that allows an organization to reduce the uncertainty about its state and behavior by means of providing a set of constraints utilized in the decision-making process involved in the maintenance of an organizational goal*. Resultantly, one of the functional requirements of PAS is associated with the capability of creating a constrained (accurate) model of the business environment that is utilized by an organization.

The modern business environment is dynamic, and the assumption of instability of the internal and external environment is advantageous when designing PAS, for such assumption will make its design more robust. The meaning of a dynamic environment from the perspective of PAS is easy to decipher, for it implies the absence of a static set of constraints and relationships between constraints that are used in creating models of the business environment used in the decision-making process. Conversely, if the assumption of stability is embedded in the design (e.g., this could be exemplified by fixed data and process models that describe constraints and the relationships between these constraints), then this will greatly limit the capability of a PAS, for any significant disturbance could render a set of constraints and their relationships obsolete and invalidate the embedded models.

What could serve as a conceptual foundation for such a system? Traditional approaches are based on functionalism, and due to their reliance on stable models, functionalist approaches do not allow for a dynamic discovery of new relevant constraints and disposal of obsolete ones. Nor do functionalist approaches allow for the dynamic adaptation and evolution of their design models. Consequently, the use of the mainstream and extended functionalist methodologies is inadequate for designing PAS. New approaches are needed.

We propose that second-order cybernetics, which emphasizes principles of autonomy, adaptation, and self-organization of CSs, could serve as a valuable vantage point from which important insights regarding the design and structure of PAS could be obtained. Because the advocated perspective is context independent, we expect the results of this study to offer equally valuable insights regarding the design of the department-, firm-, industry-, or economy-level control systems.

Cybernetics-Based Analytic Support System

We would like to offer a justification for why the principles of cybernetics could serve as a solid foundation of the structural design of PAS; we argue that cybernetics can provide a suitable foundation for the following three reasons:

- First, the domain of inquiry of cybernetics includes not only artificially engineered systems but also naturally evolving ones. Organizations exemplify such engineered, yet evolving, systems.
- Second, the subject of inquiry of cybernetics is goal-directed systems. Organizations are goal-directed systems, the survival of which is dependent on achievement of the organizational goal.

- Third, the focus of cybernetics is on the use of information, models, and control actions by goal-directed evolving systems. Organizations are such systems, and organizations actively use information, models, and control actions in order to counteract internal and external disturbances that threaten the stability of the goal-oriented behavior.

Based on this brief assessment of eligibility, the use of principles of cybernetics for designing analytic systems appears reasonable. However, cybernetics is not concerned with the structure of the analytic system, but rather, with its function. For this reason, it cannot directly provide a prescriptive blueprint of what the possible design of a system might look like. Therefore, we take a three-step indirect approach to outlining the conceptual design of PAS.

First, in step 1, we offer an overview of the general principles of cybernetic systems. Second, in step 2, we outline, based on the principles identified in step 1, a set of functionalities that a cybernetic system must possess. Finally, in step 3, we offer a mapping of the functionalities identified in step 2 to the design components that could be used in the design of PAS.

Cybernetics: An Overview

Norbert Wiener was the founder of cybernetics as a field of study of *the control and communication in the animal and the machine* (Wiener 1948); this came to be known as *first-order cybernetics*. According to first-order cybernetics, a system under study can be represented by its simplified model and perceived to be independent of its observer. Some cyberneticists felt that the emphasis in studying the systems must be placed on autonomy, self-organization, cognition, and the role of the observer in the modeling of a system; later, this movement became known as *second-order cybernetics* (Heylighen & Joslyn 2001). Being a complement, rather an alternative, to its predecessor, second-order cybernetics (Von Foerster 1960; Ashby 1962) recognizes a system under study as an agent in its own right, actively interacting with the observer. The principles of second-order cybernetics are by now so firmly embedded in the overall foundation of cybernetics that it is appropriate to discuss this subject by simply referring to it as *cybernetics*, without making a clear-cut differentiation between first- and second-order cybernetics (Heylighen & Joslyn 2001). Overall, cybernetic systems are characterized by complexity, mutuality, complementarity, evolvability, constructivity, and reflexivity (Joslyn 1992); these characteristics and their interpretations are summarized in Table 11.1.

The fundamental principles of cybernetics are selective retention, autocatalytic growth, asymmetric transitions, blind variation, recursive systems construction, selective variety, requisite knowledge and incomplete knowledge (Heylighen 1992); these principles and the interpretations of the principles are summarized in Table 11.2.

Table 11.1 General Characteristics of Cybernetic Systems

Characteristic	Interpretation of the Characteristic
Complexity	Cybernetic systems are complex structures, with many heterogeneous interacting components.
Mutuality	Components of the cybernetic system interact in parallel, cooperatively, and in real time, creating multiple simultaneous interactions among subsystems.
Complementarity	Complementarity, which is brought about by complexity and mutuality, refers to the irreducibility of the level of analysis to any one dimension.
Evolvability	Cybernetic systems tend to evolve and grow in an opportunistic manner, rather than be designed and planned in an optimal manner.
Constructivity	Cybernetic systems tend to evolve and grow in size and complexity, while historically being bound to previous states.
Reflexivity	Cybernetic systems can enter into the feedback of reflexive self-application, which may result in the reflexive phenomena of self-reference, self-modeling, self-production, and self-reproduction.

Implications of the General Principles of Cybernetic Systems for Designing the PAS

Based on general principles of cybernetics, we can derive the set of implications regarding the required functionality of PAS. The set of proposed functionalities is provided in Table 11.3.

Structural Components of PAS

The set of implications outlined previously suggests the presence of a concept that is central to a productivity-driven organization, namely, that of the *superior stable configuration*. In line with the principles of cybernetics, stability of the behavior of a goal-oriented system is associated with the presence of a successful stable configuration of the system. Given the goal of achieving a high level of efficiency of conversion of inputs into outputs, a superior stable configuration in the context of a productivity-driven organization may imply *a model of conversion of inputs into output (input–output model) characterized by a high level*

Table 11.2 General Principles of Cybernetics

Principle	Interpretation of the Principle
Selective retention	Stable configurations of the system are retained, while unstable ones are eliminated.
Autocatalytic growth	The stable configurations, which facilitate the appearance of configurations similar to themselves, will become more numerous.
Asymmetric transitions	A transition from an unstable configuration to a stable one is possible, while the transition from stable to unstable configuration is not.
Blind variation	The variation processes cannot identify in advance which of the produced variants will turn out to be selected.
Selective variety	The larger the variety of configurations a system undergoes, the larger the probability that at least one of these configurations will be selectively retained.
Recursive systems construction	BVSR (blind-variation-and-selective-retention) processes recursively construct stable systems by the recombining the stable building blocks.
Requisite variety	The larger the variety of actions available to a control system, the larger the variety of perturbations it is able to compensate.
Requisite knowledge	In order to adequately compensate perturbations, a control system must "know" which action to select from the variety of available actions.
Incomplete knowledge	The model embodied in a control system is necessarily incomplete.

of relative efficiency. Consequently, we put forward the following propositions (Table 11.4):

> Proposition 1: Stability of the organizational behavior of a productivity-driven organization is dependent on the presence of the stable input–output model.
> Proposition 2: Accomplishment of the organizational goal of a productivity-driven organization is dependent on the creation and implementation of a stable input–output model characterized by a high level of relative efficiency.
> Proposition 3: In order to control the behavior of a productivity-driven organization, PAS must be able to create and identify superior stable configurations, represented by the input–output models characterized by a high level of relative efficiency.

Table 11.3   Implications of General Principles of Cybernetics on the Functionality of PAS

Principle	Implication of the Principle with Regard to the Functionality of PAS
Selective retention	PAS must be able not only to contribute to the development of the stable organizational configurations but also to recognize them as such. For example, a successful product development process or a particularly productive organizational substructure must be identified (e.g., by using internal benchmarking?) and then retained within the organization.
Autocatalytic growth	PAS must promote the increase of the stable successful structures within an organization; this could be done through the process of organizational learning utilizing knowledge management systems.
Asymmetric transitions	PAS must be able to recognize inferior solutions in advance, possibly by means of simulation and modeling.
Blind variation	While PAS might not be able to ensure the production of only successful configurations, it must be able to identify the obviously inferior ones. This could be done by means of using what-if analysis and scenario building.
Selective variety	PAS must allow for a large variety of its own possible configurations; this could mean that PAS should be characterized by a large number of independent components.
Recursive systems construction	PAS must be able to construct stable systems by the recombination of the stable subsystems and elements, which suggests high cohesion and loose coupling of PAS components.
Requisite variety	PAS must not be constructed for one specific purpose or with a predefined functionality; instead, it must constantly be in the process of growth and development.
Requisite knowledge	PAS must be able to select from multiple available actions an appropriate response to a particular event. This may mean that PAS must have scenario-building capabilities, possibly utilizing modeling and simulations.
Incomplete knowledge	PAS must not function in a closed environment; instead, PAS must be able to interact freely with not only the competitive environment of the firm but the global environment as well.

Table 11.4 Functionality of PAS in Productivity-Driven Organizations

Functionality of PAS	Interpretation
It must contribute to the development of stable organizational configurations.	Stable configurations allow for the presence of a consistent model depicting the process of conversion of inputs into outputs by an organization, in the form of an input–output model.
It must promote an increase in stable successful structures within an organization.	Stable configurations are promoted on the basis of the effectiveness and efficiency of conversion of inputs into outputs in such a way that every distinct consistent model is characterized by the distinct level of relative efficiency of conversion of inputs into outputs.
It must be able to recognize inferior solutions in advance.	Inferior solutions represent stable configurations characterized by lower levels of effectiveness and efficiency of conversion of inputs into outputs, while superior solutions represent stable configurations characterized by higher levels of effectiveness and efficiency.
It must allow for a large variety of its own possible configurations.	A process of evaluation of the stability and quality of configurations is independent of the structure of the input–output model representing a given stable configuration; single PAS must be able to evaluate many configurations.
It must be able to construct stable systems by the recombination of stable subsystems and elements.	A process of evaluation of the stability and quality of configurations must rely on information-rich components that could be reused in new processes.
It must be able to select from multiple available actions an appropriate response to a particular event.	A process of evaluation of the stability and quality of configurations must allow for variations in inputs and outputs, as well as variations in the process of conversion itself; PAS must be able to identify not only the superior configurations but also the factors that impact the quality of configurations.
It must be able to function in the open environment.	Stable configurations must be regularly assessed and reassessed relative to the internal and external organizational environment.

Keeping the relativity of the concept of efficiency in mind, the functionality of PAS can be presented as encompassing two subsets of functionalities: internally oriented and externally oriented. The externally oriented functionality of the PAS is directed toward evaluating the external competitive environment of a productivity-driven organization, as well as identifying the differences between the current state of the organization and the states of its competitors. Internally oriented functionality, on the other hand, is directed toward optimization of the level of productivity of the organization, as well as toward identification of the factors impacting the efficiency of the input–output process. We suggest that the functionalities outlined previously could be implemented using combination of parametric and nonparametric data analytic and data mining techniques, such as data envelopment analysis (DEA), cluster analysis (CA), decision trees (DTs), neural networks (NN), association rule mining (AR), and regression analysis (RA).

Table 11.5 provides a summary of how the aforementioned components could be utilized to implement the required functionality. While the complete design of

Table 11.5 Suggested Structural Implementation of the Functionality of Cybernetic-Centered PAS

Functionality	System Requirement	Structural Components
Externally oriented	Detection of changes in the external competitive environment	Cluster analysis
	Identification of the possible factors that resulted in changes	Combination of cluster analysis and decision trees
	Identification of the relative efficiency of the organization relative to its competitors	Data envelopment analysis
	Identification of the factors associated with the differences in the relative efficiencies of the competitors	Combination of data envelopment analysis, cluster analysis, association rule mining, and decision trees
Internally oriented	Identification of the factors impacting the current level of the relative efficiency of the input–output process	Regression analysis
	Identification of the most effective ways of increasing the level of efficiency of the input–output process	Combination of data envelopment analysis and neural networks

PAS capable of analyzing the behavior of economic units proposed in this chapter is in its conceptual form, the parts of the outlined functionality have been implemented and will be demonstrated in the following chapters.

Conclusion

Any sound research study must be conducted on a solid foundation, and a defensible methodology is one of the elements attesting to the rigor of an investigation. While a single data analytic method may offer a valuable insight, such insight would be necessarily limited because it looks at the problem from a single perspective. Complex problems call for multiple perspectives to be considered, where each perspective serves not only as a contributor to a better understanding of the problem but also as a synergetic component complementing other perspectives allowing, together, for a fuller view to emerge. Thus, it is not surprising that benefits of multiple perspectives in research and practice have been noted (Harrits 2011; Poteete et al. 2010).

In the case of quantitative data analysis, a complex question is unlikely to be answered via a single method; hence, there is a need for multimethod methodologies. Prior to constructing such a methodology, however, two questions must be answered. First, what is the basis for a multimethod methodology? Second, what is the process of designing such a methodology? We believe that in this chapter, we offered an answer to the first question by showing that CST could offer theoretical support for designing multimethod methodologies. We also offered an answer to the second question—we have demonstrated that a methodology could be built on the premise of mapping the system's behavior to various parametric and nonparametric data analytic tools.

We believe that there is more to theoretically sound multimethod methodologies than a sequenced collection of analytic methods. We argue that such methodologies offer their creators the benefit of what we call *information emergence*, where multiple methods working together in synergy offer a researcher richer information than would be available if the methods were applied separately and the resulting information combined additively.

References

Ashby, W. R. (1962). Principles of the self-organizing systems. In *Int. Tracts in Computer Science and Technology and Their Applications, vol. 9: Principles of self-organization*, eds., H. Von Foerster & G. W. Zopf, Pergamon Press, Oxford, pp. 255–278.

Harrits, G. S. (2011). More than method? A discussion of paradigm differences within mixed methods research. *Journal of Mixed Methods Research*, 5(2), 150–166.

Heylighen, F. (1992). Principles of systems and cybernetics: An evolutionary perspective. In *Cybernetics and Systems*, ed. R. Trappl, World Science, Singapore, pp. 3–10.

Heylighen, F. & Joslyn, C. (2001). Cybernetics and second-order cybernetics. In *Encyclopedia of Physical Science & Technology* (3rd ed.), ed. R. A. Meyers, Academic Press, New York.

Joslyn, C. (1992). The nature of cybernetic systems. In *Principia Cybernetica Web*, eds. F. Heylighen, C. Joslyn and V. Turchin, Principia Cybernetica, Brussels. Available online at http://pespmc1.vub.ac.be/CYBSNAT.html

Poteete, A., Janssen, M., & Ostrom, E. (2010). *Working Together: Collective Action, the Commons, and Multiple Methods in Practice*. Princeton University Press, Princeton.

Samoilenko, S. (2008). Information systems fitness and risk in IS development: Insights and implications from chaos and complex systems theories. *Information Systems Frontiers*, 10(3), 281–292.

Samoilenko, S. & Osei-Bryson, K. M. (2007a). Chaos theory as a meta-theoretical perspective for IS strategy: Discussion of the insights and implications. In *Proceedings of the Southern Association for Information Systems Conference*, Jacksonville, FL, USA, May 22–23.

Samoilenko, S. & Osei-Bryson, K. M. (2007b). Increasing the discriminatory power of DEA in the presence of the sample heterogeneity with cluster analysis and decision trees. *Expert Systems with Applications*, 34(2), 1568–1581.

Samoilenko, S. & Osei-Bryson, K. M. (2010). Determining sources of relative inefficiency in heterogeneous samples: Methodology using cluster analysis, DEA and neural networks. *European Journal of Operational Research*, 206, 479–487.

Von Foerster, H. (1960). On self-organizing systems and their environments. In *Int. Tracts in Computer Science and Technology and Their Applications, Vol. 2: Self-Organizing Systems*, eds. M. C. Yovits & S. Cameron, Pergamon Press, Oxford, pp. 31–51.

Wiener, N. (1948). *Cybernetics or Control and Communication in the Animal and the Machine*. Paris, Hermann et Cie—MIT Press, Cambridge, MA.

Chapter 12

Investigation of Determinants of Total Factor Productivity: An Analysis of the Impact of Investments in Telecoms on Economic Growth in Productivity in the Context of Transition Economies

Introduction

An overall reduction in the level of investments in information and communication technologies (ICTs) is just one of the likely results brought about by current economic conditions. The depth of reduction, however, is not uniform across economies. During the 1990s, the impact of investments in ICT on the macroeconomic bottom line was especially pronounced in the United States (Oliner & Sichel 2000;

Van Ark et al. 2002; Jorgenson 2003) and, albeit to a lesser degree, in some of the Organisation for Economic Co-operation and Development (OECD) countries (Colecchia & Schreyer 2002; Van Ark et al. 2002). Because it has been established that investments in ICT contribute reliably to the macroeconomic bottom line of the developed countries (OECD 2002), the reduction in the level of investments in these economies might be minimal.

In the context of developing and transition economies (TEs), however, the outcomes of investments in ICT are mixed (Dewan & Kraemer 2000); as a result, a TE might be required to present evidence that such investments can be effectively and efficiently transformed into macroeconomic output (Heeks & Molla 2009). Moreover, despite the significant increase in the level of adoption of ICT by European countries in the period around 1998–2002, their levels of productivity, unlike those in the US, started to decline, and disparities in the macroeconomic outcomes of investments in ICT among economies became obvious (Daveri 2002). Such a reduction in growth, even among well-heeled developed countries, clearly requires TEs to demonstrate that their limited technical, financial, and human resources are not wasted (Indjikian & Siegel 2005).

From the perspective of the widely used framework of neoclassical growth accounting (Brynjolfsson & Hitt 1996), an increase in the macroeconomic bottom line, often represented by gross domestic product (GDP), may come from two sources. The first source is provided by the available levels of capital and labor, and the second source is reflected by total factor productivity (TFP). While the most straightforward way of improving a macroeconomic bottom line is simply by increasing the levels of available capital and labor, for economies with limited resources, a contribution coming from TFP is preferable. This is because TFP represents the macroeconomic growth that is not accounted for by the increase in the levels of capital and labor, and is not subject to the law of diminishing returns. Most investigations linking ICT and growth in productivity have been conducted in the context of developed countries (McGuckin et al. 1998).

The overall purpose of this research, conducted in the context of European TEs, is to identify whether the ICT sector of those economies exhibited growth in TFP (heretofore we use TFP and productivity interchangeably) associated with investments in ICT; specifically, we investigate two periods: a period of early transition (1993–2002) and a period of a later transition (2003–2008). We have decided to start our analysis in 1993 because that year provided a common starting point for TEs. Our reasoning was that it took a year from the fall of the Berlin Wall in 1991 for the process of transition to start, but year 1992 as a starting point could have favored "early starters." After that, 10 years seemed to be a reasonable period of time to allow for the changes to occur and the first results to take place. The duration of the second period was primarily determined by the availability of relevant data and the beginning in 2008 of the period of economic instability. In this investigation, we use *telecoms* as a surrogate for general ICT. Not only is it a subset of ICT, but investments in telecoms are also common to almost all economies of the world.

Overall, we aim to answer two questions: (1) Do the TEs in the sample exhibit continuous growth in TFP? (2) What are some of the factors impacting growth in TFP? The potential for investments in ICT to generate high levels of productivity growth in the context of TEs has been noted (Indjikian & Siegel 2005); however, it is not clear whether that potential has been realized. Previous investigations established that in order for investments in ICT to impact the economic bottom line, the level of investments must be above a certain threshold, and such investments must be complemented by other factors, notably, investments in human resources (Bresnahan et al. 2002; OECD 2004). In this study, we focus on the following questions that emerge from our major research questions:

1. RQ1: Does the given TE exhibit annual growth in TFP?
2. RQ2: Does the given TE exhibit continuous growth in TFP?
3. RQ3: Is there a relationship between changes in the level of investments in telecoms and changes in TFP?
4. RQ4: Is there a relationship between changes in the level of full-time telecom staff and changes in TFP?
5. RQ5: Do changes in the level of investments in telecoms and changes in the level of full-time telecom staff produce a complementary effect on changes in TFP?
6. RQ6: Is there a relationship between the level of investments in telecoms and TFP?
7. RQ7: Is there a relationship between the level of full-time telecom staff and TFP?
8. RQ8: Are the level of investments in telecoms and the level of full-time telecom staff complementary in terms of their effect on TFP?
9. RQ9: Do changes in the ratio of revenues to investments impact TFP?
10. RQ10: Does the ratio of revenues to investments have an impact on TFP?
11. RQ11: What is the dominant source of growth in TFP?

The importance of these research questions is intuitive, for answers not only will allow for a prudent allocation of limited resources (e.g., human and investment capital) but could also offer some insights regarding the more effective and efficient conversion of investments into the macroeconomic bottom line. To answer the research questions, we employ data envelopment analysis (DEA), multivariate regression (MR), and ordinary least squares regression (OLS) as our main data analytic tools. DEA is a widely used method for evaluating productivity and performance and is commonly combined with other techniques, such as cluster analysis (e.g., Shin & Sohn 2004; Hirshberg & Lye 2001), neural networks (e.g., Samoilenko & Osei-Bryson 2008a), decision trees (e.g., Samoilenko & Osei-Bryson 2007; Wu 2009), and regression analysis (e.g., Cooper & Tone 1997). We explore our research questions within the context of two time periods, 1993–2002 and 2003–2008. This permits us to assess whether there are any significant differences between the period of early transition (1993–2002) and the period of late transition (2003–2008).

Description of the Data

Data for this study were obtained from the *World Development Indicators* database, which is the World Bank's (http://web.worldbank.org) comprehensive database on development data, and the *Yearbook of Statistics*, which is published yearly by International Telecommunication Union (http://www.itu.int). The designation of TEs is applied to the countries that are in the process of transitioning from a centrally planned economy to a market-oriented economy. To minimize the heterogeneity of our sample, we identified TEs that started the transition at about the same time. Looking at the 25 countries classified as TEs in Europe and the former Soviet Union by International Monetary Fund (IMF) (2000), we selected the following 17 countries (data on the remaining was incomplete): Albania, Armenia, Azerbaijan, Belarus, Bulgaria, Czech Republic, Estonia, Hungary, Kazakhstan, Latvia, Lithuania, Moldova, Poland, Romania, Slovakia, Slovenia, and Ukraine. For the selected TEs, we were able to construct two data sets: a data set covering the period of early transition, from 1993 to 2002, and a data set spanning the period of later transition, from 2003 to 2008. However, TEs are not homogeneous in terms of development. It has been noted that TEs tend to share economic characteristics with both developed and less developed economic regions (OECD 2004).

Research Methodology

In order to answer the 11 research questions of this study, we employ a four-phase methodology supported by three data analytic tools; a summary is provided in Table 12.1.

Phase 1

To approach our research problem, we rely on the neoclassical framework of growth accounting (Solow 1957). A lack of theoretical support has been noted to be one of the serious limitations for conducting investigations assessing the impact of ICTs (Gomez & Pather 2012; Heeks 2010); by employing a widely used theoretical framework, we place this inquiry on a solid footing.

From the three inputs used by growth accounting, only capital and labor could be observed in our data, while TFP serves as a residual (often referred to as the *Solow residual*) term capturing that contribution to a macroeconomic outcome (GDP or revenues from telecoms), which is left unexplained by the changes in the

Investigation of Determinants of Total Factor Productivity ■ 147

Table 12.1 The Four-Phase Methodology of the Study

Phase #	Research Question	Technique
1	• RQ1: Does the given TE exhibit annual growth in TFP? • RQ2: Does the given TE exhibit continuous growth in TFP?	DEA, MI
2	• RQ3: Is there a relationship between changes in the level of investments in telecoms and changes in TFP? • RQ4: Is there a relationship between changes in the level of full-time telecom staff and changes in TFP? • RQ5: Do changes in the level of investments in telecoms and changes in the level of full-time telecom staff produce a complementary effect on changes in TFP? • RQ6: Is there a relationship between the level of investments in telecoms and TFP? • RQ7: Is there a relationship existing between the level of full-time telecom staff and TFP? • RQ8: Are the level of investments in telecoms and the level of full-time telecom staff complementary in terms of their effect on TFP?	MR
3	• RQ9: Do changes in the ratio of revenues to investments impact TFP? • RQ10: Does the level of the ratio of revenues to investments have an impact on TFP?	OLS
4	• RQ11: What is the dominant source of changes in TFP?	DEA, MI

levels of capital and labor. The value of TFP could be obtained by calculating the value of the Malmquist index (MI) of TFP growth (Malmquist 1953; Caves et al. 1982), which can be constructed based on the results of DEA (Färe et al. 1994).

Essentially, the approach is based on performing DEA in two points of time: let us say *Year1995* and *Year1996*. Then, for a given decision-making unit (DMU), the period of time (*Year1996–Year1995*) can be represented as the distance between the data point at time *Year1995* and the data point at time *Year1996*. For each DMU in the sample, the distance between these data points is reflective of the change in this DMU's TFP, which is represented by the MI. If the obtained value of MI is greater than 1, then the change is positive; conversely, if MI is less than 1, the change is negative. In the case of economic growth, we expect that the efficiency frontier for

Table 12.2 Input and Output Variables of the DEA Model

Input Variables of the DEA Mode	Output Variables of the DEA Model
GDP per capita (in current US$) Full-time telecommunication staff (% of total labor force) Annual telecom investment per telecom worker (in current US$) Annual telecom investment (% of GDP in current US$) Annual telecom investment per capita (in current US$) Annual telecom investment per worker (in current US$)	Total telecom services revenue per telecom worker (in current US$) Total telecom services revenue (% of GDP in current US$) Total telecom services revenue per worker (in current US$) Total telecom services revenue per capita (in current US$)

a given set of DMUs would change its position favorably over time, which will be indicated by values of MI > 1.

In the case of our investigation, we obtain nine values of MI for the first period (1993–2002) and five values of MI for the second period (2003–2008), where each value of MI serves as a measure of TFP for a given year. For the DEA part of the methodology, we have identified a model that consists of the six input and four output variables (see Table 12.2). The main goal that we pursue in performing DEA is to find out how efficient our set of TEs is in converting investment inputs into revenue outputs. Therefore, we did not include any other inputs or outputs such as those related to infrastructure, capabilities, utilization, etc. It is should be mentioned that the purpose of our DEA model is not to reflect the path by which the investments are transformed into revenues over the course of one year; rather, the intent is to depict a "fiscal efficiency" of the TEs regarding their investments in telecoms (Table 12.2).

Instead of presenting the levels of investments and revenues in absolute dollar terms, we chose to represent them in relative units. The intent in doing so is to counter the differences between TEs in terms of their size, population, and level of wealth, while representing the investments and revenues more broadly (i.e., relative to the whole population, labor force of a country, and the telecom industry).

Phase 2

The second phase is dedicated to determining the presence of statistically significant relationships between the values of TFP, investments in telecoms, and a full-time telecom workforce. While Samoilenko and Osei-Bryson (2008b) presented evidence of the relationships between investments in ICT and full-time ICT labor and GDP in the context of TEs, it has not been shown that such a link exists with regard to TFP. Furthermore, we aim to determine whether these two variables (investments and labor) are complementary in their effect on TFP. The second and third phases of our investigation employ regression analysis. In this investigation, we are interested in the presence of a statistically significant effect of the independent variables on the dependent variable. In this case, the general model of MR takes the form

$$Y = a + b_1 * X_1 + b_2 * X_2 + b_3 * X_1 X_2 + e$$

and the test for presence of a statistically significant effect amounts to testing the null hypothesis

$$H0: b_1, b_2, b_3 = 0$$

In the case of $b_1, b_2, b_3 \neq 0$, we are able to reject the null hypothesis of the study. However, because TFP represents change in productivity, we can also express the variable *annual telecom investment (in current US$)* in the form of the change in the level of investment that took place over the year; similarly, we can express the variable *full-time telecommunication staff* in the form of the change in the level of full-time telecom labor that took place over the same year. As a result, in the first MR model, we use the variables listed in Table 12.3.

Consequently, the first MR model takes the following form:

$$\begin{aligned} TFP = a &+ b_1 * \text{annual change in investments in telecoms} \\ &+ b_2 * \text{annual change in full-time telecoms staff} \\ &+ b_3 * \text{annual change in investments in telecoms} \\ &\quad * \text{annual change in full-time telecoms staff} + e \end{aligned}$$

Table 12.3 Variables in the First MR Model

Term in the First MR Model	Variable
Y1	TFP
X1	Annual change in the level of telecom investment (in current US$)
X2	Annual change in the level of full-time telecommunication staff
X1*X2	Interaction term

Table 12.4 Variables in the Second MR Model

Term in the Second MR Model	Variable
Y1	TFP
X3	Annual telecom investment (in current US$)
X4	Full-time telecommunication staff
X3*X4	Interaction Term

We also aim to determine whether the values of TFP are affected by the levels of *annual telecom investment (in current US$)* and *full-time telecommunication staff*. In the second MR model, we use the variables listed in Table 12.4.

As a result, the second MR model takes the following form:

$$\text{TFP} = a + b_4 * \text{ annual telecom investments}$$
$$+ b_5 * \text{ full-time telecoms staff}$$
$$+ b_6 * \text{ annual telecom investments}$$
$$* \text{ full-time telecoms staff} + e$$

Phase 3

There appears to be an agreement on the importance of the effectiveness and efficiency of the workforce with regards to productivity (Samoilenko & Osei-Bryson 2008a). It is not clear, however, whether there is a relationship between the capacity of the workforce to convert investments into revenues and TFP in the context of TEs. In order to investigate this issue further, we employ a proxy variable, *conversion efficiency*, which is represented by the ratio of the variable *total telecom services revenue (in current US$)* to the variable *annual telecom investment (in current US$)*, per full-time telecom worker. The use of generalized indexes (e.g., *e-readiness, network readiness, Information Technology (IT) diffusion*) and proxy variables (e.g., *IT intensity, access, connectivity*) is common in studies of the economic impact of ICT (Indjikian & Siegel 2005). The proxy variable *conversion efficiency* serves as a good intuitive representation of the state of a given economy in terms of investments in and revenues from telecoms, as well as the quality of its full-time telecom labor force. For example, given two hypothetical TEs with the values of 2 and 3 of the proxy *conversion efficiency*, one can immediately assess that a TE, ceteris paribus, can generate a higher level of revenues from the same level of investments while using the same level of workforce of another TE. Once the values of conversion efficiency were calculated for TEs in our sample, we then calculated the values

Investigation of Determinants of Total Factor Productivity ■ 151

Table 12.5 Variables in the First OLS Model

Term in the First OLS Model	Variable
Y1	TFP
X5	Change in conversion efficiency [annual change in ratios of total telecom services revenue (in current US$) to annual telecom investment (in current US$)] per full-time telecom worker

Table 12.6 Variables in the Second OLS Model

Term in the Second OLS Model	Variable
Y1	TFP
X5	Conversion efficiency [ratio of total telecom services revenue (in current US$) to annual telecom investment (in current US$)] per full-time telecom worker

of *change in conversion efficiency*, representing the annual changes in the values of conversion efficiency for all of the economies in our sample. This allows us to construct two OLS models consisting of the variables listed in Tables 12.5 and 12.6.

The first OLS model allows for determining the presence of statistically significant relationships between the values of TFP and the values of change in conversion efficiency, and utilizes the formulation of the following OLS model:

$$\text{TFP} = a + b_7 * \text{ change in conversion efficiency} + e$$

Using the second OLS model allows for determining the presence of statistically significant relationships between the values of TFP and the values of conversion efficiency, and utilizes the formulation of the following OLS model:

$$\text{TFP} = a + b_8 * \text{ conversion efficiency} + e$$

Phase 4

DEA allows an investigator not only to obtain the scores of the relative efficiency for each DMU in the sample and to calculate MI to evaluate the overall growth in productivity, but also to decompose the values of MI into two components.

The first component is represented by the *change in technology* (TC), and the second component is represented by the *change in efficiency* (EC) that took place during a period of time. Consequently, taking into consideration that TFP is represented by MI, MI = (TC + EC), we can express the possible contribution of investments in telecoms to the macroeconomic bottom line as a combination of two components:

1. Investments in telecoms → Revenues from telecoms → GDP
2. Investments in telecoms → (TC + EC) → GDP

By knowing relative values of TC and EC, we can determine whether the growth in productivity was primarily due to the technology (in the case of TC > EC) or to the efficiency (in the case of EC > TC). This will allow us to gain insights into the nature of the weak link of the overall chain of the macroeconomic impact of the investments.

Null Hypotheses of the Study

Phase 1

For the first phase of our inquiry, we formulate the following two hypotheses. Given the value of MI, representing the annual change in TFP, and the value of AMI, representing the averaged value of TFP from 1993 to 2002,

H01: For each of the 18 TEs in the sample for each year in the 1993–2002 period, MI < 1.
H02: For each of the 18 TEs in the sample for the period from 1993 to 2002, AMI <1.

Similar hypotheses were formulated for the period from 2003 to 2008. Testing of H01 allows us to identify TEs that exhibited annual growth in TFP, while testing H02 allows us to identify TEs that exhibited continuous growth in TFP over the given period.

Phase 2

Given the first MR model

$$\begin{aligned}\text{TFP} = a + b_1 &* \text{ annual change in investments in telecoms} \\ + b_2 &* \text{ annual change in full-time telecoms staff} \\ + b_3 &* \text{ annual change in investments in telecoms} \\ &* \text{ annual change in full-time telecoms staff} + e,\end{aligned}$$

we formulate the following three hypotheses:

H03: $b_1 = 0$ at $\alpha = .05$ level of significance
H04: $b_2 = 0$ at $\alpha = .05$ level of significance
H05: $b_3 = 0$ at $\alpha = .05$ level of significance

Testing H03 allows us to determine whether a given TE exhibited a statistically significant relationship between changes in the level of investments in telecoms and changes in TFP, while testing H04 allows us to determine whether a given TE exhibited a statistically significant relationship between changes in the level of full-time telecom staff and changes in TFP. Finally, testing H05 allows us to determine whether for a given TE, the changes in the level of investments in telecoms and changes in the level of full-time telecom staff produce a complementary effect on changes in TFP.

Similarly, given the second MR model

$$\text{TFP} = a + b_4 * \text{annual telecom investments} \\ + b_5 * \text{full-time telecom staff} \\ + b_6 * \text{annual telecom investments} \\ * \text{full-time telecoms staff} + \xi$$

we formulate the following three hypotheses:

H06: $b_4 = 0$ at $\alpha = .05$ level of significance
H07: $b_5 = 0$ at $\alpha = .05$ level of significance
H08: $b_6 = 0$ at $\alpha = .05$ level of significance

Testing H06 allows us to determine whether a given TE exhibited a statistically significant relationship between the level of investments in telecoms and TFP, while testing H07 allows us to determine whether such a relationship existed between the level of full-time telecom staff and TFP. Finally, testing H08 allows us to determine whether level of investments in telecoms and the level of full-time telecom staff are complementary in impacting TFP.

Phase 3

The first OLS model utilizes the following formulation:

$$\text{TFP} = a + b_7 * \text{change in conversion efficiency} + \xi$$

and is concerned with the testing of the following null hypothesis:

H09: $b_7 = 0$ at $\alpha = .05$ level of significance

Testing H09 allows us to determine whether for a given TE, changes in conversion efficiency impact TFP. The second model takes the form of

$$\text{TFP} = a + b_8 * \text{ conversion efficiency} + \xi$$

and is concerned with the testing of the following null hypothesis:

H010: $b_8 = 0$ at $\alpha = .05$ level of significance

Testing H010 allows us to determine whether for a given TE, conversion efficiency has an impact on TFP.

Phase 4

The last phase of our investigation relies on the equation

$$\text{MI} = \text{TC} + \text{EC}$$

and is concerned with testing the following hull hypothesis:

H011: TC > EC

Testing H011 allows us to determine whether the growth in productivity was primarily driven by a better technology (e.g., via acquisition of a new technology) or whether it was driven by the improvements in efficiency (e.g., via more effective utilization of the available technology).

Results and Discussion

In order to perform the DEA part of the data analysis we used the software application OnFront, version 2.02, and SAS Enterprise Miner (EM) to conduct MR and OLS analysis. The data subjected to MR and OLS were standardized prior to analysis. The results from phase 1 that are displayed Table 12.7 suggest that most TEs exhibited continuous growth both in the period of the early transition (i.e., 1993–2002) and the period of late transition (2003–2008). Further with regard to the two research questions (i.e., RQ1, RQ2) of this phase, the performance of

Table 12.7 Results of Phase 1: RQ1 and RQ2

Country	RQ1: Annual Growth in Productivity? (Y/N) 1993–2002	2003–2008	RQ2: Continuous Growth in Productivity? (Y/N) 1993–2002	2003–2008
Albania	N	N	Y	N
Armenia	N	N	Y	Y
Azerbaijan	N	N	Y	Y
Belarus	N	N	Y	Y
Bulgaria	N	N	Y	N
Czech Republic	N	Y	Y	Y
Estonia	N	N	Y	Y
Hungary	Y	N	Y	Y
Kazakhstan	N	N	Y	N
Latvia	N	N	Y	Y
Lithuania	N	N	Y	Y
Moldova	N	N	Y	Y
Poland	N	N	Y	N
Romania	N	N	Y	Y
Slovak Republic	N	N	Y	N
Slovenia	N	N	Y	Y
Ukraine	N	N	Y	Y
Yes: proportion	.06	.06	1.00	.71
Similarity	.88		.71	

most of the TEs in the late transition period was similar to what it was in the early transition period.

The results of phase 2 that are displayed in Table 12.8 suggest that the level of investments in telecoms and the level of the full-time telecom staff are not viable predictors of productivity growth.

Table 12.8 Results of Phase 2: RQ3, RQ4, and RQ5

Country	RQ3: Relationship between Changes in the Level of Investments in Telecoms and Changes in Productivity		RQ4: Relationship between Changes in the Level of Full-time Telecom Staff and Changes in Productivity		RQ5: Complementarity of Changes in the Levels of Investments in Telecoms and Full-time Telecom Staff on Changes in Productivity	
	1993–2002	2003–2008	1993–2002	2003–2008	1993–2002	2003–2008
Albania	N	N	N	N	N	N
Armenia	N	N	N	N	N	N
Azerbaijan	N	N	N	N	N	N
Belarus	Y	N	Y	N	Y	N
Bulgaria	N	N	N	N	N	N
Czech Republic	N	N	N	N	N	N
Estonia	N	N	N	N	N	N
Hungary	N	N	N	N	N	N
Kazakhstan	N	N	N	Y	N	N

(Continued)

Table 12.8 (Continued) Results of Phase 2: RQ3, RQ4, and RQ5

Country	RQ3: Relationship between Changes in the Level of Investments in Telecoms and Changes in Productivity 1993–2002	RQ3: 2003–2008	RQ4: Relationship between Changes in the Level of Full-time Telecom Staff and Changes in Productivity 1993–2002	RQ4: 2003–2008	RQ5: Complementarity of Changes in the Levels of Investments in Telecoms and Full-time Telecom Staff on Changes in Productivity 1993–2002	RQ5: 2003–2008
Latvia	N	N	N	N	N	N
Lithuania	N	Y	N	Y	N	N
Moldova	N	Y	N	Y	N	N
Poland	N	N	N	N	N	N
Romania	N	N	Y	N	N	N
Slovak Republic	N	N	N	N	N	N
Slovenia	N	N	N	N	N	N
Ukraine	N	N	N	N	N	N
Yes: proportion	.06	.12	.12	.18	.06	.00
Similarity	.82		.71		.94	

This simply means that no TE in our sample, other than Belarus, can claim that the level of investment in telecoms serves as a determinant of the growth in productivity. It would appear that 17 TEs in our sample have already achieved a threshold level of investments sufficient to produce a stream of revenue but insufficient in the absence of complementary investments to produce any significant changes in productivity. Also, no TE, other than Belarus or Romania, can demonstrate that the growth in productivity is determined by the level of the full-time telecom staff. This might mean that the rest of the 16 TEs must concentrate not on increasing the quantity of full-time workers but, rather, on increasing the efficiency and effectiveness of the utilization of investments, including increasing the quality of the existing level of their full-time workforce.

The results of phase 2 displayed in Table 12.9 identify only Kazakhstan as having an economy in which the level of investments in telecoms and the level of full-time telecom staff have a statistically significant impact on changes in productivity (H06 and H07 were rejected). These results also suggest a presence of the complementary effect of the levels of labor and capital on productivity for Kazakhstan (H08 was rejected).

The results of phase 3 displayed in Table 12.10 allow us to conclude that in the context of our set of TEs, a higher ratio of revenues to investments per full-time telecom worker does not indicate a growth in productivity and the presence of a spillover effect. However, we can observe some dissimilarity between the two periods with regard to RQ10 and RQ11. For example, for some of the TEs, the changes in the ratio may reliably serve as a predictor of changes in productivity, which means that the growth in productivity is determined by the ability of these economies to do "more with less." Based on the results of OLS, it is reasonable to conclude that at least for some of the TEs, economic growth associated with investments in telecoms is determined by the presence of a skilled full-time telecom workforce capable of efficiently and effectively converting investments into revenues.

Finally, the results of phase 4 demonstrate a major difference between the two periods with regard to the sources of the growth in productivity; while during the period from 1993 to 2002, most of the TEs in our sample exhibited growth in productivity fueled by technological changes (e.g., TC component of MI), in the 2003–2008 period, growth in productivity was based on improvements in efficiency (e.g., EC component of MI). The growth via the EC component is clearly more desirable, since macroeconomic growth is associated with more efficient utilization of the already available technology. This is in contrast to the growth via the TC component, which is indicative of the growth fuelled by the continuous acquisition of new technologies (Tables 12.11 and 12.12).

Table 12.9 Results of Phase 2: RQ6, RQ7, and RQ8

Country	RQ6: Relationship between the Level of Investments in Telecoms and TFP 1993–2002	RQ6: 2003–2008	RQ7: Relationship between the Level of Full-Time Telecom Staff and TFP 1993–2002	RQ7: 2003–2008	RQ8: Complementarity of Investments in Telecoms and Telecom Staff on TFP 1993–2002	RQ8: 2003–2008
Albania	N	N	N	N	N	N
Armenia	N	N	N	N	N	N
Azerbaijan	N	N	N	N	N	N
Belarus	N	N	N	N	N	N
Bulgaria	N	N	N	N	N	N
Czech Republic	N	N	N	N	N	N
Estonia	N	N	N	N	N	N
Hungary	N	N	N	N	N	N
Kazakhstan	Y	Y	Y	Y	Y	Y

(Continued)

Table 12.9 (Continued)   Results of Phase 2: RQ6, RQ7, and RQ8

Country	RQ6: Relationship between the Level of Investments in Telecoms and TFP		RQ7: Relationship between the Level of Full-Time Telecom Staff and TFP		RQ8: Complementarity of Investments in Telecoms and Telecom Staff on TFP	
	1993–2002	2003–2008	1993–2002	2003–2008	1993–2002	2003–2008
Latvia	Y	Y	N	N	N	N
Lithuania	N	N	N	N	N	N
Moldova	N	N	N	N	N	N
Poland	N	N	N	N	N	N
Romania	N	N	N	N	N	N
Slovak Republic	N	N	N	N	N	N
Slovenia	N	N	N	N	N	N
Ukraine	N	N	N	N	N	N
Yes: proportion	.18	.18	.12	.12	.12	.12
Similarity between periods	1.00		1.00		1.00	

Table 12.10 Results of Phase 3: RQ9 and RQ10

	RQ9: Impact of Changes in the Ratio of Revenues to Investments on TFP		RQ10: Impact of Level of the Ratio of Revenues to Investments on TFP	
Country	1993–2002	2003–2008	1993–2002	2003–2008
Albania	Y	N	N	N
Armenia	N	N	N	N
Azerbaijan	Y	N	N	Y
Belarus	N	Y	N	N
Bulgaria	N	N	N	N
Czech Republic	N	N	N	N
Estonia	N	N	N	N
Hungary	N	Y	N	N
Kazakhstan	N	Y	N	N
Latvia	N	N	N	N
Lithuania	N	Y	N	N
Moldova	N	Y	N	Y
Poland	Y	N	N	N
Romania	Y	Y	N	N
Slovak Republic	Y	Y	N	N
Slovenia	Y	N	N	N
Ukraine	Y	N	N	N
Yes: proportion	.41	.41	.00	.12
Similarity	.44		.88	

Table 12.11   Results of Phase 4: RQ11

	RQ11: Do Changes in Technology (TC) Serve as a Primary Engine of Growth of TFP?
Country	1993–2002	2003–2008
Albania	Y	N
Armenia	Y	N
Azerbaijan	Y	N
Belarus	Y	N
Bulgaria	Y	N
Czech Republic	Y	N
Estonia	Y	N
Hungary	Y	Y
Kazakhstan	Y	Y
Latvia	Y	Y
Lithuania	Y	N
Moldova	N	N
Poland	Y	N
Romania	Y	N
Slovak Republic	Y	N
Slovenia	Y	Y
Ukraine	Y	N
Yes: proportion	.94	.18
Similarity between periods	.29

Table 12.12 Comparison of Sources of Growth in Productivity: 1993–2002 versus 2003–2008

Country	Leader (L) or Follower (F)	1993–2002	2003–2008
Albania	F	TC	EC
Armenia	F	TC	EC
Azerbaijan	F	TC	EC
Belarus	F	TC	EC
Bulgaria	F	TC	EC
Czech Republic	L	TC	EC
Estonia	L	TC	EC
Hungary	L	TC	TC
Kazakhstan	F	TC	TC
Latvia	L	TC	TC
Lithuania	L	TC	EC
Moldova	F	EC	EC
Poland	L	TC	EC
Romania	F	TC	EC
Slovak Republic	L	TC	EC
Slovenia	L	TC	TC
Ukraine	F	TC	EC
EC: proportion	Overall	.06	.76
	Leaders	.00	.63
	Followers	.11	.89
Similarity between periods		.29	

Conclusion

In this study, we explored 11 research questions within the context of the early (1993–2002) and late (2003–2008) transition periods of 17 TEs. The results summarized in Table 12.13 allow us to conclude that with regard to the first nine research questions, no significant changes took place over time. For example, the results of phase 1 demonstrate that during both periods, most of the TEs in our sample exhibited annual growth in productivity.

This finding is in line with the current perspective that ICT could indeed drive economic development of emerging, transition, and developing economies

Table 12.13 Assessment of the Similarities between the Two Periods: 1993–2002 and 2003–2008

	Research Question	Similarity between Periods
RQ1	Has the TE exhibited annual growth in productivity?	.88
RQ2	Has the TE exhibited continuous growth in productivity?	.71
RQ3	Is there a relationship between changes in the level of investments in telecoms and changes in productivity?	.82
RQ4	Is there a relationship between changes in the level of full-time telecom staff and changes in productivity?	.71
RQ5	Do changes in the level of investments in telecoms and changes in the level of full-time telecom staff produce a complementary effect on changes in productivity?	.94
RQ6	Is there a relationship between the level of investments in telecoms and TFP?	1.00
RQ7	Is there a relationship between the level of full-time telecom staff and TFP?	1.00
RQ8	Are the level of investments in telecoms and the level of full-time telecom staff complementary in terms of their effect on TFP?	1.00
RQ9	Do changes in the ratio of revenues to investments impact TFP?	.88
RQ10	Does the level of the ratio of revenues to investments have an impact on TFP?	.44
RQ11	What is the dominant source of improvement in productivity? TC or EC?	.29

by impacting rates of growth (Samoilenko & Weistroffer 2010a; Ngwenyama & Morawczynski 2009). However, specific insights offered by the study are worth elaborating.

First, it would appear reasonable to expect, taking into consideration a limited economic power of TEs, that the period of late transition would be different from the period of early transition with regard to the annual growth in productivity. This is because the period of early transition could be characterized as a period of catching up in terms of the quality of technology, know-how of the workforce, and the amount of accumulated ICT-related capital. Thus, it is somewhat reasonable to expect that the growth in productivity would be inconsistent simply due to changes that TEs have gone through. After the first 10-year period, we would expect to see TEs exhibiting a more stable, consistent annual growth in productivity, for the foundation laid by the period of early transition should serve as a solid platform for an economic takeoff. However, this turned out not be the case—while most TEs indeed grew over the two periods, almost none of them grew consistently on an annual basis. Furthermore, approximately one-third of TEs did not exhibit a continuous growth in productivity over the period of late transition. This is troubling, because it was noted that TEs increased their levels of investments in telecoms in the period of late transition compared to early years (Samoilenko 2013), and yet, the increased investments did not translate into growth in productivity.

Moreover, the obtained evidence suggests that a level of investments in telecoms as well as changes in the level of investments do not serve as predictors of growth in productivity. While previous investigations determined that economies could obtain macroeconomic benefits by investing in telecoms (Samoilenko & Weistroffer 2010b) via stream of revenues, our study provides evidence that the macroeconomic benefits of investments do not come in the form of growth in productivity. This is an important insight that suggests that TEs cannot pave the road to continuing growth in productivity with investment money. Additionally, when taking into consideration the absence of the impact of the level, as well as changes in the level of the telecom workforce on growth in productivity, we can conclude that TEs are not in a position when they can grow by simply spending more on technology or by hiring more workers. The absence of the complementarity of labor and investments indicates that TEs could not grow even if they invest more in telecoms while hiring more workers. What, then, could serve as an indicator of growth?

According to the results of the data analysis, our proxy *conversion efficiency* does not serve as a predictor of growth in productivity. The implication is simple, if counterintuitive—if one TE has a superior ratio of revenues to investments per telecom worker compared with another TE, that does not imply that the economy with a higher ratio exhibited a stronger growth in productivity. While it was reasonable to expect that if an economy has a high revenue-to-investments ratio, the higher ratio is due to a high level of productivity, this expectation turned out false. It is possible that while the impact of investments on revenues is straightforward, the impact of investments on the productivity component of the macroeconomic bottom line is indirect, being mediated by intermediate precursors and targets (Samoilenko 2014).

However, it appears that the change in conversion efficiency does serve as a predictor of growth in productivity for some economies. This is an important finding, because this means that some TEs could manipulate their growth in productivity by monitoring revenues-to-investments ratio. Simply put, it is possible to grow by continuously doing more with less, by employing a smaller number of highly skilled workers who are more efficient in utilization of investments in telecoms.

The results provided by the last phase of our inquiry are in line with the doing-more-with-less interpretation—we have determined that the sources of growth in productivity have shifted over time. If in the early stage of transition, the growth was primarily driven by investments in technology, the later period of transition had growth in productivity that was driven by the increases in efficiency of utilization of the available technology. However, keeping in mind that growth in productivity is comprised of two components—change in technology and change in efficiency—the best route to continuous growth in productivity is via balanced approach (Samoilenko 2013), where both components contribute to growth.

The results raise questions that could serve as directions for future inquiries. First, what are some of the complementary factors that allow investments in telecoms to impact productivity growth? Second, what are some of the ways of improving the efficiency of conversion of investments in telecoms into revenues? Third, what is the optimal revenue-to-investment ratio per telecom worker that indicates the need for the expansion of the full-time workforce?

One of the limitations of our inquiry is associated with the availability of data; it has been noted that in the context of emerging market economies, "researchers face sampling and data collection problems" (Hoskisson et al. 2000). Thus, this limitation seems to be fairly common for such a context.

Acknowledgment

Material in this chapter previously appeared in "Investigation of determinants of total factor productivity: An analysis of the impact of investments in telecoms on economic growth in productivity in the context of transition economies," *International Journal of Technology Diffusion* 5:1, 26–43.

References

Bresnahan, T. F., Brynjolfsson, E., & Hitt, L. M. (2002). Information technology, workplace organization, and the demand for skilled labor: Firm level evidence. *Quarterly Journal of Economics*, 117, 339–376.

Brynjolfsson, E., & Hitt, L. M. (1996). Paradox lost: Firm level evidence on returns to information systems spending. *Management Science*, 42, 541–558.

Caves, D., Christensen, L., & Diewert, W. (1982). Multilateral comparisons of output, input and productivity using superlative index numbers. *Economic Journal*, 92, 73–86.

Colecchia, A. & Schreyer, P. (2002). ICT investment and economic growth in the 1990s: Is the United States a unique case? A comparative study of nine OECD countries. *Review of Economic Dynamics*, 5, 408–442.

Cooper, W. W. & Tone, K. (1997). Measures of inefficiency in data envelopment analysis and frontier estimation. *European Journal of Operational Research*, 99(1), 72–88.

Daveri, F. (2002). The new economy in Europe, 1992–2001. *Oxford Review of Economic Policy*, 18(3), 345–362.

Dewan, S. & Kraemer, K. (2000). Information technology and productivity: Evidence from country level data. *Management Science (Special Issue on the Information Industries)*, 46(4), 548–562.

Färe, R., Grosskopf, S., Norris, M., & Zhang, Z. (1994). Productivity growth, technical progress, and efficiency change in industrialized countries. *American Economic Review*, 84(1), 66–83.

Gomez, R. & Pather, S. (2012). ICT evaluation: Are we asking the right questions? *The Electronic Journal of Information Systems in Developing Countries*, 50, 1–14.

Heeks, R. (2010). Do information and communication technologies (ICTs) contribute to development? *Journal of International Development*, 22, 625–640.

Heeks, R. & Molla, A. (2009). Impact Assessment of ICT-for-Development Projects: A Compendium of Approaches, IDPM Development Informatics Working Paper No. 36. Available online at http://www.sed.manchester.ac.uk/idpm/research/publications/wp/di/index.htm. Manchester.

Hirschberg, J. G., & Lye, J. N. (2001). Clustering in a data envelopment analysis using bootstrapped efficiency scores. Department of Economics, University of Melbourne, Melbourne, Australia.

Hoskisson, R., Eden, L., Lau, C., & Wright, M. (2000). Strategy in emerging economies. *Academy of Management Journal*, 43(3): 249–267.

IMF (2000). Transition Economies: An IMF Perspective on Progress and Prospects. Available online at http://www.imf.org/external/np/exr/ib/2000/110300.htm#I, retrieved May 20, 2009.

Indjikian, R. & Siegel, D. (2005). The impact of investments in IT on economic performance: Implications for developing countries. *World Development*, 33(5), 681–700.

Jorgenson, D. W. (2003). Information technology and the G7 economies. Conference on Digital Transformations: ICT's Impact on Productivity, London Business School, London.

Malmquist, S. (1953). Index numbers and indifference curves. *Trabajos de Estatistica*, 4(1), 209–242.

McGuckin, R. H., Streitwieser, M. L., & Doms, M. (1998). The effect of technology use on productivity growth. *Economics of Innovation and New Technology*, 7, 1–27.

Ngwenyama O. & Morawczynski, O. (2009). Factors affecting ICT expansion in emerging economies: An analysis of ICT infrastructure expansion in five Latin American countries. *Information Technology for Development*, 15(4), 237–258.

OECD (2002). *OECD Information Technology Outlook: ICTs and the Information Economy.* OECD, Paris.

OECD (2004). *DAC Network on Poverty Reduction: ICTs and Economic Growth in Developing Countries.* OECD, Paris.

OECD (2005). The contribution of ICTs to pro-poor growth: No. 379. *OECD Papers*, 5(1), 59–72.

Oliner, S. D. & Sichel, D. E. (2000). The resurgence of growth in the late 1990's: Is information technology the story? *Journal of Economic Perspectives*, 4(14), 3–22.

Samoilenko, S. (2013). Investigating factors associated with the spillover effect of investments in telecoms: Do some transition economies pay too much for too little? *Information Technology for Development*, 19(1), 40–61.

Samoilenko, S. (2014). Investigating the impact of investments in telecoms on microeconomic outcomes: Conceptual framework and empirical investigation in the context of transition economies. *Information Technology for Development*, 20(3), 251–273.

Samoilenko, S. & Osei-Bryson, K. M. (2008a). Strategies for telecoms to improve efficiency in the production of revenues: An empirical investigation in the context of transition economies. *Journal of Global Information Technology Management*, 11(4), 59–79.

Samoilenko, S. & Osei-Bryson, K. M. (2008b). An exploration of the effects of the interaction between ICT and labor force on economic growth in transitional economies. *International Journal of Production Economics*, 115, 471–481.

Samoilenko, S. & Weistroffer, H. R. (2010a). Improving the relative efficiency of revenue generation from ICT in transition economies: A product life cycle approach. *Information Technology for Development*, 16(4), 279–303.

Samoilenko, S. & Weistroffer, H. R. (2010b). Spillover Effect of Telecom Investments on Technological Advancement and Efficiency Improvement in Transition Economies. In Proceedings of the SIG GlobDev 3rd Annual Workshop ICT in Global Development; Saint Louis, Missouri, USA, December 12, 2010.

Shin, H. W., & Sohn, S. Y. (2004). Multi-attribute scoring method for mobile telecommunication subscribers. Expert Systems with Applications, 26(3), 363–368.

Solow, R. M. (1957). Technical change and the aggregate production function. *The Review of Economics and Statistics*, 39(3), 312–320. The MIT Press.

Van Ark, B., Inklaar, R., & McGuckin, R. (2002). "Changing Gear"—Productivity, ICT and Services Industries: Europe and the United States, No. 02-02, Economics Program Working Papers, The Conference Board, Economics Program, http://econpapers.repec.org/RePEc:cnf:wpaper:0202.

Chapter 13

Human Development and Macroeconomic Returns within the Context of Investments in Telecoms: An Exploration of Transition Economies

Introduction

The macroeconomic impact of investments in information and communication technology (ICT) is a well-researched topic (OECD 2005a,b,c; IMF 2001; Samoilenko & Osei-Bryson 2008a,b) within a relatively homogenous context of developed economies (Lam & Lam 2005; Madden & Savage 1999; Dunne et al. 2004; Siegel 1997) but a notably underresearched one in a more diverse context of developing, emerging, least developed, and transition economies (TEs) (Roztocki & Weistroffer 2008; Hoskisson et al. 2000). Because developed countries share a common set of important social, economic, and political characteristics (Ngwenyama & Morawczynski 2009), the findings of the studies conducted in the settings of the developed economies can be easily generalized and the results of the investigations in the form of the easily adoptable best practices and lessons learned shared by the peer developed economies. However, the heterogeneity of other contexts (Roztocki &

Weistroffer 2008; Hoskisson et al. 2000) precludes straightforward transfer of practical insights and policy-making knowledge between the rest of the economies that have yet to obtain the spectacular results from investments in ICT (Arcelus & Arocena 2000; Barro & Sala-i-Martin 1995; Sala-i-Martin 1996). Fortunately, the context of TEs offers an attractive research setting for investigators studying the impact of investments in ICT on the macroeconomic bottom line of the developing, emerging, and least developed countries (Samoilenko 2008), for it has been noted that TEs share characteristics of developed and less developed economies of the world (OECD 2004).

This study is part of our program of research (see Table 13.1) on the impact of investments in ICT on productivity, particularly within the context of TEs. Here we are concerned about the impact of human development, as measured by the Human Development Index (HDI), on macroeconomic outcomes and total factor productivity (TFP). Our study involves the following research questions:

1. RQ1: Is there is a statistically significant relationship between HDI and gross domestic product (GDP)?
2. RQ2: Is there is a statistically significant relationship between HDI and revenues from telecoms?
3. RQ3: Is there is a statistically significant relationship between HDI and TFP?

Our motivation for addressing these three questions is that they relate to a broader research question: *What are the socioeconomic factors that impact macroeconomic returns (e.g., GDP, revenues from telecoms, TFP) within the context of investments in telecoms?* Here we limit our focus of socioeconomic factors to those relating to human development as measured by the HDI.

Research questions RQ1–RQ3 will be explored within the context of the efficient *leaders* and the less efficient *followers* subgroups of the TEs that were identified in our previous studies (see Table 13.2). It should be noted that while other researchers have inquired into the relationship between investments in ICT and various measures of social and economic development (Bollou 2006; Ngwenyama & Morawczynski, 2009), including HDI (Ngwenyama et al. 2006), no investigations to our knowledge have been conducted to inquire into the possible relationship between HDI and the macroeconomic impact of investments in ICT.

We base our inquiry on the framework of neoclassical growth accounting and utilize data envelopment analysis (DEA) and multivariate regression (MR) to conduct the analysis of the data, which was provided by the previous inquiry of Samoilenko and Osei-Bryson (2008a) and the Human Development Report (UN, 2009). A major reason for using DEA in this study is to compute the TFP values based on the Malmquist index (MI), which was originally suggested by Malmquist (1953). Caves et al. (1982) defined the MI of TFP growth. Later, Färe et al. (1994) demonstrated that the MI could be constructed based on the results of DEA. Since DEA relative efficiency scores are calculated for each point in time t (e.g., year 1993),

Human Development and Returns in Telecom Investment ■ 171

Table 13.1 Previous Results of Samoilenko and Osei-Bryson's Research Program on IT and Productivity

Study	Findings	Follow-Up Question
Samoilenko (2008)	The study identified some of the general factors contributing to the differences in the levels of efficiency of utilization of investment in telecoms between the more efficient group of TEs (the leaders) and the less efficient group (the followers).	Is the difference in the levels of efficiency of utilization of investment in telecoms between the leaders and the followers due to the differences in the levels of investments, or is it due to the differences in the efficiency of the processes of conversion of investments into revenues?
Samoilenko and Osei-Bryson (2008a)	The results indicate that the followers are able to obtain the higher levels of revenues from telecoms not because of the higher levels of investments in telecoms but because of the leaders' more efficient processes of conversion of investments into revenues.	Is there a significant complementarity effect between the levels of investments in telecoms and full-time telecom labor that is impacting the levels of revenues from telecoms? Is there a similar discrepancy between the leaders and the followers with regard to the impact of investments in telecoms on TFP?
Samoilenko and Osei-Bryson (2008b)	The investigation identified the presence of a statistically significant complementarity effect of the levels of investments and labor on the levels of revenues from telecoms only in the case of the leaders; for the followers, the effect was not statistically significant.	Is there a similar complementarity effect of the levels of labor and investments on TFP?
Samoilenko and Osei-Bryson (2010)	The study proposed and tested a methodology allowing for relating "white-box" components, such as investments in telecoms and telecom labor, to the "black-box" component in the form of TFP. Results indicate the presence of the relationship between investments and labor and TFP for the leaders only.	What are some of the factors impacting the presence of the relationship between investments in telecoms and TFP?

Table 13.2 Groups Sample of 18 TEs

Subgroup	Membership of the Group
Leaders	Czech Republic, Estonia, Hungary, Latvia, Lithuania, Poland, Slovenia, Slovakia
Followers	Albania, Armenia, Azerbaijan, Bulgaria, Kazakhstan, Kyrgyzstan, Moldova, Romania, Ukraine

Source: Samoilenko, S. and Osei-Bryson, K. M., Linking Investments in Telecoms and Productivity Growth in the Context of Transition Economies within the Framework of Neoclassical Growth Accounting: Solving Endogeneity Problem with Structural Equation Modeling, in Proceedings of 18th European Conference on Information Systems, Pretoria, South Africa, June 6–9, 2010.

for a given decision-making unit it is possible to calculate the change in relative efficiency scores between any pair of consecutive points in time t and $t + 1$ (e.g., year 1993 and year 1994). The calculated value of change in the scores will represent the MI and reflect TFP. We present our inquiry as follows. The next section of the paper provides an overview of the theoretical framework and states the research questions of the study. Then we present results of the data analysis, followed by a discussion of the findings. We conclude the paper with an overview of the contribution, directions for further inquiries, and limitations of the inquiry.

Background

Theoretically, there is no obvious reason why developed economies can obtain outstanding macroeconomic benefits from investments in ICT, while less developed economies cannot (Madden & Savage 1998; Eggleston et al. 2002). According to a well-established framework of neoclassical growth accounting, which is widely used in both contexts (Oliner & Sichel 2000; Schreyer 2000; Daveri 2000; Jorgenson & Stiroh 2000; Whelan 2000; Hernando & Nunez 2002), the macroeconomic benefit of investments could come from two sources. If the macroeconomic bottom line is represented by GDP, then the first source is represented by the stream of revenues that is generated from investments in ICT (UN ICT Task Force Report 2005; WT/ICT Development Report 2006), and the second source is represented by the outcome of the spillover effect of investments in ICT—a contribution to GDP that is not directly associated with investments (Samoilenko & Osei-Bryson 2010). It is this investment independence of the second source, commonly referred to as total factor productivity, that makes it a highly attractive target in the quest of improving the macroeconomic impact of investments in ICT, for within the neoclassical framework, TFP is essentially free.

Previous inquiries into the macroeconomic impacts of investments in ICT along these two routes yielded some important insights. It was determined that the "investments-to-revenues" model works well only if a threshold level of ICT capital infrastructure has been developed (*The Economist* 2004), and then, on top of the developed infrastructure, if the level of investments is high enough (Oliner & Sichel 2000; Jorgenson 2001; Jorgenson & Stiroh 2000). Keeping in mind the resource-intensive nature of ICT, investigators inquired into the factors complementary to investments in ICT (Kraemer & Dedrick 2001; Pohjola 2002) that could produce a synergistic effect on the macroeconomic bottom line; the state of the full-time ICT workforce was determined to be one such complementary factors (Samoilenko & Osei-Bryson 2008b). Notably, researchers determined that it is not the quantity but, rather, the quality of the full-time ICT workforce that plays an important role not only in converting a stream of investments in ICT into revenues (Samoilenko & Osei-Bryson 2008a), but also in achieving a spillover effect of investments that is captured by TFP (Samoilenko & Osei-Bryson 2010). Based on the results of these studies suggesting the importance of an effective and efficient ICT workforce to the macroeconomic bottom line, investigators proposed that workforce development programs may offer a new route allowing for better leveraging the impact of investment in ICT. Overall, taking into consideration the well-established insights regarding the significance of such factors as the level of investments in ICT, quality of the ICT workforce, and presence of complementary investments for achieving the macroeconomic impact of investments in ICT, it appears that a basic *push*-side model (see Figure 13.1) of the macroeconomic success of investments in ICT could be outlined. However, due to the consumer-oriented nature of ICT, at least a

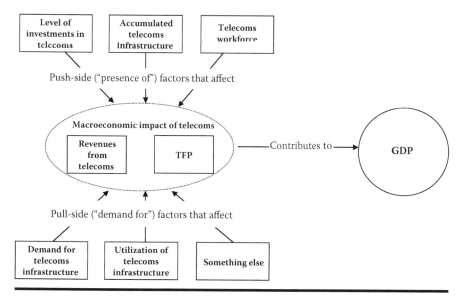

Figure 13.1 Macroeconomic impact of investments in telecoms: current insights.

portion of investments will be directed toward producing products and/or services for customer consumption. Taking this into consideration, it is only reasonable to suggest that some efforts of the researchers should be directed toward the development of the *pull* side of the model of the macroeconomic success of investments in ICT, for it is a consumer demand for ICT products and services that, at least in part, is reflected in the stream of revenues and drives the level of investments.

Recent investigations of the impact of investments in telecoms (a subset of investments in ICT) in the context of TEs identified that the better-developed TEs (the leaders) with a higher level of investments in telecoms and a more productive workforce do demonstrate a relationship between investments in telecoms and macroeconomic growth, while the less developed TEs (the followers) do not (Samoilenko & Osei-Bryson 2010). This disparity can be easily explained by the aforementioned push model of investments in ICT, where the main reasons for the failure of the followers to achieve the macroeconomic impact of investments in telecoms can be traced to the insufficient level of investments in telecoms and the inefficient telecom workforce. The investigators also provided evidence that in the case of the followers, the state of telecom infrastructure and the utilization of telecom infrastructure serve as factors affecting the level of investments in telecoms (Samoilenko & Osei-Bryson 2010). This means that in the case of the less developed TEs (i.e., the followers), a rudimentary pull model (see Figure 13.1) of investments in ICT may include such factors as insufficiently developed infrastructure and unsatisfied demand for services that rely on the utilization of that infrastructure.

However, the same investigation found no evidence that in the case of the better-developed TEs (the leaders) the level of investments in telecoms was associated with the state of telecom infrastructure or the utilization of telecom infrastructure (Samoilenko & Osei-Bryson 2010). The implication of this finding is interesting, for this tells us that in the case of the less developed TEs, investments in telecoms are probably driven by the structural and functional deficiencies of their telecom infrastructure, but once the infrastructure is sufficiently developed, as in the case of the leaders, something else drives the investments and, consequently, impacts the macroeconomic bottom line. The importance of knowing the answer to this question is intuitive, for regardless of the context, if a given economy is to progress, then it is bound at some point to sufficiently develop its infrastructure and to satisfy a basic customer demand associated with the utilization of the infrastructure, thus ending up in the situation where something else is driving the investments and impacting the economic growth. It is only reasonable to assume the benefit of knowing what this "something else" is in advance.

As stated earlier, our motivation for addressing these three questions (see Figure 13.2 for an illustration) is that they relate to a broader research question: *what are the socioeconomic factors that impact macroeconomic returns (e.g., GDP, revenues from telecoms, TFP) within the context of investments in telecoms?* In order to begin an inquiry into this undoubtedly multidimensional, complex problem, we propose investigating a role that an overall socioeconomic development of economies, as it is represented

Human Development and Returns in Telecom Investment ■ 175

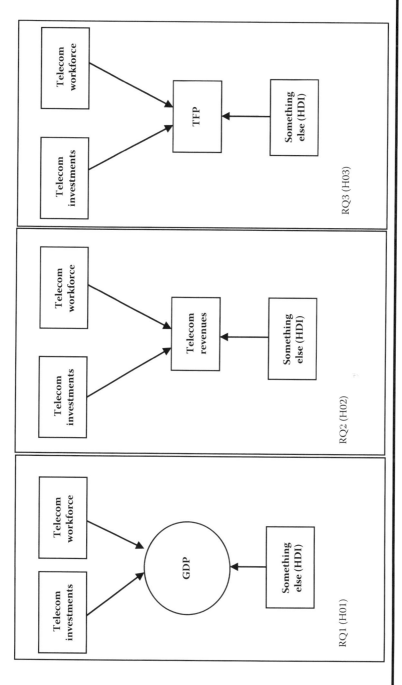

Figure 13.2 Domain of the investigation of the current study.

and measured by the United Nations (UN) HDI (UN 1990), plays in impacting the macroeconomic outcomes of investments in ICT. The reasoning behind using HDI as a possible indicator of a macroeconomic impact of investments in ICT is an intuitive one: an increase in the value of HDI for a given economy indicates improvements in the areas of education and standards of living (Depotis 2005; Neumayer 2001; Sagar & Najam 1998), and such increase may fuel the consumer demand for high-margin, less infrastructure-dependent products and services offered by ICT. We understand, however, that HDI is imperfect as a measure of socioeconomic development (Paehlke 2003; Cahill 2005; Schimmack 2008) and suggest that our inquiry serves as a springboard for future studies that may consider a wider and more precise spectrum of variables representing the degree of socioeconomic development.

Theoretical Framework and Research Questions of the Study

In this study, we expand the usual formulation of the neoclassical production function

$$Y = f(A, K, L) \tag{13.1}$$

where Y = output (most often in the form of GDP), A = TFP, K = capital stock, and L = quantity of labor/size of labor force by including HDI as another independent variable and denoting it as *HDI*. Consequently, the neoclassical production function allows us to relate HDI, *ICT capital*, *ICT labor*, and Y in the following fashion:

$$Y = f(\text{TFP, ICT capital, ICT labor, HDI}) \tag{13.2}$$

We are going to use Equation 13.2 to generate three research models: the first two models are used to explore the relationship between HDI and the macroeconomic output variables GDP and *revenues from ICT*, respectively, while our third model is used to explore whether HDI, the "something else," determines TFP, the presence of the spillover effect. Our three research models are expressed as follows:

$$\text{GDP} = \beta_0 + \beta_1 * \text{ICT capital} + \beta_2 * \text{ICT labor} + \beta_{H \to G} \text{HDI} + \xi_1 \tag{13.3}$$

$$\text{Revenues from ICT} = \beta_{10} + \beta_{11} * \text{ICT capital} + \beta_{12} * \text{ICT labor} + \beta_{H \to R} \text{HDI} + \xi_2 \tag{13.4}$$

$$\text{TFP} = \beta_{20} + \beta_{21} * \text{ICT capital} + \beta_{22} * \text{ICT labor} + \beta_{H \to G} \text{HDI} + \xi_3 \tag{13.5}$$

We will use the variable *annual investments in telecoms* as a proxy for ICT capital, and the variable *full-time telecom staff* as a proxy for ICT labor.

Exploration of our research questions will involve testing the following hypotheses:

1. H01: There is no statistically significant relationship between HDI and GDP ($\beta_{H \to G} = 0$). Given our interest in exploring differences between the leaders and followers, we will test this null hypothesis separately for both the leaders and followers.
2. H02: There is no statistically significant relationship between HDI and revenues from telecoms ($\beta_{H \to R} = 0$). Similarly to H01, hypothesis H02 will be tested separately for both the leaders and followers.
3. H03: There is no statistically significant relationship between HDI and TFP ($\beta_{H \to T} = 0$). Similarly to H01, hypothesis H03 will be tested separately for both the leaders and followers.

Overview of the Data

In this investigation, we utilize a data set on 18 TEs spanning the period from 1993 to 2002 that was used in a previous study of Samoilenko & Osei-Bryson (2010). The original data were obtained from the World Development Indicators database of the World Bank (http://web.worldbank.org) and the *Yearbook of Statistics* (2004; http://www.itu.int/ITU-D/ict/publications) of the *International Telecommunication Union* (*ITU*; http://www.itu.int). The values of the HDI index were obtained from the Human Development Report (UN 2009). Most of the studies inquiring into the macroeconomic impact of investments in ICT either analyze chronological time series (e.g., Ngwenyama et al. 2006) or point-in-time (UNDP 2004) data.

For the purposes of the study, we decided to concentrate on a single year, year 2000. Table 13.3 displays the mean values for relevant variables (e.g., GDP, MI,

Table 13.3 Mean Values of the Variables for *Leaders* and *Followers* Groups

Variable	Leaders	Followers
Annual investments in telecoms (current US$)	$529,724,490.19	$138,103,505.57
Annual revenues from telecoms (current US$)	$1,841,045,788.05	$365,197,999.15
GDP (current US$)	$44,653,918,142.86	$14,690,637,125.00
Number of full-time telecom staff	18,647.43	34,168.88
HDI	0.85	0.76
MI	1.13	1.32

HDI, full-time telecom staff, investments in telecoms, revenues from telecoms) for the *leaders* and *followers* groups of our sample.

Results of the Data Analysis

Our results suggest that while for the *leaders* group, HDI has a statistically significant impact on GDP, this relationship does not hold for the *followers* group. Similarly, our results suggest that while for the *leaders* group, HDI has a statistically significant impact on TFP, this relationship does not hold for the *followers* group. It should be noted that in a previous study (Samoilenko & Osei-Bryson 2010), we found that while for the *leaders* group, *ICT capitalization* has a statistically significant impact on TFP, this relationship does not hold for the *followers* group (Tables 13.4 and 13.6). Interestingly, our current results also suggest that with regard to the impact of HDI on revenues from ICT, there is no difference between the *leaders* and the *followers* groups (Tables 13.5).

Discussion and Conclusion

The outcomes of the tests of H01 offer evidence that in the case of the leaders, the levels of telecom labor and HDI do serve as statistically significant predictors of GDP, while in the case of the followers, it is the levels of capital investment and labor that have statistically significant impacts on GDP. It is somewhat not surprising that the level of telecom labor is statistically significant in this regard for both settings, for it is an ICT workforce that serves as a "caretaker" of the capital investments. The significance of the level of capital investments in telecoms for the followers suggests that this group, unlike the leaders, could increase its GDP by engaging in a straightforward "white-box" process of simply investing more in telecoms and hiring more of telecoms staff. At this point, the level of socioeconomic development of the followers simply does not appear to be an important factor affecting their macroeconomic bottom line.

The results of the data analysis also suggest that HDI is not one of the determinants of the level of ICT-based revenues from telecoms for either group of TEs. In the case of the leaders, however, full-time telecom staff does have an impact on the level of ICT-based revenues, while in the case of the followers, it is the level of investments in telecoms that is a factor affecting the level of ICT-based revenues. This evidence provides support to the preliminary conclusion that the leaders and the followers are, indeed, at different stages with regard to their respective states of telecom development, and if in the case of the followers, an increase in the level of telecom revenues requires an increase in the level of telecom investments (i.e., the followers do not invest enough), in the case of the leaders, it is an efficient conversion of investments into revenues performed by telecom staff (e.g., smaller number

Table 13.4 Impact of HDI on GDP

H01: HDI Has No Statistically Significant Impact on GDP at 5% Level of Significance (LoS)

Group	Parameter	Estimate	t value	Pr > |t|	Adj. R^2	Test of H01
Followers	Investments in telecoms	43.8591	4.36	0.0121	.8882	Accepted
	Full-time telecom staff	137725	2.86	0.0461		
	HDI	5.543E10	1.22	**0.2902**		
Leaders	Investments in telecoms	−8.4288	−1.77	0.1741	.9981	Rejected
	Full-time telecom staff	2524472	25.62	0.0001		
	HDI	2.069E11	4.21	**0.0244**		

Table 13.5 Impact of HDI on Revenues from ICT

H02: HDI Has No Statistically Significant Impact on Revenues from ICT at 5% LoS

Group	Parameter	Estimate	t value	$Pr > |t|$	Adj. R^2	Test of H02
Followers	Investments in telecoms	4.5696	4.61	0.0058	.8563	Accepted
	Full-time telecom staff	788.1	0.09	0.9313		
	HDI	2.7168E9	0.31	**0.7661**		
Leaders	Investments in telecoms	0.9064	0.88	0.4302	.9330	Accepted
	Full-time telecom staff	89071.7	4.00	0.0162		
	HDI	−3.747E9	−0.32	**0.7685**		

Table 13.6 Impact of HDI on TFP

H03: HDI Has No Statistically Significant Impact on TFP at 5% LoS

Group	Parameter	Estimate	t value	$Pr > |t|$	Adj. R^2	Test of H03
Followers	Investments in telecoms	5.56E–10	0.45	0.6763	.2832	Accepted
	Full-time telecom staff	–6.6E–6	–1.11	0.3279		
	HDI	–2.5284	–0.45	**0.6746**		
Leaders	Investments in telecoms	–611E–12	–5.33	0.0129	.8523	Rejected
	Full-time telecom staff	0.000010	4.26	0.0237		
	HDI	6.6821	5.64	**0.0110**		

of workers handling greater quantity of investment inputs) that matters more than a simple increase in investment inputs.

The most interesting insight, however, is provided by the results of testing of H03; while in the case of the leaders HDI, investments in telecoms, and telecom staff all appear to determine TFP, none of the variables seem to impact TFP in the case of the followers. This suggests not only that the presence of the spillover effect is dependent on multiple factors but also that the countries in the *followers* group are simply not ready yet at the socioeconomic level where they can reap the macroeconomic benefit of investments in telecoms beyond the investments-to-revenues model. Overall, the results of testing of the null hypotheses of this study provide some important insights into the increasing complexity of the process of obtaining the macroeconomic impact from investments in ICT. While obtaining the stream of revenues from investments in telecoms seems to be, while not a cheap, a fairly straightforward undertaking, achieving the spillover effect from investments in telecoms appears to be an incomparably more complex process requiring many more variables working together.

Where do the findings of our study fit within the existing body of knowledge? Previously, it was reported that in the context of the least developed economies, investments in ICT and components of HDI (namely, education and health care) serve as predictors of GDP growth (Ngwenyama & Morawczynski 2007) and that, overall, there is a high positive correlation between ICT indicators and HDI (UNDP 2004). While this evidence is in line with the findings of our investigation, it is still not clear, even conceptually, what type of a mechanism exists that allows for macroeconomic growth to provide some sort of a feedback to the push- and pull-side factors of our model, specifically, HDI and investments in telecoms. The importance of this follow-up question is fundamental, for the answer will allow us to explain the sources of growth in the values of indicators that impact the macroeconomic bottom line, thus providing us with a model reminiscent of a close-loop second-order cybernetic system showing a negative feedback–type mechanism, devoid of conceptual "miracles" and "black holes." We intend to address this question in a future study.

Acknowledgment

Material in this chapter previously appeared in "Human development and macroeconomic returns within the context of investments in telecoms: An exploration of transition economies," *Journal of Information Technology for Development* 22:4, 550–561.

References

Arcelus, F. J. & Arocena, P. (2000). Convergence and productive efficiency in fourteen OECD countries: A non-parametric frontier approach. *International Journal of Production Economics*, 66(2), 105–117.

Barro, R. & Sala-i-Martin, X. (1995). *Economic Growth*. McGraw Hill, Boston, MA.

Bollou, F. (2006). ICT Infrastructure Expansion in Sub-Saharan Africa: An analysis Of Six West African Countries from 1995–2002. *Electronic Journal on Information Systems in Developing Countries*, 26(5), 1–16.

Bollou, F., Ngwenyama, O. & Morawczynski, O. (2006). The impact of investments in ICT, health and education on development: A DEA analysis of five African countries from 1993–1999. *Proceedings of the 14th European Conference on Information Systems*, pp. 1043–1055.

Cahill, M. (2005). Is the human development index redundant? *Eastern Economic Journal*, 31(1), 1–5.

Caves, D., Christensen, L., & Diewert, W. (1982). The economic theory of index numbers and the measurement of input, output, and productivity. *Econometrica*, 50, 1393–1414.

Daveri, F. (2000). Is Growth an Information Technology Story in Europe too? EPRU Working Paper Series 00-12, Economic Policy Research Unit (EPRU), University of Copenhagen. Department of Economics.

Depotis, D. (2005). Measuring human development via data envelopment analysis: The case of Asia and the Pacific. *Omega: International Journal of Management Science*, (33), 385–390.

Dunne, T., Foster, L., Haltiwanger, J., & Troske, K. R. (2004). Wages and productivity dispersion in US manufacturing: The role of computer investment. *Journal of Labor Economics*, 22(2), 397–430.

Eggleston, K., Jensen, R., & Zeckhauser, R. (2002). Information and Communication Technologies, Markets and Economic Development. Discussion Papers Series, Department of Economics, Tufts University 0203, Department of Economics, Tufts University.

Färe, R., Grosskopf, S., Norris, M., & Zhang, Z. (1994). Productivity growth, technological progress, and efficiency in industrialized countries. *American Economic Review*, 84, 374–380.

Hernando, I. & Nunez, S. (2002). The Contribution of ICT to Economic Activity: A Growth Accounting Exercise with Spanish Firm-Level Data. Banco de España Working Papers 0203, Banco de España.

Hoskisson, R., Eden, L., Lau, C., & Wright, M. (2000). Strategy in emerging economies. *Academy of Management Journal*, 43(3), 249–267.

IMF (2001). *International Financial Statistics*. IMF, Washington, DC.

Jorgenson, D. W. (2001). Information technology and the US economy. *American Economic Review*, 91(1), 1–32.

Jorgenson, D. W. & Stiroh, K. J. (2000). US economic growth in the new millennium. *Brooking Papers on Economic Activity*, 1, 125–211.

Kraemer, K. L. & Dedrick, J. (2001). Information technology and economic development: Results and policy implications of cross-country studies. In *Information Technology, Productivity, and Economic Growth*, ed. M. Pohjola, Oxford University Press, Oxford.

Lam, P.-L. & Lam, T. (2005). Total factor productivity measures for Hong Kong telephone. *Telecommunications Policy*, 29(1), 53–69.

Madden, G. & Savage, S. (1998). CEE telecommunications investment and economic growth. *Information Economics and Policy*, 10(2), 173–195.

Madden, G. & Savage, S. J. (1999). Telecommunications productivity, catch-up and innovation. *Telecommunications Policy*, 23(1), 65–81.

Malmquist, S. (1953). Index numbers and indifference surfaces. *Trabajos de Estatistica*, 4, 209–242.
Neumayer, E. (2001). The human development index—A constructive proposal. *Ecological Economics*, 39, 101–114.
Ngwenyama, O., Andoh-Baidoo, F. K., Bollou, F., & Morawczynski, O. (2006). Is there a relationship between ICT, health, education and development? An empirical analysis of five West African countries from 1997–2003. *EJISDC: The Electronic Journal on Information Systems in Developing Countries*, 23(5), 1–11.
Ngwenyama, O. & Morawczynski, O. (2007). Unraveling the impact of investments in ICT, education and health on development: An analysis of archival data of five West African countries using regression splines. *The Electronic Journal on Information Systems in Developing Countries*, 29(5), 1–15.
Ngwenyama, O. & Morawczynski, O. (2009). Factors affecting ICT expansion in emerging economies: An analysis of ICT infrastructure expansion in five Latin American countries. *Information Technology for Development*, 15(4), 237–258.
OECD (2004). *DAC Network on Poverty Reduction: ICTs and Economic Growth in Developing Countries*. OECD, Paris.
OECD (2005a). Good practice paper on ICTs for economic growth and poverty reduction. *The DAC Journal*, 6(3), 1–69.
OECD (2005b). Background paper: The contribution of ICTs to pro-poor growth: No. 384. *OECD Papers*, 5(2), 15–52.
OECD (2005c). The contribution of ICTs to pro-poor growth: No. 379. *OECD Papers*, 5(1), 59–72.
Oliner, S. D. & Sichel, D. E. (2000). The resurgence of growth in the late 1990s: Is information technology the story? *Journal of Economic Perspectives*, 14(4), 3–22.
Paehlke, R. (2003). *Democracy's Dilemma: Environment, Social Equity, and the Global Economy*. MIT Press, Cambridge, MA.
Pohjola, M. (2002). New Economy in Growth and Development. WIDER Discussion Paper 2002/67, United Nations University World Institute for Development Economics Research (UNU/WIDER). Helsinki, Finland.
Roztocki, N. & Weistroffer, H. (2008). Editorial preface: Information technology in transition economies. *Journal of Global Information Technology Management*, 11(4), 2–9.
Sagar, A. & Najam, A. (1998). The human development index: A critical review. *Ecological Economics*, 25, 249–264.
Sala-i-Martin, X. (1996). The classical approach to convergence analysis. *Economic Journal*, 106(4), 1019–1036.
Samoilenko, S. (2008). Contributing factors to information technology investment utilization in transition economies: An empirical investigation. *Information Technology for Development*, 14(1), 52–75.
Samoilenko, S. & Osei-Bryson, K. M. (2008a). Strategies for telecoms to improve efficiency in the production of revenues: An empirical investigation in the context of transition economies. *Journal of Global Information Technology Management*, 11(4), 56–75.
Samoilenko, S. & Osei-Bryson, K. M. (2008b). An exploration of the effects of the interaction between ICT and labor force on economic growth in transitional economies. *International Journal of Production Economics*, 115(2), 471–481.

Samoilenko, S. & Osei-Bryson, K. M. (2010). Linking Investments in Telecoms and Productivity Growth in the Context of Transition Economies within the Framework of Neoclassical Growth Accounting: Solving Endogeneity Problem with Structural Equation Modeling. In Proceedings of 18th European Conference on Information Systems, Pretoria, South Africa, June 6th–9th, 2010.

Schimmack, U. (2008). The structure of subjective well-being. In *The Science of Subjective Well-Being*, eds. M. Eid & R. Larsen, The Guilford Press, New York, pp. 97–123.

Schreyer, P. (2000). The Contribution of Information and Communication Technology to Output Growth a Study of the G7 Countries. OECD Science, Technology and Industry Working Papers 2000/2, OECD, Directorate for Science, Technology and Industry.

Siegel, D. (1997). The impact of computers on manufacturing productivity growth: A multiple-indicators, multiple-causes approach. *The Review of Economics and Statistics*, 79(1), 68–78.

Solow, R. (1957). Technical change and the aggregate production function. *Review of Economics and Statistics*, 39(3), 312–320.

The Economist (2004). Reaping the benefits of ICT: Europe's productivity challenge. A report from the Economist Intelligence Unit. Available online at http://graphics.eiu.com/files/ad_pdfs/MICROSOFT_FINAL.pdf

UN (1990). The Human Development Report 1990. UNDP HDR. Available online at http://hdr.undp.org/en/reports/global/hdr1990/

UN (2009). The Human Development Report 2009. UNDP HDR. Available online at http://hdr.undp.org/en/reports/global/hdr2009/

UN ICT Task Force Report (2005). Innovation and Investment: Information and Communication Technologies and the Millennium Development Goals. Report Prepared for the United Nations ICT Task Force in Support of the Science, Technologies and Innovation Task Force of the United Nations Millennium Project. Available online at www.unmillenniumproject.org/documents/Innovation_InvestmentMaster.pdf

UNDP (United Nations Development Program) (2004) *ICT and Human Development: Towards Building a Composite Index for Asia Realizing the Millennium*. Elsevier, New Delhi.

Whelan, K. (2000). Computers, obsolescence, and productivity. Federal Reserve Board Finance and Economics Discussion Series 2000-06, Board of Governors of the Federal Reserve System (USA).

WT/ICT Development Report (2006). Measuring ICT for social and economic development. International Telecommunication Union's World Telecommunication/ICT Development Report, 8th edition. Available online at http://www.itu.int/dms_pub/itu.../D-IND-WTDR-2006-SUM-PDF-E.pdf

Yearbook of Statistics (2004). Telecommunication Services Chronological Time Series 1993–2002. ITU Telecommunication Development Bureau (BDT), International Telecommunication Union.

Chapter 14

The Spillover Effects of Investments in Telecoms: Insights from Transition Economies

Introduction

The stream of research dedicated to investigating the relationship between investments in information and communication technologies (ICTs) and their macroeconomic outcomes is by now well established (OECD 2005a,b,c; IMF 2001; Samoilenko & Osei-Bryson 2008a,b). Not all settings, however, have received the equal attention of investigators, as the overwhelming majority of studies have been conducted in the context of developed countries (Lam & Lam 2005; Madden & Savage 1999; Dunne et al. 2004; Siegel 1997). Resultantly, while the accumulated evidence of the positive impact of investments in ICT on the economies of developed countries is by now ample, the research concerning the effects of investments in ICT on developing, emerging, and transition economies (TEs) is still scarce, which led to a call for conducting additional substantive research beyond the context of developed economies (OECD 2004).

One of the commonly noted reasons for the scarcity of studies lies in the limited availability of the reliable time series data (Hoskisson et al. 2000); another reason is associated with the absence of the clear-cut taxonomy that differentiates emerging, developing, and TE (Samoilenko 2008). For example, the United Nations noted that no agreed-upon criteria present for categorizing either developed or developing countries (http://unstats.un.org/unsd/cdb/cdb_dict_xrxx.asp?def_code=491), and Hoskisson et al. (2000) warned that the term *emerging economy* may mean different things to different researchers, for there exists no established list of countries that agreed to be called emerging economies. There exists, however, a common definition of TEs as economies that are in transition from centralized planning system to a free market economy (Roztocki & Weistroffer 2008).

The usefulness of in-depth case studies of selected countries (Palvia 2006) in this context is often limited, for while clearly being distinct from developed economies, the emerging, developing, and TEs (Roztocki & Weistroffer 2008; Hoskisson et al. 2000) do not represent a homogenous group. Consequently, researchers have been encouraged to perform group analyses at various levels (Bagchi & Kirs 2009) and to conduct longitudinal studies (Palvia 2006; Hoskisson et al. 2000) in order to obtain a broader perspective on the relationship between investments in ICT and their economic outcomes. While investigators are called to expand research in the area of global IT management (Palvia 2006), they face a challenge associated with the general scarcity of sound theory-driven approaches, methods, and methodologies allowing for identification of the relevant economy-specific antecedents and their business and IT strategy consequents (Palvia 2006).

Despite the inherent heterogeneity of the context of emerging, developing, and TEs, which limits generalizability of the results of studies, it was noted that the context of TEs is advantageous in this regard (Samoilenko 2008) for it is composed of a subgroup (i.e., *Leaders* subgroup of TEs) that can be considered to share economic characteristics with developed economic regions and a subgroup (i.e., *Followers* subgroup of TEs) that can be considered to share economic characteristics with less developed economic regions (OECD 2004). As such, analysis of differences between TEs can serve as a useful surrogate for direct analysis to explore differences in success perception of IT between developing/emerging and developed economies. Thus, while the data that we use in this study are directly related to TEs, our exploration can provide insights that are useful for understanding similarities and differences between developed and developing/emerging economies with regards to the perception of IT success.

In order to explore the similar paths along which investments in ICT can impact the macroeconomic outcomes in both settings, in this study, we utilize the research framework of neoclassical growth accounting that is widely used in information systems (IS) research (McGuckin & Stiroh 2002; Brynjolfsson & Hitt 1996). Within the context of this framework, an increase in the macroeconomic bottom line (e.g., GDP) can come from two sources. The first source is represented by the "white-box" components, such as the available levels of capital (e.g., investments in ICT) and labor (e.g., ICT workforce). The origins of the "white-box" component are clear-cut and transparent. The second source is reflected by total factor productivity (TFP), a "black-box" component origin, the composition of which is less clear.

Resultantly, if we are to conceptualize the growth in GDP as a function of capital investments, labor, and "something else," then within the framework of neoclassical growth accounting, the term "something else" is represented by TFP, where it serves as the residual (often referred to as *Solow's residual*) term capturing that contribution to GDP that is left unexplained by the inputs of capital and labor. Of the three inputs used by the growth accounting model, only capital and labor are empirically observable, while the values for TFP must be derived computationally. Based on the idea of the productivity index, originally suggested by Malmquist (1953), Caves et al. (1982) defined the Malmquist index (MI) of TFP growth. Later, Färe et al. (1994) demonstrated that the MI could be constructed based on the results of data envelopment analysis (DEA). Let us recall that DEA calculates the scores of the relative efficiency of decision-making units (DMUs; e.g., TEs in the case of our study) relative to the efficient frontier, which "envelops" the data set. While DEA scores are calculated for a given point in time t (e.g., year 1993), an investigator can conduct DEA at two points in time t and $t + 1$ (e.g., year 1993 and year 1994) and observe a change in the associated scores of the relative efficiency for each DMU over that period. The calculated value of change in the scores will represent the MI and reflect TFP.

The most straightforward way of improving a macroeconomic bottom line is clearly by increasing the contribution of such "white-box" components as the levels of available capital and labor. However, a contribution coming from a "black-box" component, TFP, is preferable because it represents the macroeconomic growth that is not accounted for by any "white-box" resource-intensive and potentially scarce components that are subject to a law of diminishing returns. It will be clearly advantageous for economic growth if a contribution of any "white-box" component produces externalities that manifest themselves in the form of economic growth that "spills over" its own contribution. For example, we can easily argue the economic value of investments in ICT by demonstrating that investments in ICT contribute in a "white-box" straightforward fashion to the growth in GDP via revenues from ICT. But the economic value of investments in ICT will greatly increase if we can argue that investments in ICT not only impact growth in GDP in a straightforward fashion but also produce spillover effect on other components, such as TFP. This spillover effect of investments in ICT will represent the contribution to economic growth that is, within the framework of neoclassical growth accounting, free for it is not accounted for in the original investments and is "spilled over" in the form of externalities of investments. Unfortunately, it is much easier to establish a link between investments in ICT and GDP (e.g., via revenues from ICT) than between investments in ICT and TFP. Unsurprisingly, scarcity of scientific evidence is particularly noticeable in regard to establishing a link between investments in ICT and TFP, as well as identifying circumstances that may impact it.

This study is part of our research program (see Table 14.1) on the impact of investments in ICT on productivity, particularly within the context of TEs. Its purpose is twofold. First, we propose a research model that allows for not only linking

Table 14.1 Previous Results of Samoilenko and Osei-Bryson's Research Program on IT and Productivity

Study	Findings	Follow-Up Question
Samoilenko (2008)	The study identified some of the general factors contributing to the differences in the levels of efficiency of utilization of investment in telecoms between the more efficient group of TEs (the leaders) and the less efficient group (the followers).	Is the difference in the levels of efficiency of utilization of investment in telecoms between the leaders and the followers due to the differences in the levels of investments, or is it due to the differences in the efficiency of the processes of conversion of investments into revenues?
Samoilenko and Osei-Bryson (2008a)	The results indicate that the followers are able to obtain the higher levels of revenues from telecoms not because of the higher levels of investments in telecoms but because of the leaders' more efficient processes of conversion of investments into revenues.	Is there a significant complementarity effect between the levels of investments in telecoms and full-time telecom labor that is impacting the levels of revenues from telecoms? Is there a similar discrepancy between the leaders and the followers in regard to the impact of investments in telecoms on TFP?
Samoilenko and Osei-Bryson (2008b)	The investigation identified the presence of a statistically significant complementarity effect of the levels of investments and labor on the levels of revenues from telecoms only in the case of the leaders; for the followers, the effect was not statistically significant.	Is there a similar complementarity effect of the levels of labor and investments on TFP?
Samoilenko and Osei-Bryson (2010)	The study proposed and tested a methodology allowing for relating "white-box" components, such as investments in telecoms and telecom labor, to the "black-box" component in the form of TFP. Results indicate the presence of the relationship between investments and labor and TFP for the leaders only.	What are some of the factors impacting the presence of the relationship between investments in telecoms and TFP?

investments in ICT and TFP but also linking and determining the effect of several ICT-related factors on the impact of investments in ICT. Our research model is an extension of the approach of Samoilenko and Osei-Bryson (2010), and is consistent with the assumptions of the framework of neoclassical growth accounting. Second, we apply the proposed model to the context of TEs by focusing on the following research questions:

- RQ1: Do TEs achieve a spillover effect from investments in ICT that is manifested in relationship between investments in telecoms and TFP?
- RQ2: Is the state of ICT in TEs impacted by the availability and the utilization of the ICT infrastructure?
- RQ3: Is the availability of ICT in TEs impacted by the state of economy?

We present our inquiry in the following sequence. First, we outline the research problem of our investigation in more detail. Then, we provide an overview of the component technique supporting the proposed solution, as well as offer a justification for the chosen technique. Overview of the data used in our study will be presented next, followed by the results of the data analysis. Discussion of the results and a brief conclusion follow.

Research Problem of the Study

The framework of the growth accounting has been widely used to estimate contribution of ICT to the macroeconomic bottom line in the context of developed and developing countries (Oliner & Sichel 2000; Schreyer 2000; Daveri 2000; Jorgenson & Stiroh 2000; Whelan 2000; Hernando & Nunez 2002). However, from the perspective of this theoretical framework, economic growth (TFP) is exogenous to the production function, and, as such, it cannot be directly explained by the variables endogenous to the function (e.g., investments in ICT of ICT-related labor).

Samoilenko and Osei-Bryson (2010) proposed a comprehensive three-step methodology allowing for relating investments in ICT to GDP and TFP within the framework of neoclassical growth accounting. Their methodology is described in Table 14.2 and illustrated in Figure 14.1.

While that approach allows for testing of the presence of the relationship between investments in ICT and TFP, it does not allow for progressing to a higher vantage point and gaining insights regarding some of the economic factors that may impact the presence of the relationship. In the current study, we concentrate on expanding step 3 of the methodology of Samoilenko and Osei-Bryson (2010) and propose a research model depicted in Figure 14.2.

We use the following operational definitions for the constructs included in the research model:

Table 14.2 Steps of Comprehensive Methodology of Samoilenko and Osei-Bryson (2010)

Step	Technique	Purpose	Outcome
Step 1	Data envelopment analysis	Obtain the values of MI	Values of TFP
Step 2	Multivariate regression analysis	Test the presence of the relationship between capital investments in ICT, ICT labor, and GDP	Strength of the relationship between the "white-box" independent variables and the dependent variable
Step 3	Structural equation modeling	Test the presence of the indirect/mediated relationship between investments in ICT and TFP	Strength of the indirect/mediated relationship between the "white-box" independent variable and the "black-box" error term

The Spillover Effects of Investments in Telecoms ■ 193

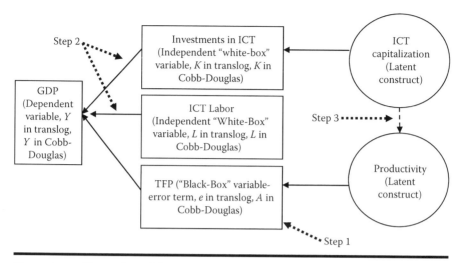

Figure 14.1 Illustration of the methodology of Samoilenko and Osei-Bryson (2010).

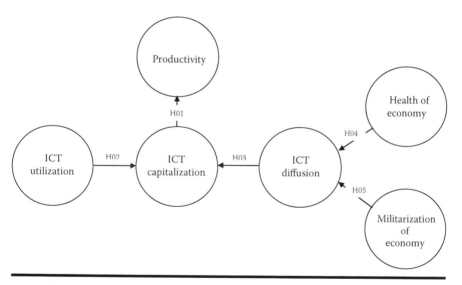

Figure 14.2 The research model of the study.

- *ICT Capitalization* is the fiscal state of ICT reflecting relationships between ICT-related investments, revenues, and labor in reference to the overall state of the economy.
- *Productivity* is an annual change in macroeconomic bottom line that is not directly associated with the changes in investments in labor.
- *ICT Utilization* is the degree of the utilization of the available ICT-related infrastructure.

- *ICT Diffusion* is the state of the available ICT-related infrastructure.
- *Health of Economy* is the fiscal state reflecting participation of an economy in international trade.
- *Militarization of Economy* is the state of an economy reflecting the fiscal and human resources dedicated to noncivilian purposes.

This research model is used to explore our three research questions by means of the following associated null hypotheses:

1. RQ1: Do TEs achieve a spillover effect from investments in ICT that is manifested in relationship between investments in telecoms and TFP?
    - H01: There is no statistically significant relationship between *ICT Capitalization* and *Productivity*.
2. RQ2: Is the state of ICT in TEs impacted by the availability and the utilization of ICT infrastructure?
    - H02: There is no statistically significant relationship between *ICT Utilization* and *ICT Capitalization*.
    - H03: There is no statistically significant relationship between *ICT Diffusion* and *ICT Capitalization*.
3. RQ3: Is the availability of ICT in TEs impacted by the state of economy?
    - H04: There is no statistically significant relationship between the *ICT Diffusion* and the *Health of Economy*.
    - H05: There is no statistically significant relationship between the *ICT Diffusion* and the *Militarization of Economy*.

We now offer a discussion on the relationships between the constructs of our research model that are reflected in our hypotheses.

- It has been previously recognized that the overall accumulated level of investments in ICT, which must be above a certain threshold to impact a macroeconomic bottom line (Dewan & Kraemer 2000; Piatkowski 2002), can be represented in terms of two components. The first component is the existing accumulated capital (e.g., in the form of existing capital infrastructure), which is in our model represented by the construct *ICT Diffusion*. The second component is the additional level of capital investments that is added annually, which is represented in our model by the construct *ICT Capitalization*. Thus, the constructs *ICT Capitalization* and *ICT Diffusion*, taken together, provide a representation of the state of ICT capital of an economy.
- The relationships between the constructs *ICT Capitalization, ICT Utilization*, and *ICT Diffusion* are intended to represent a testable assumption that investments in ICT in TEs are impacted by the level of existing ICT infrastructure and the demand for ICT services. Given variations in consumer demand for different levels of ICT services, it seems reasonable to expect that the

relationship depicted in our research model will hold for the poorer, less developed TEs and will not hold for the relatively wealthier TEs that are more similar to developed economies.
- The construct *Health of Economy* is intended to represent the relative level of wealth and globalization of a given economy. Again, keeping in mind the insufficient level of existing ICT infrastructure in the context of TEs, we expect an association between the constructs *Health of Economy* and *ICT Diffusion*, such that wealthier TEs will have better ICT infrastructure, and poorer TEs have worse ICT infrastructure. However, we will not expect this relationship to hold in the context of developed economies that have already sufficiently developed their ICT infrastructure.
- Finally, the construct *Militarization of Economy* is intended to represent expenditures of TEs that are not associated with the civilian needs. Taking into consideration that TEs have a lower level of economic development than developed economies, we expect that higher levels of spending on noncivilian needs would be associated with the less developed ICT infrastructure.

The reader should note that we do not claim that our research model offers a detailed description of the process of how investments in ICT contribute to the unexplained growth in the form of TFP. Rather, the purpose of our model is to serve as a vehicle of understanding whether the spillover effect from investments in ICT indeed takes place, as well as to serve as a means for gaining insights into factors that may affect the presence of the spillover effect.

The approach that we chose to test our research model is structural equation modeling (SEM) as implemented in the partial least squares (PLS) method (e.g., Chin 1995, 1998a; Gefen & Straub 2005; Chin & Hubona 2006). It should be noted that the PLS method (Wold 1966) is one of the least restrictive models that impose minimal demands on measurement scales, size, and residual distribution of the sample (Chin 1995, 1998a; Barclay et al. 1995). Clearly, keeping in mind a sample of convenience representing the context of TEs, the choice of the PLS method in our investigation is appropriate. Additionally, because the use of the PLS method is advisable under the condition of low theoretical knowledge, and in cases where research models are at the early stage of development (Jöreskog & Wold 1982; Barclay et al. 1995; Chin 1995), our choice to implement SEM with PLS is justified also. For a more detailed overview on the PLS, we would like to direct the interested reader to the works of Gefen and Straub (2005), Gefen et al. (2000), and Chin and Hubona (2006).

Overview of the Data

To clearly outline the general context of the current inquiry, we use a commonly accepted definition of TE as economies that are in transition from centralized planning system to a free market economy (Roztocki & Weistroffer 2008). In particular, we concentrate on the group of 18 ex-communist (Bagchi & Kirs 2009) *transition economies in Europe and the former Soviet Union* (IMF 2001). The Eurasian TEs (such as those in our sample) exhibit regional commonalities and share multiple cultural traits (Deichmann et al. 2006), thus representing a relatively homogenous group of economies in this regard and making them suitable for a group analysis (e.g., Beilock & Dimitrova 2003; Dimitrova & Beilock 2005). Consequently, in this study, we assume that the impact of cultural differences (Palvia et al. 2002; Watson et al. 1997) is controlled for.

In this investigation, we utilize a time-series data set that on 18 TEs spanning the period from 1993 to 2002 was used in our previous studies (Samoilenko 2008). The data were obtained from the *World Development Indicators* database (web .worldbank.org/WBSITE/EXTERNAL/DATASTATISTICS) and the Yearbook of Statistics (2004) (http://www.itu.int/ITU-D/ict/publications) *of International Telecommunication Union* (ITU; http://www.itu.int). Previously, however, Samoilenko and Osei-Bryson (2007) determined that over the 10-year period, from 1993 to 2002, this set of 18 TEs contained two groups that clearly differed in their levels of investments and revenues from ICT, as well as in terms of the efficiency of the utilization of the investments in ICT (Samoilenko & Osei-Bryson, 2007). Consequently, this sample of 18 TEs can be represented in terms of two clusters (see Table 14A.1): the more efficient group was labeled the *Leaders* and the less efficient group the *Followers* (Samoilenko & Osei-Bryson 2007). Furthermore, it was established that the *Leaders* differ from the *Followers* in many aspects, such as those related to militarization of the economy, quality of human resources (i.e., education and health), and level of sociotechnical development (Samoilenko 2008). Moreover, to gain an additional insight into the nature of the differences between two groups, Samoilenko and Osei-Bryson (2008) considered the results previously noted by Andonova and Diaz-Serrano (2007) regarding the possible impact of political institutions on the diffusion of telecommunication technologies and compared the *Leaders* and the *Followers* in terms of the levels of political risks. The results demonstrated that the *Leaders* differ favorably from the *Followers* in this regard also, possibly by offering a greater degree of political commitment for investor protection (Andonova & Diaz-Serrano 2007). Consequently, given the heterogeneity in our set of TEs and our interest in using an analysis of TEs as a useful surrogate for direct analysis to explore differences in success perception of IT between developing/emerging and developed countries, we can expand our original research questions to include three sets of null hypotheses: one set considering the *Leaders* and *Followers* together, the second set considering only the *Leaders*, and the third set considering only the *Followers*.

In order to reduce the bias associated with the heterogeneity of the sample, we needed to represent all 18 TEs in such a way that the difference in geographical size, population, economic wealth, and so on would be countered. Thus, the constructs of our research model were represented by the measures described in Table 14.3.

Table 14.3 Measures of the Research Model

Measure	Source Variables	Representation	Latent Construct
TFP	MI	Annual change in productivity	Productivity
RatioGDPtoInvestment	1. GDP per capita (in current US$) 2. Annual Telecom Investment per capita (in current US$)	Ratio of GDP per capita to Annual Telecom Investment per capita	ICT Capitalization
RatioProductivity	1. Annual Total Revenue from Telecoms (% of GDP) 2. Annual Investments in Telecoms (% of GDP)	Ratio of Annual Total Revenue from Telecoms to Annual Investments in Telecoms	ICT Capitalization
RatioStafftoInvestment	1. Full-time Telecom Staff 2. Annual Investment in Telecoms (in current US$)	Ratio of Full-time Telecom Staff to the Annual Investment in Telecoms	ICT Capitalization
TelephoneMainlines	Telephone Mainlines (per 1000 people)	Telephone Mainlines (per 1000 people)	ICT Diffusion
MobilePhones	Mobile Phones (per 1000 people)	Mobile Phones (per 1000 people)	ICT Diffusion
FixedMobilePhoneSubscr	Fixed Line and Mobile Phone Subscribers (per 1000 people)	Fixed Line and Mobile Phone Subscribers (per 1000 people)	ICT Diffusion

(Continued)

Table 14.3 (Continued) Measures of the Research Model

Measure	Source Variables	Representation	Latent Construct
InterTelecomOutgoing	International Telecom, Outgoing Traffic (in minutes per subscriber)	International Telecom, Outgoing Traffic (in minutes per subscriber)	ICT Utilization
RevenuePerSubscriber	1. Annual Total Revenue from Telecoms (% of GDP) 2. Fixed Line and Mobile Phone Subscribers (per 1000 people)	Ratio of Annual Total Revenue from Telecoms to the total number of Fixed Line and Mobile Phone Subscribers	ICT Utilization
Exports%ofGDP	Exports of Goods and Services (% of GDP)	The value of all goods and other market services provided to the rest of the world as a share of GDP	Health of Economy
Trade%ofGDP	Trade (% of GDP)	The sum of exports and imports of goods and services measured as a share of GDP	Health of Economy
RatioMilExpToMilPers	1. Total Annual Military Expenditure (in current US$) 2. Total Size Military Force	Ratio of Total Annual Military Expenditure to the Total Size of Military Force	Militarization of Economy
MilExpPerCapita	1. Total Annual Military Expenditure (in current US$) 2. Total Population	Ratio of Total Annual Military Expenditure to the Total Population	Militarization of Economy

The important difference between the construct *Productivity* and the other five constructs is that the measures representing the five constructs are provided by the data, while the measure for *Productivity* is computationally derived using DEA.

We assume that if we can demonstrate the presence of a statistically significant relationship between the constructs of our research model, then we would be able to infer the presence of the indirect relationship between the factors affecting investments in ICT, as well as the relationship between investments in ICT and TFP. However, this can only be done if we demonstrate that the measures that we use to represent our constructs are valid and reliable. We next present the results of our data analysis.

Results of Data Analysis

Preliminary Data Analysis

First, we used the SPSS package to conduct an exploratory principal component analysis (PCA) in order to determine whether the measures we chose to represent our constructs demonstrate a specific pattern of loadings, align in the same direction, and load together on the same principal component. There are two latent constructs in our research model; therefore, we requested two components to be extracted. We also requested the results of the Kaiser–Meyer–Olkin (KMO) test of sampling adequacy and Bartlett's test of sphericity to be included in the output as these two measures are commonly used to determine whether or not a data set could be successfully analyzed using factor analysis. In order for our data set to pass these two tests, KMO value must be above 0.5, and Bartlett's test value must be less than 0.05 (Bollen & Long 1993). Results of the analysis produced a KMO value of 0.661 and a Bartlett's test value of 0.000; thus, we conclude that our data set passed the two tests and is suitable for PCA.

Next, we performed five-component PCA, choosing only values above 0.7 to be displayed. The sixth construct, *Productivity*, is represented by a single calculated measure; thus, it was not included in PCA. We also requested the most common rotation option, *varimax*, in order to obtain an easy to interpret solution, where each of our measures would be maximally associated with a single construct.

Based on the results of the output presented in Table 14.4, we conclude that our measures are fit for representing their respective constructs. At this point, we could continue our inquiry and perform PLS analysis, the results of which are presented next.

PLS Analysis: Steps, Procedures, and Results

Assessment of the Measurement Model

Investigators have an option of choosing among many available packages to conduct a PLS analysis; PLS-G (Chin 1998b) was used in this study. We assess the

Table 14.4 Results of PCA (Rotated Component Matrix)

Measure	ICT Capitalization	Militarization of Economy	Health of Economy	ICT Diffusion	ICT Utilization
RevenuePerSubscriber					.876
InterTelecomOutgoing					.941
Fixed&MobilePhoneSubscr				.823	
MobilePhones				.932	
TelephoneMainlines				.872	
ProductivityRatio	.909				
RatioGDPInvestPerCapita	.972				
RatioFtTW/ICTinvest	.922				
RatioMilExp/MilPers		.837			
MilExpPerCapita		.890			
Exports%ofGDP			.944		
Trade%ofGDP			.964		

adequacy of our measurement model by evaluating the following three criteria: (1) the reliability of the individual items and their constructs, (2) the convergent validity of the measures representing each construct, and (3) the discriminant validity of the measures (Hulland 1999).

A commonly accepted test of reliability of individual items consists of assessment of the loadings of the measures on their construct, while the assessment of the reliability of the constructs is performed by evaluating the composite reliability of the constructs. Results of the assessment presented in Table 14.5 demonstrate that our research model passed the test of composite reliability assessment, for the measures of the internal consistency (*Composite reliability* column) are higher than suggested by Nunnally (1978), which is a baseline of 0.7, and shared by each construct, and its measured variance [*Average Variance Extracted (AVE)* column] is significantly higher than the suggested value of 0.5 (Rivard & Huff 1988).

We examined the reliability of the individual measures next.

The values of the loadings of the measures provided in Table 14.6 indicate that our research model also passed the second test of assessment of the reliability of the individual items. Individual loadings of all items are greater than 0.8, which indicates that the measures and the construct share the significant amount of variance.

Convergent validity of the measures is assessed through the evaluation of the measure of internal consistency (Fornell & Larcker 1981). First, we look at the magnitude and significance of the *t*-values for the loadings of each of the individual items. Additionally, the process of evaluation involves assessment of the loadings of the measures on their own constructs, where it is expected that the measures representing a construct would exhibit high loadings on that construct and low loadings on all the other constructs in the model. The results displayed in Table 14.7 demonstrate that all *t*-values for all measures of the five constructs are significant, which indicates that the model passed the first test of the convergent validity.

Table 14.5 Assessment of Reliability of Constructs

Construct	Composite Reliability	AVE	Squared Root of AVE
Productivity	1.000	1.000	1.000
Militarization of Economy	0.982	0.964	0.9818
Health of Economy	0.990	0.979	0.9894
ICT Diffusion	0.946	0.855	0.9246
ICT Capitalization	0.968	0.909	0.9534
ICT Utilization	0.883	0.791	0.8893

202 ■ Theoretical Research Frameworks Using Multiple Methods

Table 14.6 Assessment of Reliability of Measures

Measure	Loading	Communality
MI	1.0000	1.0000
RatioMilExp/MilPers	0.9838	0.9678
MilExpPerCapita	0.9802	0.9607
Exports%ofFDP	0.9906	0.9814
Trade%ofGDP	0.9886	0.9774
FixedMobilePhoneSubscr	0.9912	0.9825
MobilePhone	0.8884	0.7893
TelephoneMainlines	0.8905	0.7930
RatioGDPInvestPerCapita	0.9910	0.9822
ProductivityRatio	0.9591	0.9199
RatioFtTW/ICTInvest	0.9082	0.8249
InterTelecomOutgoing	0.9578	0.9173
RevenuePerSubscriber	0.8151	0.6644

Table 14.7 Assessment of Convergent Validity

Measure	t-Value
RatioMilExp/MilPers	318.0371
MilExpPerCapita	241.0288
Exports%ofFDP	799.7814
Trade%ofGDP	555.8673
FixedMobilePhoneSubscr	482.3323
MobilePhone	49.9610
TelephoneMainlines	84.5551
RatioGDPInvestPerCapita	206.0844
ProductivityRatio	31.5843
RatioFtTW/ICTInvest	21.9873
InterTelecomOutgoing	6.5862
RevenuePerSubscriber	5.4178

In order to further assess convergent validity, we obtained the values of the loadings by following a method outlined by Chin and Hubona (2006), which is as accurate as and more efficient than the one offered by Gefen and Straub (2005). Results provided in Table 14.8 demonstrate that all measures present in our research model share a lot of variance with their own construct, which indicates high convergent validity. At the same time, we note that no measure loads highly on any other construct but its own, which indicates high discriminant validity.

One of the suggested ways (e.g., Fornell & Larcker 1981) of determining discriminant validity in PLS is by assessing the average variance that is shared by a construct and the construct's measures. This measure is provided by the PLS-Graph output as *Average Variance Extracted* (AVE). The commonly accepted practice is to substitute the diagonal elements of the correlation matrix that includes the correlations between the model's constructs with the square root of AVE, and then to compare the substituted values with the values of the off-diagonal elements. The adequacy of the discriminant validity is demonstrated if the diagonal elements of the matrix are greater than the off-diagonal elements (Hulland 1999). The results of the final test of convergent and discriminant validity of our research model are presented in Table 14.9.

The results of the assessment provided in Table 14.9 allow us to conclude that our research model successfully passed the last test of discriminant validity, and we can therefore proceed further with the assessment of the structural model.

PLS Analysis: Assessment of the Structural Model

The process of the assessment of the structural model involves testing the significance of the hypothesized relationships between the constructs specified in the research model. By running PLS-Graph analysis, we observe the path coefficients between the constructs in the model. The significance of the path coefficients is evaluated by running a bootstrapping procedure, which yields *T*-values for each path; the significance level of the path is established using a two-tailed *t*-distribution table. The result of the assessment of the structural model is presented in Table 14.10.

Discussion of the Results

The results of our inquiry allow us to answer the research questions we formulated in the beginning of the paper; the answers, and the possible interpretations, are provided in Table 14.11.

Our data analysis demonstrates that while for the *Leaders* group there is a statistically significant relationship between the constructs *Productivity* and *ICT Capitalization*, for the *Followers* group, existence of the corresponding relationship

Table 14.8 Assessment of Convergent and Discriminant Validity

Measure	Productivity	Mil_of_Ec	H_of_Ec	ICT_Diff	ICT_Cap	ICT_Util
Fixed&MobilePhoneSubscr	-0.05	0.72	0.54	**0.99**	-0.24	0.14
InterTelecomOutgoing	0.04	0.20	-0.11	0.05	0.32	**0.96**
MobilePhones	-0.02	0.62	0.38	**0.89**	-0.13	0.20
TelephoneMainlines	-0.09	0.69	0.65	**0.89**	-0.36	0.02
ProductivityRatio	0.37	-0.18	-0.30	-0.24	**0.96**	0.39
MI	**1.00**	-0.09	-0.01	-0.06	0.37	0.04
RatioGDPInvestPerCapita	0.38	-0.19	-0.26	-0.26	**0.99**	0.29
RatioFtTW/ICTinvest	0.30	-0.25	-0.23	-0.31	**0.91**	0.13
RatioMilExp/MilPers	-0.08	**0.98**	0.35	0.76	-0.21	0.33
MilExpPerCapita	-0.10	**0.98**	0.27	0.69	-0.20	0.36
Exports%ofGDP	-0.01	0.36	**0.99**	0.60	-0.30	-0.12
Trade%ofGDP	-0.01	0.27	**0.99**	0.55	-0.25	-0.10
RevenuePerSubscriber	0.03	0.57	-0.09	0.23	0.16	**0.82**

Table 14.9 Assessment of Discriminant Validity of the Research Model

Construct	Variance (Original On-Diagonal Values Replaced with Square Root of AVE)					
Productivity	**1.000**					
Militarization of Economy	−0.087	**0.9818**				
Health of Economy	−0.012	0.318	**0.9894**			
ICT Diffusion	−0.059	0.736	0.581	**0.9246**		
ICT Capitalization	0.371	−0.211	−0.281	−0.279	**0.9534**	
ICT Utilization	0.044	0.354	−0.111	0.120	0.295	**0.8893**

Table 14.10 Strengths of Structural Path for Constructs in the Research Model

Group of TEs	t-Value	Significance (at $p < 0.05$)	Structural Path	Test of the H0
Complete Set	1.8633	Not significant	ICT Capitalization to Productivity	H01 accepted
The Followers	1.8218	Not significant		H01 accepted
The Leaders	2.2881	Significant		H01 rejected
Complete Set	7.7878	Significant	ICT Diffusion to ICT Capitalization	H02 rejected
The Followers	3.5353	Significant		H02 rejected
The Leaders	1.1696	Not significant		H02 accepted
Complete Set	2.7958	Significant	ICT Utilization to ICT Capitalization	H03 rejected
The Followers	2.8019	Significant		H03 rejected
The Leaders	1.7954	Not significant		H03 accepted
Complete Set	8.0871	Significant	Health of Economy to ICT Diffusion	H04 rejected
The Followers	7.1869	Significant		H04 rejected
The Leaders	6.4556	Significant		H04 rejected
Complete Set	11.4635	Significant	Militarization of Economy to ICT Diffusion	H05 rejected
The Followers	7.2093	Significant		H05 rejected
The Leaders	5.8360	Significant		H05 rejected

Table 14.11 Answers to the Research Questions of the Study

Research Question	Answer within the Context	Possible Interpretation
Do TEs achieve a spillover effect from investments in ICT that is manifested in the relationship between investments in telecoms and TFP?	There is no evidence that the less efficient subset of TEs, the Followers, achieve a spillover effect from investments in ICT. However, there is evidence that the more efficient Leaders are able to demonstrate a spillover effect from investments in ICT.	Only the more efficient TEs are able to demonstrate the link between investments in ICT and growth in productivity.
Is the capitalization of ICT in TEs impacted by the availability and the utilization of ICT infrastructure?	There is evidence that in the case of the less efficient TEs, the Followers, the availability and the utilization of ICT infrastructure have an impact on the capitalization of ICT. However, there is no evidence that the capitalization of ICT of the Leaders is impacted by the availability and the utilization of ICT infrastructure.	Only in the case of less efficient TEs investments in ICT appear to be driven by the limited ICT infrastructure and the utilization of ICT.
Is the availability of ICT in TEs impacted by the state of the economy?	There is evidence that in the case of the TEs, regardless of their level of efficiency, the state of economy has an impact on the availability of ICT.	Regardless of the level of relative efficiency of TE, the level of the existing ICT infrastructure is associated with the level of economic development, and not associated with the level of military spending.

is not supported by the data. Let us place the findings of this study within the broader context of previous investigations, which determined that

- The *Leaders* have higher averaged levels of investments and revenues from telecoms, as well as the higher averaged level of GDP, than the *Followers* (Samoilenko & Osei-Bryson 2007).
- The *Leaders* have higher levels of economic development and of accumulated ICT capital, as well as a higher level of sociotechnical development, than the *Followers* (Samoilenko 2008).
- The *Leaders* have a positive effect of complementarity of investments in telecoms and investments in telecoms on macroeconomic bottom line, while the *Followers* have a negative effect (Samoilenko & Osei-Bryson 2008a).
- The *Leaders* have a higher level of relative efficiency of utilization of investments and a higher level of efficiency of the production of revenues from telecoms than the *Followers* (Samoilenko & Osei-Bryson 2008b).

While it appears to be clear what determines the higher level of revenues from telecoms and a greater impact of investments in telecoms on macroeconomic bottom line in the case of the *Leaders* vs. the *Followers*, the previous studies did not shed any light on the relationship between investments in telecoms and TFP. This investigation, however, demonstrates that the differences between the *Leaders* and the *Followers* extend to the existence of the relationship between investments in telecoms and TFP also. Furthermore, the proposed research model in this investigation also allows us to inquire into the possible nature of the differences that allow the *Leaders* to achieve a spillover effect from investments in ICT, but preclude the *Followers* from achieving the same result.

In order to understand the underlying factors that are possibly responsible for the difference between two subgroups of the sample, we can compare the *Leaders* and the *Followers* in terms of the averaged values of the indicators that were used to represent the latent constructs in the current investigation; the results are summarized in Table 14.12.

The results of the comparison of the measures of the constructs suggest that the *Leaders* and the *Followers* do not significantly differ in terms of the corresponding averaged values of the measure *MI*, representing annual productivity growth reflected by the construct *Productivity*. But this is, figuratively speaking, just the tip of the iceberg, for the two groups do significantly differ in terms of the values of the measures representing the rest of the constructs. In regard to the relationship between the constructs *Productivity* and *ICT Capitalization*, these differences could be interpreted as follows: on average, ceteris paribus,

- The *Leaders* invest in telecoms in excess of seven times more than the *Followers*.
- For a given level of investment, the *Leaders* generate three and a half times the revenues generated by *Followers*.
- For a given level of investment, the *Leaders* employ less than 1/34 of the number of full-time telecom workers than the *Followers*.

Table 14.12 Comparison of the Leaders and the Followers

Construct	Measure	Average Value Leaders	Average Value Followers	Difference Magnitude	Difference Percentage
Productivity	MI	1.23	1.20	0.980	98%
ICT Capitalization	RatioGDPtoInvestment	100.4142029	732.8558883	7.298	729.83%
	RatioProductivity	3.03785	10.70187128	3.522	352.28%
	RatioStaffToInvestment	0.056913043	1.940048936	34.088	3408.79%
ICT Diffusion	FixedMobilePhoneSubscr	497.1483696	183.0814947	0.3683	36.83%
	MobilePhone	199.4960159	25.62736383	0.128	12.85%
	TelephoneMainlines	298.9945275	157.4541394	0.527	52.66%
ICT Utilization	InterTelecomOutgoing	131.283	95.30649	0.726	72.60%
	RevenuePerSubscriber	310.3536348	133.6965596	0.431	43.08%
Health of Economy	Exports%ofGDP	53.98767536	40.18138723	0.744	74.43%
	Trade%ofGDP	112.4362188	91.56905957	0.814	81.44%
Militarization of Economy	RatioMilExp/MilPers	12422.13742	2350.772684	0.189	18.92%
	MilExpPerCapita	77.57529565	19.5409883	0.252	25.19%

In summary, the *Leaders* invest more per capita in telecoms, with a greater efficiency and effectiveness, while using fewer full-time telecom workers in comparison to the *Followers*. Based on the insights offered by this study, it is reasonable to suggest that in order for the *Followers* to establish the link between investments in ICT and growth in productivity, they should develop and pursue policies that concentrate on bringing the levels of values of the measures used in this study closer to the levels of the *Leaders*. The impact of the changes could be investigated based on the simulation approach suggested earlier by Samoilenko and Osei-Bryson (2008a).

However, we must also take into consideration that for the *Leaders*, the construct *ICT Capitalization* "resides" in a totally different environment than it does for the *Followers*. The results of the data analysis suggest that only in the case of the less efficient *Followers* the deficiencies in the existing ICT infrastructure and the demand for its utilization have an impact on *ICT Capitalization*. This may suggest that the *Leaders* have already achieved a level of development where the growth in ICT is fueled not by a basic demand for products and services, but by the demand for added value (e.g., iPhone apps vs. cell phone service). In the case of the *Followers*, as Table 14.12 demonstrates, we can observe clear relative deficiency in the levels of ICT-related infrastructure as compared to *the Leaders*, and even greater discrepancy in regard to the level of utilization of the ICT infrastructure. This may suggest that the *Leaders* do not base their growth on the simple increase in volume of commodity-like generic products and services, but instead face the more specific customer demand and tougher competition, which forces the *Leaders* to work smarter and, resultantly, do more with less.

It is somewhat not surprising that *Health of Economy* turned out to be associated with *ICT Diffusion*, for we expect wealthier economies to have better ICT infrastructure. What is surprising, as Table 14.12 illustrates, is a clearly significant difference in this regard between the *Leaders* and the *Followers*. If, as our research model suggests, the spillover effect of investments in ICT is associated with saturation of ICT-related infrastructure, which, in turn, is associated with the level of economic development, then the unfortunate insight is that the *Followers* may have a long way to go before obtaining the same type of benefits from investments in ICT as the *Leaders* were able to obtain.

Unexpectedly, results also suggest that regardless of the level of economic development, *Militarization of Economy* appears to be associated with the state of existing ICT infrastructure. Also surprising is that *Militarization of Economy* is the construct representing the smallest difference between the *Leaders* and the *Followers*; this may suggest that the *Followers* spend too much on military and have too many people serving in military relative to their level of economic development.

Limitations of the Study

It should be noted that in this study we were often limited by the absence of data on all potentially relevant variables for all countries in the sample for the period covered by the sample. This unavailability of data affected how some of our constructs were defined. For example, we defined *Health of Economy* as the *fiscal state reflecting participation of an economy in international trade*. We note that it could be reasonably argued that the construct *Health of Economy* should also include variables such as *FDI, portfolio flows, and income payments*. So one concern may be that in the absence of all relevant data, is it more appropriate to use the term *Trade* instead of *Health of Economy*? Certainly it can be argued that in the era of globalization, for most countries, particularly those of the developing world, the strength of the economy is often perceived as being largely determined by the extent to which the country participates in international trade. The level of imports is related to the purchasing power of the given economy, which can be considered to be related to the wealth and perceived strength of the given economy. The level of exports can be considered to be related to the relative competitiveness of the productive capacity of the economy. Similarly, our use of the term *ICT Diffusion* may appear to be overly broad since the corresponding data mainly refer to telecom diffusion. Yet while it is true that telecoms are not the only form of ICT, in many developing and transition economies, the telecom category of ICT is the most pervasive form of ICT, being diffused and utilized through all sectors of the given society for both economic and personal activities. Furthermore, *Partnership on Measuring ICT for Development* lists multiple telecoms-related variables (e.g., fixed telephone lines, mobile cellular telephone subscriptions, fixed Internet subscribers, etc.) to construct a core list of ICT indicators on ICT infrastructure and access (ITU 2010). It should also be acknowledged that like this study, many other studies (e.g., Bagchi & Udo 2007; Bollou & Ngwenyama 2008) have used data on telecoms as a useful though imperfect surrogate for ICTs within the context of developing and transition economies.

Conclusion

In this study, we presented a new methodology for testing the relationship between investments in ICT and TFP that is consistent with the framework of neoclassical growth accounting. Furthermore, we used this methodology to conduct direct exploration of TEs and indirect exploration of similarities and differences between developing/emerging and developed countries with regard to the perception of IT success. To adequately and appropriately address the complicating issue of endogeneity that arises, our methodology involves the use of multiple methods: data envelopment analysis, multivariate regression analysis, and SEM. Given the fact that insufficient attention has been given to IT and productivity research in developing and transition economies, we applied this methodology to 18 transition economies. Specifically, we explored the hypothesis that for a given sample of TEs, there exists no statistically discernible relationship between the capital investments in ICT and TFP. Major contributions of this research are the new methodology and the results that followed from its application to TEs, particularly insights that they offer that are relevant to understanding the differences and similarities between developed vs. developing/emerging economies with regard to factors that are relevant to the impact of IT on productivity.

Acknowledgment

Material in this chapter previously appeared in "The spillover effect of investments in telecoms: Insights from transition economies," *Journal of Information Technology for Development* 17:3, 213–233.

References

Andonova, V. & Diaz-Serrano, L. (2007). Political Institutions and the Development of Telecommunications, IZA Discussion Papers 2569, Institute for the Study of Labor (IZA).

Bagchi, K. & Kirs, P. (2009). Group analysis at regional levels can be meaningful in global IS research. *Journal of Global Information Technology Management*, 12(4), 1–5.

Bagchi, K. & Udo, G. (2007). Empirically testing factors that drive ICT adoption in Africa and OECD set of nations. *Issues in Information Systems*, 8(2), 45–52.

Barclay, D., Thompson, R., & Higgins, C. (1995). The partial least squares (PLS) approach to causal modeling: Personal computer adoption and use an illustration. *Technology Studies*, 2(2), 285–309.

Beilock, R. & Dimitrova, D. (2003). An exploratory model of inter-country Internet diffusion. *Telecommunication Policy*, 27(3–4), 237–252.

Bollen, K. & Long, S. (eds.) (1993). *Testing Structural Equation Models.* Sage Focus Editions, Vol. 154. Sage Publications, Inc., Thousand Oaks, CA.

Bollou, F. & Ngwenyama, O. (2008). Are ICT investments paying off in Africa? An analysis of total factor productivity in six West African countries from 1995 to 2002. *Information Technology for Development*, 14(4), 294–307.

Brynjolfsson, E. & Hitt, L. M. (1996). Paradox lost: Firm level evidence on returns to information systems spending. *Management Science*, 42, 541–558.

Caves, D., Christensen, L., & Diewert, W. (1982). The economic theory of index numbers and the measurement of input, output, and productivity. *Econometrica,* 50, 1393–1414.

Chin, W. W. (1995). Partial least squares is to LISREL as principal components analysis is to common factor analysis. *Technology Studies*, 2, 315–319.

Chin, W. W. (1998a). The partial least squares approach to structural equation modeling. In *Modern Methods for Business Research,* ed. G. A. Marcoulides, Lawrence Erlbaum Associates, Mahwah, NJ, pp. 295–336.

Chin, W. W. (1998b). PLS-Graph (Version 03.00 build 1126).

Chin, W. & Hubona, G. S. (2006). *Structural Equation Modeling (SEM) Using PLS-Graph Software: A Tool for Quantitative Researchers Analyzing Path-Based Models.* In Proceedings of the 12th Americas Conference on Information Systems, August 2006.

Daveri, F. (2000). *Is Growth an Information Technology Story in Europe too?* EPRU Working Paper Series 00-12, Economic Policy Research Unit (EPRU), University of Copenhagen. Department of Economics.

Dimitrova, D. & Beilock, R. (2005). Where freedom matters: Internet adoption among the former socialist countries. *The International Journal for Communication Studies*, 67(2), 173–187.

Dunne, T., Foster, L., Haltiwanger, J., & Troske, K. R. (2004). Wages and productivity dispersion in US manufacturing: The role of computer investment. *Journal of Labor Economics*, 22(2), 397–430.

Deichmann, J., Eshghi, B., Haughton, D., Masnghetti, M., Sayek, S., & Topi, L. (2006). Exploring break-points and interaction effects among predictors of the international digital divide. *Journal of Global Information Technology Management*, 9(4), 47–71.

Dewan, S. & Kraemer, K. (2000). Information technology and productivity: Evidence from country level data. *Management Science (Special Issue on the Information Industries)*, 46(4), 548–562.

Färe, R., Grosskopf, S., Norris, M., & Zhang, Z. (1994). Productivity growth, technological progress, and efficiency in industrialized countries. *American Economic Review*, 84, 374–380.

Fornell, C. & Larcker, D. F. (1981). Evaluating structural equation models with unobservable variables and measurement error. *Journal of Marketing Research*, 18, 39–50.

Gefen, D., Straub, D., & Boudreau, M. (2000). Structural equation modeling techniques and regression: Guidelines for research practice. *Communications of the AIS*, 7(7), 1–78.

Gefen, D. & Straub, D. W. (2005). A practical guide to factorial validity using PLS-Graph: Tutorial and annotated example. *Communications of the AIS*, 16(5), 91–109.

Hernando, I. & Nunez, S. (2002). *The Contribution of ICT to Economic Activity: A Growth Accounting Exercise with Spanish Firm-Level Data*. Banco de España Working Papers 0203, Banco de España.

Hoskisson, R., Eden, L., Lau, C., & Wright, M. (2000). Strategy in emerging economies. *Academy of Management Journal*, 43(3), 249–267.

Hulland, J. (1999). Use of partial least squares (PLS) in strategic management research: A review of four recent studies. *Strategic Management Journal*, 20(2), 195–204.

IMF (2001). *International Financial Statistics*. IMF, Washington, DC.

ITU (2010). *Core ICT Indicators 2010*. International Telecommunication Union, Geneva, Switzerland. Available online at http://www.itu.int/ITU-D/ict/partnership/material/Core%20ICT%20Indicators%202010.pdf

Jöreskog, K.G. & Wold, H. (1982). The ML and PLS techniques for modeling with latent variables: Historical and comparative aspects. In *Systems Under Indirect Observation: Causality, Structure, Prediction (Vol. I)*, eds. H. Wold and K. Jöreskog, North-Holland, Amsterdam.

Jorgenson, D.W. & Stiroh, K. J. (2000). US economic growth in the new millennium. *Brooking Papers on Economic Activity*, 1, 125–211.

Lam, P.-L. & Lam, T. (2005). Total factor productivity measures for Hong Kong telephone. *Telecommunications Policy*, 29(1), 53–69.

Madden, G. & Savage, S. J. (1999). Telecommunications productivity, catch-up and innovation. *Telecommunications Policy*, 23(1), 65–81.

Malmquist, S. (1953). Index numbers and indifference surfaces. *Trabajos de Estatistica*, 4, 209–242.

McGuckin, R. H. & Stiroh, K. J. (2002). Computers and productivity: Are aggregation effects important? *Economic Inquiry*, 40(1), 42–59.

Nunnally, J. C. (1978). *Psychometric Theory* (2nd ed.). McGraw-Hill, New York.

OECD (2004). *DAC Network on Poverty Reduction: ICTs and Economic Growth in Developing Countries*. OECD, Paris.

OECD (2005a). Good practice paper on ICTs for economic growth and poverty reduction. *The DAC Journal 2005*, 6(3).

OECD (2005b). Background paper: The contribution of ICTs to pro-poor growth: No. 384. *OECD Papers*, 5(2), 15–52.

OECD (2005c). The contribution of ICTs to pro-poor growth: No. 379. *OECD Papers*, 5(1), 59–72.

Oliner, S. D. & Sichel, D. E. (2000). The resurgence of growth in the late 1990s: Is information technology the story? *Journal of Economic Perspectives*, 14(4), 3–22.

Palvia, C., Palvia, S., & Whitworth, J. (2002). Global information technology: A meta analysis of key issues. *Information and Management*, 39, 403–414.

Palvia, P. (2006). Key IS management issues: Need for an international research program. *Journal of Global Information Technology Management*, 9(2), 1–4.

Piatkowski, M. (2002). *The 'New Economy' and Economic Growth in Transition Economies*. WIDER Discussion Paper No. 2002/63. WIDER, Helsinki.

Rivard, S. & Huff, S. (1988). Factors of success for end-user computing. *Communications of ACM*, 31(5), 552–561.

Roztocki, N. & Weistroffer, H. (2008). Editorial preface: Information technology in transition economies. *Journal of Global Information Technology Management*, 11(4), 2–9.

Samoilenko, S. (2008). Contributing factors to information technology investment utilization in transition economies: An empirical investigation. *Information Technology for Development*, 14(1), 52–75.

Samoilenko, S. & Osei-Bryson, K. M. (2007). Increasing the discriminatory power of DEA in the presence of the sample heterogeneity with cluster analysis and decision trees. *Expert Systems with Applications*, 34(2), 1568–1581.

Samoilenko, S. & Osei-Bryson, K. M. (2008a). Strategies for telecoms to improve efficiency in the production of revenues: An empirical investigation in the context of transition economies. *Journal of Global Information Technology Management*, 11(4), 56–75.

Samoilenko, S. & Osei-Bryson, K. M. (2008b). An exploration of the effects of the interaction between ICT and labor force on economic growth in transitional economies. *International Journal of Production Economics*, 115(2), 471–481.

Samoilenko, S. & Osei-Bryson, K. M. (2010). *Linking Investments in Telecoms and Productivity Growth in the Context of Transition Economies Within the Framework of Neoclassical Growth Accounting: Solving Endogeneity Problem with Structural Equation Modeling*. In Proceedings of 18th European Conference on Information Systems, Pretoria, South Africa, June 6th–9th, 2010.

Schreyer, P. (2000). *The Contribution of Information and Communication Technology to Output Growth: A Study of the G7 Countries*. OECD Science, Technology and Industry Working Papers 2000/2, OECD, Directorate for Science, Technology and Industry.

Siegel, D. (1997). The impact of computers on manufacturing productivity growth: A multiple-indicators, multiple-causes approach. *The Review of Economics and Statistics*, 79(1), 68–78.

Watson, R., Kelly, G., Galliers, R., & Brancheau, J. (1997). Key issues in information systems management: An international perspective. *Journal of Management Information Systems*, 13(4), 91–115.

Whelan, K. (2000). *Computers, obsolescence, and productivity.* Federal Reserve Board. Finance and Economics Discussion Series 2000-06, Board of Governors of the Federal Reserve System (U.S.).

Wold, H. (1966). Estimation of principal components and related models by iterative least squares. In *Multivariate Analysis,* ed. P. R. Krishnaiah, Academic Press, New York, pp. 391–420.

Yearbook of Statistics (2004). *Telecommunication Services Chronological Time Series 1993–2002.* ITU Telecommunication Development Bureau (BDT), International Telecommunication Union.

Appendix

Table 14A.1 18 Transition Economies: The *Leaders* and the *Followers*

The Leaders	The Followers
Bulgaria (2002), Czech Rep (1993–2002), Estonia (1994–2002), Hungary (1993–2002), Latvia (1994, 1995, 1997–2002), Lithuania (1999–2002), Poland (1993–2002), Slovenia (1993–2002), Slovakia (1995–1998, 2000–2002)	Albania (1993–2002), Armenia (1993–2002), Azerbaijan (1993–2002), Belarus (1993–2002), Bulgaria (1993–2001), Estonia (1993), Kazakhstan (1993–2002), Kyrgyz Rep (1993–2002), Latvia (1993, 1996), Lithuania (1993–1998), Moldova (1993–2002), Romania (1993–2002), Slovakia (1993,1994, 1999), Ukraine (1993–2002)

Chapter 15

Investigating Factors Associated with the Spillover Effect of Investments in Telecoms: Do Some Transition Economies Pay Too Much for Too Little?

Introduction

It has been noted that while the stream of research investigating the relationship between investments in information and communication technologies (ICT) and their macroeconomic outcomes is well established (OECD 2005a,b,c; IMF 2001; Samoilenko & Osei-Bryson 2008a,b), the majority of studies have been conducted in the context of developed economies (Lam & Lam 2005; Madden & Savage 1999; Dunne et al. 2004; Siegel 1997). Unfortunately, the inherent relative heterogeneity of emerging, developing, and transition economies (TEs) complicates the application of insights offered by the studies of developed countries to their settings (Roztocki & Weistroffer 2008, 2011; Hoskisson et al. 2000). Consequently, policy and decision makers of TEs often face unique managerial challenges that do not lend themselves to be addressed by the techniques, best practices, and tools available to developed economies (Arcelus & Arocena 2000; Barro & Sala-i-Martin 1995; Sala-i-Martin 1996). Furthermore, managers of the developed world seeking to penetrate new markets and to establish a global presence for their firms may find themselves inadequately prepared for dealing with the specificity of the new contexts. Fortunately, it was acknowledged that the context of the countries that are in transition from centralized planning system to a free market economy, commonly designated as *transition economies* (Roztocki & Weistroffer 2008, 2011), offers an advantageous setting for conducting generalizable research (Samoilenko 2008). The feature that makes the environment of TEs attractive to researchers is their "economic duality," where TEs share some economic characteristics with developed countries, while also having economic traits that are more in line with less developed regions (OECD 2004). This is an important advantage, because on the one hand, previous investigations provided compelling evidence that ICT expansion has led to robust returns and economic growth in developed economies (OECD 2005a; Oliner & Sichel 2002; Jalava & Pohjola 2002), while on the other hand, the research conducted in the setting of emerging, developing, and TEs reveals that investments in ICT impact the macroeconomic bottom line at a significantly lower level (Dewan & Kraemer 2000; Pohjola 2001; Piatkowski 2003). Given this disparity of outcomes, TEs can serve as a bridge spanning the divide separating developed and less developed countries, and offer a platform from which much needed investigations (OECD 2004) could be conducted and the findings generalized to a broader pool of economies. As a result, investigations conducted in the context of TEs are not only of scholarly but also of practical importance, for their results may serve as a useful addition to the otherwise limited managerial toolkit of decision makers of emerging, developing, and TEs.

The overall objective of the current investigation is to gain a greater understanding of the impact of investments in ICT on the macroeconomic bottom line of TEs. To fulfill this objective, we aim to, first, identify factors that are associated with the presence of the impact of the investments on macroeconomic growth, and, second, assess changes that took place in regard to the impact over time. More specifically, in the current study, we focus on investments in telecoms (a subset of ICT) within the context of 18 ex-communist (Bagchi & Kirs 2009) *Transition Economies in Europe* and the *Former Soviet Union* (IMF 2001) composed of a group of the more efficient *Leaders* and a group of the less efficient *Followers* (Samoilenko & Osei-Bryson 2010). We achieve the research objective by following a multistep methodology allowing for accomplishing two primary goals. The first goal of our study is to identify a set of criteria relevant to investments in telecoms that differentiate the "higher macroeconomic impact" TEs (the *Leaders*) from the "lower macroeconomic impact" TEs (the *Followers*). The second goal is to assess change that took place over time in the level of macroeconomic performance of the investments of the laggards and to inquire whether the *Followers* have been pursuing empirically justifiable strategies directed toward improving the level of the impact of investments in telecoms relative to the *Leaders*.

Previously, Samoilenko and Weistroffer (2010a) utilized structural equation modeling (SEM) implemented with partial least squares (PLS) to conduct a test for significance of the relationship between investments in telecoms and two components of macroeconomic growth, the one resulting from change in technology (TC) and the other associated with a change in efficiency (EC). The analysis was performed using the data set spanning the period from 1993 to 2002. The authors determined that while both groups of TEs (e.g., the *Leaders* and the *Followers*) exhibited the presence of the relationship between investments in telecoms and TC, the relationship between investments in telecoms and EC was present only in the case of the *Leaders* (Samoilenko & Weistroffer 2010a). We further develop the findings of the investigation of Samoilenko and Weistroffer (2010a) by employing decision trees (DTs) to identify some of the variables that are associated with and could be significant to the presence of the relationship between investments in telecoms and different components of growth in productivity. DT is a nonparametric data analytic tool that uses an algorithm-based induction model to partition a data set into multiple subsets, most commonly for the purposes of classification and prediction. There are multiple methods allowing for identifying variables of discriminatory value, but in this study we use classification DT to identify a variable or a set of variables that differentiate two groups of TEs the most. Taking into consideration that the top-level splits starting at the root node of the tree (which represents the complete data set) are the most important for the partitioning, the corresponding splitting criteria will point at the dimensions that differentiate the two groups of TEs. Based on the outcome of DT induction, we derive a set of differentiating factors that, if used as prescriptive measures, may allow for improving the impact of investments in telecoms on the macroeconomic bottom line.

To accomplish the second goal, we analyze the data set representing the same 18 TEs, but spanning the period from 2003 to 2008, with the purpose of identifying some of the changes that actually took place, and then comparing them with the prescribed ones. We empirically evaluate the outcome of the comparison by conducting data envelopment analysis (DEA). DEA is a nonparametric method commonly used for the purpose of measuring the efficiency of decision-making units (DMUs). The DMUs in the set, which in the case of this study are represented by TEs, are ranked based on the calculated scores of the relative efficiency, where the highest ranking DMUs are considered to be relatively efficient. There could be multiple relatively efficient DMUs in the set; thus, the data set will become "enveloped" by the efficient frontier consisting of the relatively efficient DMUs. If the position of the efficient frontier changes over time, then we can capture this change to assess the impact in terms of the efficiency of the production of revenues and utilization of investments in telecoms.

By accomplishing these research goals, we aim to contribute to a better understanding of the role and the impact of information technology in the context of TEs.

Specifically, we will gain better insights into managerial challenges associated with effective and efficient conversion of investments in telecoms into macroeconomic outcomes, and will offer a theoretically sound methodology allowing for deriving descriptive and prescriptive strategies geared toward addressing those challenges. Additionally, we will identify some of the changes in regard to telecoms-related factors that took place over the period from 1993 to 2008 within the context of TEs. Also, we will be able to evaluate the relative impact of the identified changes on the macroeconomic impact of telecoms in the context of TEs. We present our investigation in the following sequence: after providing an overview of the background of this study, we present the theoretical foundation of the inquiry. Then we describe the data and the research methodology used in our investigation. This is followed by the presentation and discussion of the results of the data analysis. A conclusion and the overview of the limitations of the study are provided at the end of the paper.

Background of the Study

Regardless of the setting, whether a developed, a transition, or a less developed economy, there are two obvious and interrelated routes by which investments in ICT could impact a macroeconomic bottom line. The first route is by providing a return on investments in the form of a stream of revenues that contributes to the overall level of GDP (UN ICT Task Force Report 2005; WT/ICT Development Report 2006). The issue of improving the level of performance of investments in ICT along this route primarily deals with determining whether the level of investments is high enough for impacting the macroeconomic bottom line (Jorgenson & Stiroh 2000; Oliner & Sichel 2000; Council of Economic Advisors 2001; Jorgenson 2001, determining whether the investments are utilized efficiently (Samoilenko & Osei-Bryson 2008a), and identifying complementary investments that could provide a synergistic effect on the level of revenues (Kraemer & Dedrick 2001; Pohjola 2002). We must note that complementary factors are not limited to internal to the economy domestic investments, but could also include such external factors as foreign direct investments (Gholami et al. 2006). Samoilenko and Osei-Bryson (2008a) investigated the efficiency of the production of revenues from investments in telecoms in the context of TEs and determined that the subgroup with a higher level of investments was also the group with a higher level of revenues. However, the authors presented evidence that the lower level of revenues of the group of TEs with the lower levels of investments was not due to insufficient levels of investments, but due to the inefficiencies of the process of conversion of investments into revenues. Previously, Samoilenko and Osei-Bryson (2007) also presented evidence that the complementarities of investments in telecoms and full-time telecom staff play an important role in the process of revenue generation, and that those TEs that do not exhibit a complementary effect of investments and labor generate, ceteris paribus, lower levels of revenue from telecoms than the TEs that do. These findings are in agreement with a consensus that in order to impact the macroeconomic bottom line via a stream of revenues, investments in ICT must be made at a sufficiently high level, must be utilized efficiently, and must be accompanied by complementary investments (Dewan & Kraemer 2000; Kraemer & Dedrick 2001; Pohjola 2002; Piatkowski 2002; OECD 2004; Samoilenko & Osei-Bryson 2007, 2008a,b).

The second route by which investments in ICT could impact the macroeconomic bottom line is via the *spillover effect*, where the impact of investments is manifested indirectly, in the form of externalities beyond the stream of revenue. Given an option, it is much more beneficial to choose the second route to the first, simply because the presence of the spillover effect allows investments to impact the macroeconomic bottom line in two ways: first, as a stream of revenues, and second, as "something else" of unexplained provenance that appears to be free. It appears to be free simply because no part of the investments is actually allocated in advance for obtaining the spillover effect; this is in contrast to revenues that are never free in the sense that they always require allocation of investments upfront. Consequently, if an economy allocates resources to impact its macroeconomic bottom line (e.g., GDP) via investment in ICT, then the impact could come in the form of revenues from ICT, or revenues from ICT accompanied by the spillover effect of investments in ICT, but not the spillover effect by itself. In other words, no investments could be made directly and exclusively into externalities bypassing revenue producing "internalities."

Is the spillover effect worth pursuing? While the "investments to revenues" route is affected by the law of diminishing returns, the "spillover" route is not. Consequently, the attractiveness of the second route is fairly obvious, especially for relatively weak economies with limited investment resources. Recently, Samoilenko and Osei-Bryson (2010) outlined and tested a methodology for determining the presence of the spillover effect of investments in telecoms. The findings indicate that the relatively more efficient TEs do demonstrate the relationship between investment in telecoms and growth in productivity, thus providing the evidence for the presence of the spillover effect of investments in telecoms. However, one important issue remained open. Namely, the authors investigated the presence of the spillover effect of investments in telecoms by means of investigating the relationship between investments in telecoms and *overall* growth in productivity. The important point to consider is that overall productivity growth, as commonly depicted in existing research (Khouja 1995; Shao & Lin 2001; Daßler et al. 2002; Chen & Zhu 2004; Samoilenko & Osei-Bryson 2008a,b), is a composite of two parts, *change in efficiency* and *change in technology*, and that it is possible for an economy to exhibit an overall economic growth that is driven by only one component (Samoilenko & Weistroffer 2010a). Later, Samoilenko and Weistroffer (2010a) presented evidence that TEs could base the economic growth solely on the change in the technology component (e.g., by investing in cutting edge technology), while having no, or even negative, growth in the change of the efficiency component (e.g., the productivity of the workforce deteriorates due to the inability to keep up with the technology, possibly caused by a sharp learning curve).

Undoubtedly, evidence that investments do, or do not, produce the identifiable spillover effect is of importance to foreign and domestic managers considering investments in the economy. However, such evidence does not offer a great deal of information in regard to a possible course of action to a decision maker tasked with the responsibility of improving the performance of investments. Moreover, even if investigators obtain evidence of disparity of TEs in regard to the presence of the relationship between investments and a specific component of growth in productivity (Samoilenko & Weistroffer 2010a), decision makers could still benefit greatly from insights as to why TEs differ in this regard. What is required then is to complement the existing findings with the additional insights provided by investigations conducted from a lower level of granularity. If a policy maker in TE obtains evidence that investments in ICT have been driving technological change (TC) at the expense of improvements in ICT workforce efficiency, then his/her strategy of improving the macroeconomic impact of investments can be formulated with more precision, and the limited resources allocated with more efficiency. In the absence of such insights, policy and decision makers will not be equipped with the empirical data necessary to decide if additional investments should be directed toward the quality of the technology or a quantity of the labor, or whether the process of the conversion of investments into revenues should be improved by means of purchasing a better technology versus by means of implementing workforce development programs.

Theoretical Foundation and Research Questions of the Study

A common formulation of neoclassical production function is presented as

$$Y = f(A, K, L) \tag{15.1}$$

where
Y = measure of economic output (most often in the form of GDP),
K = measure of capital, an endogenous variable explaining part of Y,
L = measure of labor, also an endogenous variable explaining part of Y, and
A = total factor productivity (TFP), an exogenous, unexplained by the endogenous components, part of Y.

Based on Equation 15.1, growth accounting uses Cobb–Douglas production function:

$$Y = A * K^\alpha * L^\beta, \tag{15.2}$$

where α and β are constants determined by the production technology.

By taking the logarithm, we obtain the following formulation:

$$\log Y = \log A + \alpha \log K + \beta \log L \tag{15.3}$$

Considering that A is a residual that can be expressed as error term e, the following general form for Equation 15.3, given Equation 15.1, is offered:

$$\log Y = \beta_0 + \beta_{1*} \log K + \beta_{2*} \log L + e. \tag{15.4}$$

In the context of our investigation, Y is represented by GDP, K is represented by the level of investments in telecoms, and L is represented by the quantity of a full-time telecom staff of a given TE. Given the ability to calculate the value of TFP using MI, investigators can also obtain the value of e, as well as the values of its components, change in efficiency (EC) and change in technology (TC). Resultantly, Equation 15.1 can be presented as

$$Y = f(A_{EC} + A_{TC}, K, L), \tag{15.5}$$

and the value of the error term in Equation 15.4 can be rewritten as

$$e = e_{EC} + e_{TC},$$

where
 e_{EC} = EC component of MI
 e_{TC} = TC component of MI

As a result, Equation 15.4 can be presented as

$$\log Y = \beta_0 + \beta_{1*} \log K + \beta_{2*} \log L + e_{EC} + e_{TC}. \tag{15.6}$$

Investigators interested in the impact of K on e_{EC} or e_{TC} cannot relate the variables directly, for this would violate the basic assumption of the independence of the endogenous variable and the error term; furthermore, TFP is an exogenous variable representing an unexplained component of growth within the neoclassical production function and thus cannot be directly related to an endogenous variable K.

The methodology of Samoilenko and Osei-Bryson (2010) relied on SEM to resolve this issue (commonly referred to as *endogeneity problem*) and to allow researchers to relate K and e indirectly via latent constructs *Productivity* and *ICT Capitalization*. This methodology was extended by Samoilenko and Weistroffer

(2010a) to allow for relating K to two components of e, namely, e_{EC} ($K{\rightarrow}EC$) and e_{TC} ($K{\rightarrow}TC$).

As it was mentioned earlier, Samoilenko and Weistroffer (2010a) identified that while the *Leaders* and the *Followers* demonstrated the existence of the relationship between investments in telecoms and the growth in productivity driven by TC ($K{\rightarrow}TC$), only the *Leaders* subset of TEs exhibited the relationship between investments in telecoms and the growth in productivity driven by change in efficiency ($K{\rightarrow}EC$). The authors also stated that the *Leaders* have a higher level of investments in telecoms and a lower level of full-time telecom workforce relative to the *Followers* (Samoilenko & Weistroffer 2010a); this claim, however, was not substantiated by a rigorous application of any established data analytic method. Consequently, a question of what differentiates economies exhibiting the spillover effect of investments in telecoms remains open and corresponds to the first research question of the current study, as formulated below.

> RQ1: What are some of the factors that differentiate a group of TEs that exhibit relationship between investments in telecoms and TC ($K{\rightarrow}TC$), from the group that exhibit relationship between investments in telecoms and EC ($K{\rightarrow}EC$)?

To answer RQ1, we will conduct DT analysis to identify the variable, or a set of variables, that is used in partitioning the overall sample of TEs into two subsets, one set characterized by $K{\rightarrow}TC$ and another characterized by $K{\rightarrow}EC$. We must note that this partitioning does not imply that one group cannot have both types of relationship present simultaneously; quite the opposite—we already know that the *Leaders* exhibit the presence of both $K{\rightarrow}TC$ and $K{\rightarrow}EC$; instead, it does imply that in order for our two groups to differ in regard to the presence of the relationship, one group (e.g., the *Followers*) must be characterized by the presence of only one type of the relationship, either $K{\rightarrow}TC$ or $K{\rightarrow}EC$. Furthermore, it is possible to use a similar type of partitioning even if both groups exhibit the presence of $K{\rightarrow}TC$ and $K{\rightarrow}EC$, but one group is dominant in terms of one type of the relationship. Accommodation of such situation may require changing the differentiating criterion (e.g., ratio, percentage, etc.), but still will produce a logical partitioning that could be subjected to DT analysis.

Knowing some of the factors that differentiate the group of $K{\rightarrow}TC$ from the group of $K{\rightarrow}EC$ over the period from 1993 to 2002, we can use those factors as a proxy for the differences between $K{\rightarrow}TC$ and $K{\rightarrow}EC$ and inquire whether the same factors are still responsible for differentiating the two groups over the period from 2003 to 2008, or whether the two groups became more homogenous in that regard. This allows us to formulate our second research question, as follows.

> RQ2: What are the changes that took place over the period from 1993 to 2008 in regard to the values of factors that differentiated the group $K{\rightarrow}TC$ from the group $K{\rightarrow}EC$ over the period of time from 1993 to 2002?

In order to answer RQ2, we will assess the two groups of TEs in regard to the values of the variable that was selected to perform a top-level split by DT analysis. The assessment will involve comparison of the averaged values across two groups for 1993–2002 and 2003–2008 periods.

It is only reasonable to assume that not only some changes in regard to that variable indeed took place, but also that the changes impacted the level of the macroeconomic performance of the TEs. Consequently, keeping in mind that the differences between the group of $K{\rightarrow}TC$ and the group of $K{\rightarrow}EC$ were represented by the differences in the value of the proxy, we can formulate our last research question, as follows:

> RQ3: What is the impact of the change in regard to the values of differentiating the group $K{\rightarrow}TC$ and $K{\rightarrow}EC$ factors that took place over the period from 1993 to 2008 on the level of macroeconomic performance of investments in telecoms in the context of TEs?

In order to answer RQ3, we will perform DEA to calculate the scores of the relative efficiencies for both groups ($K{\rightarrow}TC$ and $K{\rightarrow}EC$) for the period from 2003 to 2008, and then compare the calculated scores with the scores for the period from 1993 to 2002.

At this point, we can formulate the following null hypotheses corresponding to RQ1, RQ2, and RQ3:

- H01: The DT analysis will not produce a top-level split allowing for separating the sample into the two subsets, where each subset will contain at least 95% of the members of each subset.
- H02: The DT analysis will not produce a top-level split of the 2003–2008 data set using the same variable that was used for the top-level split of the 1993–2002 data set.
- H03: There is no statistically significant difference between the averaged scores of the relative efficiency of the two groups for the period 1993–2002 and 2003–2008.

Overview of the Data

In this investigation, we utilize two sequential time-series data sets on 18 TEs. The first data set, which covers the period from 1993 to 2002, was used in the previous study of Samoilenko and Weistroffer (2010a), while the second data set, which covers the period from 2003 to 2008, has not been previously analyzed. The first period intended to represent an earlier stage of transition, which started in 1991 with the onset of the collapse of the Soviet Union (many countries of our data set were members of the Soviet Union). We begin with the year 1993 mainly for the reason of the availability of

Table 15.1 Two Subgroups Comprising the Sample of 18 TEs

Subgroup	General Membership of the Subgroup
The *Leaders*	Czech Rep, Estonia, Hungary, Latvia, Lithuania, Poland, Slovenia, Slovakia
The *Followers*	Albania, Armenia, Azerbaijan, Belarus, Bulgaria, Estonia, Kazakhstan, Kyrgyz Rep., Moldova, Romania, Ukraine

the data; 1993 was the earliest year that allowed us to obtain the data on most of the TEs in Europe and the former Soviet Union. The second period intended to represent a later stage of transition; the year 2008 was chosen as an end of the period due to the availability of the data. The data for both sets were obtained from World Development Indicators (WDI) database of the World Bank and the Yearbook of Statistics (2009) of the International Telecommunication Union (ITU). The complete membership of the sample of 18 TEs is represented in terms of two clusters (see Table 15.1): the more efficient *Leaders* also characterized by a higher level of investments and revenues from telecoms, and the less efficient *Followers* with a lower level of investments and revenues from telecoms (Samoilenko & Osei-Bryson 2010).

The original research model of Samoilenko and Osei-Weistroffer (2010a) reflected three latent variables, *ICTCap*, *EC*, and *TC*, and two links, *ICTCap→EC*, and *ICTCap→TC*; the details are provided in Table 15.2.

The next section of the paper is dedicated to an overview of the methodology that we employ to answer the research questions of this study.

Research Methodology

Previous investigations offered evidence of the differences between the *Leaders* and the *Followers* not only in regard to the general presence of the spillover effect (Samoilenko & Osei-Bryson 2010) but also to the link between investments in telecoms and change in efficiency component of the effect (Samoilenko & Weistroffer 2010a). However, this evidence is insufficient for gaining an understanding of the reasons for the differences between the two groups of TEs. The methodology used in the current investigation should not be considered an extension of the methodology of Samoilenko and Osei-Bryson (2010) or Samoilenko and Weistroffer (2010a). And while in the context of this investigation we use findings of Samoilenko and Weistroffer (2010a) as an input to our inquiry, there is no reason why another data set, partitioned in a criterion-based manner, could not serve as an input to the proposed methodology.

Our methodology relies on DT induction to generate decision rules that could provide insights into some of the factors responsible for the heterogeneity of the sample of 18 TEs in regard to the presence of the relationship between constructs *ICT Capitalization* and *Productivity Driven by Change in Technology (ICTCap→TC)* and

Table 15.2 Measures of the Current Research Model

Measure	Source Variables	Representation	Latent Construct
TFP	MI	Annual change in productivity	Productivity driven by change in technology ("TC")
Technical change component of TFP	TC component of MI	Annual change in productivity driven by change in technology	Productivity driven by change in technology ("TC")
TFP	MI	Annual change in productivity	Productivity driven by change in efficiency ("EC")
Efficiency change component of TFP	EC component of MI	Annual change in productivity driven by change in efficiency	Productivity driven by change in efficiency ("EC")
RatioGDPtoInvestment	1. GDP per capita (in current US$) 2. Annual Telecom Investment per capita (in current US$)	Ratio of GDP per capita to Annual Telecom Investment per capita.	ICT Capitalization ("ICTCap")
RatioProductivity	1. Annual Total Revenue from Telecoms (% of GDP) 2. Annual Investments in Telecoms (% of GDP)	Ratio of Annual Total revenue from Telecoms to Annual investments in Telecoms	ICT Capitalization ("ICTCap")
RatioStafftoInvestment	1. Full-time telecom staff 2. Annual investment in Telecoms (in current US$)	Ratio of Full-time telecom staff to the Annual investment in telecoms	ICT Capitalization ("ICTCap")

Source: Based on Samoilenko, S., & Weistroffer, H.R. Spillover Effect of Telecom Investments on Technological Advancement and Efficiency Improvement in Transition Economies. In *Proceedings of the SIG GlobDev 3rd Annual Workshop ICT in Global Development*; Saint Louis, Missouri, USA. December 12, 2010.

ICT Capitalization and *Productivity Driven by Change in Efficiency (ICTCap→EC)* in terms of the measures of the construct *ICT Capitalization*.

This will require the inclusion of a composite target variable (let us say *Group&RelationshipExistence*) that identifies a group within the sample of 18 TEs and the presence of the relationships identified by Samoilenko and Weistroffer (2010a); we will use a categorical variable to indicate whether the relationship between the two constructs is present. Consequently, because there are only three possible scenarios in regard to the relationship between the constructs, the domain of values for the *RelationshipExistence* part of our target categorical variable can be represented as follows:

1. Only presence of the relationship *ICTCap→TC*
2. Only presence of the relationship *ICTCap→EC*
3. Presence of the relationships *ICTCap→TC* and *ICTCap→EC*

If we take into consideration the heterogeneity of the sample, where according to Samoilenko and Osei-Bryson (2010) there are two subgroups, then the *Group* part of our target categorical variable can be represented as *Leaders* or *Followers*; thus, we end up with the following complete domain of values for our target categorical variable *Group&RelationshipExistence*: *Leaders1, Leaders2, Leaders3, Followers1, Followers2,* and *Followers3*. The representation of domain of values is not important; thus, to simplify the actual coding, we chose binary "0" and "1" to represent the values of the target variable.

After the target variable *Group&RelationshipExistence* is added to the original dataset, the next step is to perform DT induction to generate a model that describes the differences between the two groups in terms of the indicators of the latent variable *ICT Capitalization*. In the next step of our methodology, we will use generated decision rules to identify the top-level split variable that was selected for partitioning of the complete set of 18 TEs into the two subsets. From the theoretical perspective, the variable selected for the top-level split allows for the greatest information gain via reduction of entropy in the data set. Consequently, the top-split variable is the most significant for partitioning of the sample, and the values of the top-split variable reflect the nature of the heterogeneity of the two resulting subsets. However, while it is possible that over time the heterogeneity of two subsets decreased, it is also possible that no appropriate actions have been undertaken by the *Followers* and the heterogeneity in regard to the values of the top-level split variable is still present. Consequently, we will use a 2003–2008 data set to inquire into the changes that took place, as well as assess the impact of the changes.

Once the changes (or an absence of thereof) in regard to the values of the top-level split variable have been identified, we will conduct DEA to inquire into the changes that took place over time in regard to the level of performance of the two groups of TEs. In order to assess the changes, we will obtain the scores of the relative efficiency for the two groups for the 2003–2008 period; the scores will then be compared to the scores for the 1993–2002 period. As a result, we will be able to obtain some important insights regarding the relative changes in the level of performance of two groups of TEs that took place over the period of 16 years, from 1993 to 2008. The methodology of the investigation is depicted in Figure 15.1.

The proposed methodology is structured as the following sequence of steps:

Step 1—Assign the appropriate values to the target variable *Group&RelationshipExistence* for the *Leaders* and the *Followers*; the results of the study of Samoilenko and Weistroffer (2010a) serve as an input to this step.

Step 2—Combine the *Leaders* and the *Followers* subsets of the sample with the target variable *Group&RelationshipExistence* into a single combined data set.

Step 3—Conduct DT analysis to obtain decision rules according to which the combined data set can be partitioned along the values of the target variable.

Step 4—Conduct DT analysis of the 2003–2008 data set in order to determine the presence of the changes in the values of the top-level split variable of the 1993–2002 data set.

Step 5—Compare 1993–2002 and 2003–2008 data sets in terms of the values of the indicators that represented the latent constructs in SEM.

Step 6—Conduct DEA to calculate the scores of the relative efficiencies of the two groups of TEs for the periods 1993–2002 and 2003–2008.

Step 7—Assess the changes in the scores of the relative efficiencies of the two groups of TEs by comparing the results of DEA for the period 1993–2002 with the results for 2003–2008.

We provide a summary of the research questions with the corresponding null hypotheses, data analytic tools utilized for testing the hypotheses, and the steps of the study in Table 15.3.

At this point, we can restate RQ1, RQ2, and RQ3 in the form of the following null hypotheses:

- H01: Given the sample of 18 TEs consisting of the groups *ICTCap →TC* and *ICTCap →EC*, the DT analysis will not produce a top-level split allowing for separating the sample into the two subsets, where each subset will contain at least 95% of the members of each group.
- H02: Given the sample of 18 TEs consisting of the groups *ICTCap→TC* and *ICTCap →EC*, the DT analysis will not produce a top-level split of the 2003–2008 data set using the same variable that was used for the top-level split of the 1993–2002 data set.
- H03: Given the sample of 18 TEs consisting of the groups *ICT Cap→TC* and *ICT Cap →EC*, there is no statistically significant difference between the averaged scores of the relative efficiency of the two groups for the period 1993–2002 and 2003–2008.

We next present the results of the data analysis.

228 ■ *Theoretical Research Frameworks Using Multiple Methods*

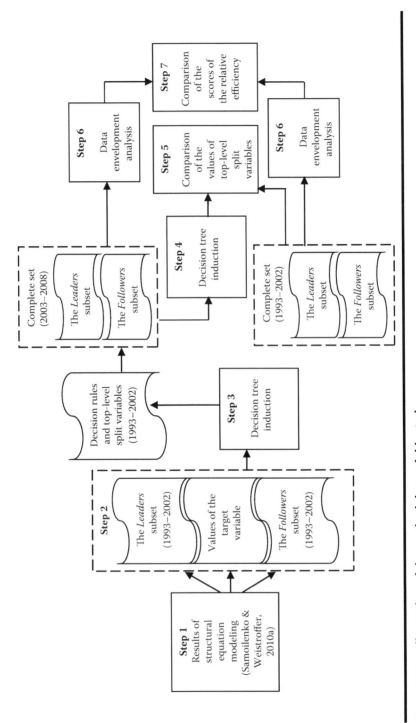

Figure 15.1 Illustration of the methodology of this study.

Table 15.3 Summary of the Steps of the Methodology

Research Question	Method	Steps	Null Hypothesis
RQ1: What are some of the factors that differentiate a group of TEs that exhibit relationship between investments in telecoms and TC (K→TC) from the group that exhibits relationship between investments in telecoms and EC (K→EC)?	DT	1, 2, 3	H01
RQ2: What are some of the changes that took place in regard to the values of differentiating factors over the period of time 1993–2002 and 2003–2008?	DT	4, 5	H02
RQ3: What is the impact of identified changes in the values of differentiating factors on the level of macroeconomic performance of investments in telecoms in the context of TEs?	DEA	6, 7	H03

Results of the Data Analysis

RQ1: Steps 1, 2, and 3

Samoilenko and Weistroffer (2010a) determined that there is no difference between the *Leaders* and the *Followers* in regard to the presence of the relationship *ICTCap→TC*. However, the authors identified the differences between the *Leaders* and the *Followers* in regard to the presence of the relationship *ICTCap→EC*. Consequently, we assign to the target variable *Group&RelationshipExistence* categorical values of *Leaders3* and *Followers1*. This allows us to proceed with Step 2 of the methodology and to combine the two subsets of the sample, the *Leaders* and the *Followers*, with the target variable *Group&RelationshipExistence* into a single combined data set. We performed DT analysis using SAS Enterprise Miner (EM) data mining software package. Once the data were imported into the EM library, we specified *Group&RelationshipExistence* variable as a target and ran the analysis. After the first DT model was generated, we reset the variable used in the first split to "don't use" and re-ran the analysis. Altogether, two single-split models were generated, giving us some insights regarding the differences between the *Leaders* and the *Followers*. Resultantly, we are able to test H01 in regard to two measures constructed by different variables. We provide the results of DT analysis, expressed in the form of the decision rules, in Table 15.4.

Table 15.4 Decision Rules Generated by DT Analysis of 1993–2002 Data Set

Model	Decision Rules	H01 Testing Criterion	Test of H01

1st model	IF ***RatioStaffToInvestment*** < 0.0925
	THEN
	    Leaders3: **98.5%**
	    Followers1: 1.5%
	IF 0.0925 <= ***RatioStaffToInvestment***
	THEN
	    Leaders3: 2.1%
	    Followers1: **97.9%**
		To reject H01 *if* the population of each of the two branches produced by the 1st split contains 95% or more members of the same group.
		1st branch contains 98.5% of the *Leaders* (**98.5% > 95%**)
		2nd branch contains 97.9% of the *Followers* (**97.9% > 95%**)
			H01 rejected

2nd model	IF 174.6695 >= ***RatioGDPtoInvestment***
	THEN
	    Leaders3: 3.4%
	    Followers1: 96.6%
	IF ***RatioGDPtoInvestment*** < 174.6695
	THEN
	    Leaders3: **67.3%**
	    Followers1: 32.7%
		To reject H01 *if* the population of each of the two branches produced by the 2nd split contains 95% or more members of the same group.
		1st branch contains 96.6% of the *Followers* (**96.6% > 95%**)
		2nd branch contains 67.3% of the *Leaders* (**67.3% < 95%**)
			H01 fail to reject

Results of the analysis allow us to reject H01 corresponding to our RQ1. According to the first split, the *Leaders* clearly differ from the *Followers* in terms of the values of the variable *RatioStafftoInvestment*, which is a ratio of the number of full-time telecom employees to the annual investments in telecoms. The simplest interpretation of this split is that given the same level of the investment, the *Leaders* have fewer telecom employees than the *Followers*.

The second model created a split based on the value of the variable *RatioGDPtoInvestment*, which is the ratio of the *GDP per capita* to *annual investment in telecom, per capita*. Similarly to the results of the first model, we can identify significant differences between the two groups. Thus, based on the conducted DT analysis, we conclude that the *Leaders* do differ from the *Followers* in terms of the inputs.

RQ2: Steps 4 and 5

The purpose of Step 4 is to determine whether the same variable that was chosen for the top-level split of 1993–2002 data set will also be chosen for the 2003–2008 data set. We do not need to perform the DT analysis of the 1993–2002 data set because the target variable introduced into the original data set during Step 1 served as a proxy to separate the *Leaders* subset from the *Followers*. Consequently, to compare the results for the period from 1993 to 2002, we refer to the decision rules and the split variables to Table 15.4. The results of DT analysis of the 2003–2008 data set are presented in Table 15.5.

A comparison of the findings summarized in Tables 15.4 and 15.5 demonstrates that not only is a different variable chosen for the top-level split, but also that the corresponding split fails to isolate 95% of the population of either group into the subsets produced by the split. At this point, we can compare the two data sets in terms of the values of the indicators that represented the latent constructs in SEM.

The results, presented in Table 15.6, clearly indicate a significant decrease in the heterogeneity of the two groups relative to the values of the variables.

We also decided to compare the two data sets in terms of the values of such relevant variables, as the levels of investments, revenues, and full-time telecom labor (Table 15.7).

Surprisingly, the comparison demonstrates that the *Followers* not only closed the gap separating them from the *Leaders* in regard to the values of those variables, but also started to lead in terms of the levels of investments and revenues.

RQ3: Steps 6 and 7

The last two steps of the methodology are dedicated to assessing the differences between the *Leaders* and the *Followers* in regard to their levels of relative efficiency of utilization of investments and production of revenues. To accomplish this goal, we conducted DEA to calculate the scores of the relative efficiencies of the two

Table 15.5 Decision Rules Generated by DT Analysis of 2003–2008 Data Set

Model	Decision Rules	H02 Testing Criterion	Test of H02

1st model	IF ***RatioProductivity*** < 4.7266
	THEN
	  0 : 81.4%
	  1 : 18.6%
		To reject H02 *if* the DT analysis will not produce a top-level split of the 2003–2008 data set using the same variable that was used for the top-level split of the 1993–2002 data set.
		(**81.4% < 95%**)
			H02 fail to reject

2nd model	IF 329.9735 <= ***RatioGDPtoInvestment***
	AND 4.7266 <= ***RatioProductivity***
	THEN
	  0 : 76.5%
	  1 : 23.5%
		To reject H02 *if* the DT analysis will not produce a top-level split of the 2003–2008 data set using the same variable that was used for the top-level split of the 1993–2002 data set.
		(**76.5% < 95%**)
			H02 fail to reject

Table 15.6 Comparisons of the Values of the Indicators, 1993–2002 vs. 2003–2008

Period	Indicator	The Leaders	The Followers	Difference
1993–2002	ProductivityRatio	3.40	10.70	68.22%
	RatioGDPtoInvestment	100.41	732.86	86.30%
	RatioStafftoInvestment	0.06	1.94	96.91%
2003–2008	ProductivityRatio	8.67	8.68	0.12%
	RatioGDPtoInvestment	218.70	237.17	7.79%
	RatioStafftoInvestment	0.98	0.98	0.00%

Table 15.7 Comparisons of the Levels of Investments, Revenues, and Labor, 1993–2002 vs. 2003–2008

Period	Variable	The Leaders	The Followers	Difference
1993–2002	Full-time telecom staff (% of total labor force)	0.4295	0.4434	3.13%
	Annual telecom investment (% of GDP in current US$)	1.1027	0.5997	−83.88%
	Total telecom revenue (% of GDP in current US$)	2.8431	2.0422	−39.22%
2003–2008	Full-time telecom staff (% of total labor force)	0.4793	0.4689	−2.22%
	Annual telecom investment (% of GDP in current US$)	0.5694	0.9870	42.31%
	Total telecom revenue (% of GDP in current US$)	4.00	4.41	9.30%

groups of TEs for the periods 1993–2002 and 2003–2008. We conducted the analysis under assumptions of constant return to scale (CRS), variable return to scale (VRS), and non-increasing return to scale (NIRS) in order to assess the performance of TEs across all the stages of the product life cycle (PLC) curve. Then we assessed the changes in the scores of the relative efficiencies of the two groups of TEs by comparing the averaged scores for the period 1993–2002 with the averaged scores for 2003–2008. The results of the assessment are provided in

Table 15A.1 in the Appendix. The comparison suggests that while there was no statistically significant difference between the levels of relative efficiency of utilization of investments during 1993–2002, the difference became significant during 2003–2008.

The results also suggest that in regard to the level of the relative efficiency of the production of revenues from telecoms, the difference between the *Leaders* and the *Followers* was significant during 1993–2002 and remains significant during 2003–2008.

The results presented in Figure 15.2 demonstrate that, overall, the *Followers* remain less efficient than the *Leaders* in regard to the utilization of resources; there are some minor improvements in some stages (e.g., *Introduction* and *Growth*), but also minor deteriorations in others (e.g., *Maturity*, *Saturation*, and *Decline*).

In general, the changes seem to be not significant in regards to the efficiency of resource utilization. A different picture emerges if we take a look at Figure 15.3. Not only do the *Followers* remain less efficient than the *Leaders* in regard to the production of revenues, but also, overall, there is a significant drop in the levels of the relative efficiency across all stages of the PLC curve.

According to Samoilenko and Weistroffer (2010b), this pattern of relative inefficiencies suggests that the *Followers* can derive greater macroeconomic benefit from improving their level of efficiency of the process of conversion of investments into revenues, then from further increasing their level of investments.

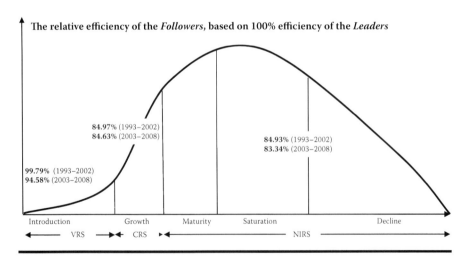

Figure 15.2 Relative efficiency of resource utilization: the *Followers* vs. the *Leaders*.

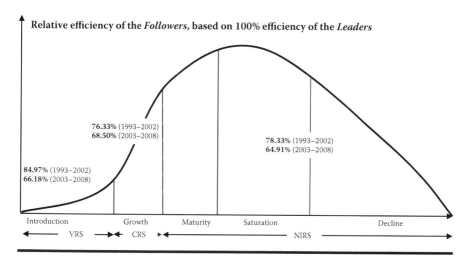

Figure 15.3 Relative efficiency of revenue production: the *Followers* vs. the *Leaders*.

Discussion of the Results

The findings of our investigation support the point that the impact of investments of telecoms in TEs on macroeconomic bottom line is not an issue that can be addressed in a one-shot fashion by a "silver bullet" solution in the form of increased level of investments. Instead, decision and policy makers in TEs must approach the subject in a step-by-step fashion: first, gradually increasing the level of investments; second, resolving the problem of identifying and dedicating resources to complementary investments; and then, finally, proceeding in accordance to the findings of the detailed analysis of contribution of investments to the macroeconomic growth. It is at the last stage of the outlined process that the results of our inquiry could be of benefit to decision and policy makers, for the methodology and insights of this investigation allow for not only pinpointing the area of growth where investments do show the impact but also determining the area of no impact where work needs to be done.

Previous studies illustrated the notion that economies can obtain some degree of a macroeconomic benefit simply by continuing to invest in the latest technology (Samoilenko & Weistroffer 2010a). The evidence of the link indicates that there exists a reliable, but not a very effective and efficient outcome of investments allocated for purposes of economic development. This is because the macroeconomic impact of a better technology, by itself, is limited. Unsurprisingly, the relationship between investments and the TC component of productivity growth holds across the sample of 18 TEs (Samoilenko & Weistroffer 2010a). This is evidence that in order to purchase a better technology, an economy does not have to invest heavily in telecoms, or be relatively efficient in utilization of investments or production of revenues. Moreover, it seems that as high-tech products become cheaper, even economies that invest relatively less in telecoms (e.g., the *Followers*) than their wealthier counterparts (e.g., the *Leaders*) can afford to grow by means of, literally, being pulled by the continuously advancing technological frontier.

Unfortunately, as evidence indicates, the impact of investments in telecoms on overall macroeconomic growth cannot be achieved only by means of growth in technology. The *Followers* do exhibit a relationship between investments in telecoms and TC (Samoilenko & Weistroffer 2010a), but not between investments in telecoms and TFP (Samoilenko & Osei-Bryson 2010), of which TC is only a component. There is a straightforward implication of the finding that only the *Leaders* exhibit the presence of the relationship between investments in telecoms and EC (Samoilenko & Weistroffer 2010a). While economies do not have to be effective and efficient in order to invest in better technology, they do have to be effective and efficient in order to utilize it to the full extent. Therefore, money can buy infrastructure, but efficient and effective utilization of the infrastructure does not automatically come along as a part of a package deal. Even if the purchase, transfer, and adoption of a new technology can take place almost immediately, there is a requisite adjustment period associated with the development of the ability to utilize the technology effectively and efficiently. This is not entirely unlike the scenario associated with the implementation of enterprise systems, where for a period of time (e.g., shakedown phase) the use of an advanced technology brings about reduction in productivity and deterioration of business performance.

These insights suggest the importance of TEs having an empirically justified and balanced investment policy, which possibly aims at not having strength in one area (e.g., latest technology paired with the obsolete skill set of workers), but rather at not having weaknesses in any (e.g., utilized technology is matched by the existing skill set of workers). We can express these insights by introducing the following two policies: a *balanced policy of investments in telecoms* and an *unbalanced policy of investments in telecoms*. We define a balanced investment policy as *a policy that allows a TE to demonstrate the presence of the relationship between investments in telecoms and overall growth in productivity*, and the unbalanced investment policy as *a policy that allows a TE to demonstrate the presence of the relationship between investments in telecoms and one of the components (TC or EC) of macroeconomic growth in productivity*. Simply put, unlike an unbalanced policy, a balanced investment policy should result in the presence of the spillover effect driven by both components of TFP. It would appear reasonable to conjecture that the investment policy of the *Leaders* was balanced, and the investment policy of the *Followers* was unbalanced. Still, the question of what differentiates the *Leaders* from the *Followers* "under the hood" in regard to their respective macroeconomics outcomes of investments in telecoms remains open. The answer to this question will allow us to gain insights into how an unbalanced investment policy could be transformed into a balanced one. Clearly, economies must invest in better technology on a regular basis, and judging by the relationship of investments in telecoms and change in technology, both groups do. However, while both groups supply their engines of economic growth with the gasoline of investments, the *Leaders* accelerate much faster than the *Followers*.

We used DT analysis to answer the question of why and to inquire into the differences between the *Leaders* and the *Followers* that may offer some insights regarding the factors that impact relationship between investments in telecoms and EC component of growth in productivity. Results of DT analysis demonstrated that while a relatively low level of investments in telecoms is negatively associated with the relationship between investments and efficiency-based growth in productivity, a relatively high level of investments does not serve as a determinant of the link between the two. This finding can be interpreted as follows: *a relatively high level of investments is a necessary but not a sufficient condition for the presence of the spillover effect from investments in telecoms.* While this finding is important, clearly, policy makers of the *Followers* will not be able to formulate useful investment strategies if armed solely with this interpretation, for such strategies simply will lead to a relative homogeneity of the levels of investments across TEs. And, as previous investigations demonstrated, an increase in the level of investments in ICT means just that and does not imply any other benefits. However, the results of DT analysis also indicated that the full-time telecom staff of the *Leaders*, ceteris paribus, is able to handle larger investments by smaller number of workers than the staff of the *Followers*. This is an important finding, for the results clearly indicate that the *Followers* have a higher ratio of full-time workers to investments than the *Leaders* do, and that the value of this ratio can be successfully used to separate one group from the other.

Consequently, if the value of this ratio can be used to separate a group of TEs with the presence of the spillover effect from the group of TEs without, then the policy makers of the *Followers* should promote strategies directed at bringing their ratio levels in line with those of the *Leaders*. Because the measure *RatioStafftoInvestment* is represented by a ratio of *Full-time telecom staff* to the *Annual investment in telecoms*, then in order to lower the value of the ratio to the level of the *Leaders*, the *Followers* could follow two routes: first, to increase the level of investments while keeping the level of the workforce constant, and second, to decrease the level of the full-time workforce while keeping the level of investments constant. Of course, a more complex and risky strategy is possible, where an increase in the level of investments is complemented, simultaneously, with a decrease in the level of the workforce. However, neither the simple increase in the level of investments nor the layoff of full-time workers presents a very effective solution in a long run. A gradual policy of lowering the hiring rate of full-time workers while not decreasing the level of investments appears more reasonable. We must acknowledge, however, that the increase in the workforce development investments ought to be considered a prerequisite for the successful implementation, or even consideration, of such policy.

This conjecture is in line with the findings of previous studies, for it was noted that expansion in ICT usually leads to significant TCs that bring about consequent increase in demand for skilled technology workers (Siegel 1997; Bresnahan et al. 2002). Previous studies also demonstrated that the better skilled and equipped workforce is more likely to achieve higher rates of ICT-related innovation and increased productivity (OECD 2004). Thus, managers, policy makers, and decision makers face the challenges of how to stimulate appropriate development of the workforce and how to define requirements for human capital that will complement TCs in ICT (Indjikian & Siegel 2005). Failing to address these challenges is not a good option, for the scarcity of technically specialized labor is seen as hindering the rate of ICT-related innovation and productivity improvements (Bollou 2006; OECD 2004; Ngwenyama & Morawczynski 2009).

However, the results of the data analysis suggest that avoidance is the route that was chosen by the *Followers*. At first glance, the situation of the *Followers* has significantly improved, for they appear to differ much less from the *Leaders* during the 2003–2008 period than they did during the earlier period of transition, that from 1993 to 2002. Furthermore, the *Followers* significantly increased their levels of investments, revenues, and full-time labor up to the point where now they surpass the *Leaders* in terms of revenues and investments relative to the GDP levels. The data clearly suggest that during 2003–2008, the *Followers* have bridged the "investments and revenues" divide that used to separate them so clearly from the *Leaders* during the period from 1993 to 2002. All seemed to indicate that the *Followers* are on the right track, until the results of DEA brought the unfortunate news regarding the levels of the relative efficiency of the *Followers*. In reality, the *Followers* did not improve much (if at all) in terms of the efficiency of the utilization of investments; moreover, they also exhibited a significant drop in efficiency of production of revenues relative to the *Leaders*.

The information provided in Table 15.7 offers an interesting vantage point for evaluating the path that the *Followers* have traveled from 2003 to 2008: given approximately the same levels of full-time telecom workforce, the *Followers* invest 40% more to receive only 10% more in revenues than the *Leaders* do. This significant increase in the levels of investments is the reason why the variable *RatioStafftoInvestment* does not serve as a factor differentiating two groups any longer. Thus, it seems only reasonable to suggest that the primary strategy of the *Followers* directed at improving the macroeconomic impact of investments in telecoms should be focused on improving the efficiencies of investments utilization and revenue production. According to Figure 15.3, the *Followers* receive 70 cents in revenues where the Leaders make a dollar; with such a low level of efficiency, the cost of macroeconomic impact of investments in telecoms becomes prohibitively expensive, for the *Followers*, literally, "pay too much for too little."

Overall, the results of our study allow us to answer the research questions of this inquiry as follows:

Answer to RQ1:
> *Those TEs that exhibit a relationship between investments in telecoms and the growth in productivity driven by the change in efficiency (the Leaders) have a higher level of investments in telecoms and lower level of full-time telecom workforce relative to the TEs that do not (the Followers).*

Answer to RQ2:
> *A significant reduction in relative heterogeneity between the Leaders and the Followers took place over the 1993 to 2008 period in regard to the values of the factors that used to differentiate two groups during the earlier stage of transition (1993–2002); most notably, the Followers dramatically increased their level of investments in telecoms.*

Answer to RQ3:
> *No reduction in relative heterogeneity between the Leaders and the Followers took place over the 1993 to 2008 period in regard to the level of macroeconomic performance of investments in telecoms; moreover, there was a significant increase in the relative heterogeneity in regard to the levels of the relative efficiency of the production of revenues.*

The answers provided to the research questions of the study allow us to consider the two goals of the investigation to be achieved; we report on the accomplishment of the goals as follows:

Goal 1:
> *A set of criteria relevant to investments in telecoms that differentiate the "higher macroeconomic impact" TEs (the Leaders) from the "lower macroeconomic impact" TEs (the Followers) must include factors allowing to gauge not only levels of investments, revenues, and workforce but also factors allowing to assess the levels of efficiency of investment utilization and revenue production.*

Goal 2:
> *The Followers have not been pursuing empirically justifiable strategies directed toward improving the level of the impact of investments in telecoms relative to the Leaders; the results of the study suggest that the Followers concentrated on increasing the levels of investments, revenues, and workforce at the expense of improvements in the levels of efficiency of the utilization of investments and production of revenues.*

Limitations of the Study and Future Research

This study has several limitations. The first limitation is associated with presenting the construct *ICT Capitalization* in a standalone fashion, without considering other relevant constructs and factors (*The Economist* 2004) that affect the impact of ICT (Baliamoune-Lutz 2003), such as state of civil infrastructure (Henderson et al. 2002; Kessides 2004), strength of political institutions (Dimitrova & Beilock 2005), and others. As a result of this limitation, the current study is better viewed as a contribution to and a component of a larger research program within Information and Communication Technologies for Development (ICT4D) stream. The second limitation is associated with the measures representing our constructs *Productivity Driven by Change in Technology* and *Productivity Driven by Change in Efficiency*; we feel that while the measures used in this study are valid and reliable, the complexity of the latent construct calls for additional measures. Consequently, more studies needed to identify and validate a set of factors and variables that could be used to represent, consistent with a theoretical framework, the two constructs in a more comprehensive fashion. The third limitation is associated with the limited conceptual model of the analysis, which lacks constructs that allow for a wider picture of the economic environment and for investigating circumstances under which spillover effect takes place. Finally, our focus on investments in telecoms adds another limitation to our investigation, for while investments in telecoms do represent a subset of the overall investments in ICT, the telecom sector may not necessarily be representative of the overall ICT sector of an economy.

Conclusion

The results of our investigation can be of use to the research and practitioner communities alike. From the theoretical standpoint, the contribution of this investigation is in the area of theory building and in providing additional insights for the gradual development of the framework outlining the success factors of ICT4D. From the methodological standpoint, we have offered a theoretically sound methodology that allows for inquiring into the impact of investments in ICT on the macroeconomic bottom line from the lower level of granularity than was previously done. Consequently, a more detailed picture is available to decision and policy makers focused on the area of increasing the impact of investments in ICT. From the practical standpoint, one of the contributions of our study is in identifying a route for a more effective and efficient allocation of investments for the purposes of ICT4D. Another practical contribution of this research is providing laggard economies with insights that can be used to formulate empirically justifiable descriptive and prescriptive strategies directed at improving the impact of investments in ICT on the macroeconomic bottom line. The results of our study also allow for gaining better insights into managerial challenges associated with effective and efficient conversion of investments in ICT into macroeconomic outcomes.

Acknowledgment

Material in this chapter previously appeared in "Investigating factors associated with the spillover effect of investments in telecoms: Do some transition economies pay too much for too little?" *Journal of Information Technology for Development* 19:1, 40–61.

References

Arcelus, F. J. & Arocena, P. (2000). Convergence and productive efficiency in fourteen OECD countries: A non-parametric frontier approach. *International Journal of Production Economics*, 66(2), 105–117.

Bagchi, K. & Kirs, P. (2009). Group analysis at regional levels can be meaningful in global IS research. *Journal of Global Information Technology Management*, 12(4), 1–5.

Baliamoune-Lutz, M. (2003). An analysis of the effects and determinants of ICT diffusion in developing countries. *Journal of Information Technology for Development*, 10, 151–169.

Barro, R. & Sala-i-Martin, X. (1995). *Economic Growth*. McGraw Hill, Boston, MA.

Bollou, F. (2006). ICT infrastructure expansion in sub-Saharan Africa: An analysis of six West African countries from 1995 to 2002. *Electronic Journal on Information Systems in Developing Countries*, 26, 1–16.

Bresnahan, T. F., Brynjolfsson, E., & Hitt, L. M. (2002). Information technology, workplace organization, and the demand for skilled labor: Firm level evidence. *Quarterly Journal of Economics*, 117, 339–376.

Chen, Y. & Zhu, J. (2004). Measuring information technology's indirect impact on firm performance. *Information Technology and Management*, 5(1–2), 9–22.

Daßler, T., Parker, D., & Saal, D. (2002). Economic performance in European telecommunications, 1978–1998: A comparative study. *European Business Review*, 14(3), 194–209.

Dewan, S. & Kraemer, K. (2000). Information technology and productivity: Evidence from country level data. *Management Science (Special Issue on the Information Industries)*, 46(4), 548–562.

Dimitrova, D. & Beilock, R. (2005).Where freedom matters: Internet adoption among the former socialist countries. *The International Journal for Communication Studies*, 67(2), 173–187.

Dunne, T., Foster, L., Haltiwanger, J., & Troske, K. R. (2004). Wages and productivity dispersion in US manufacturing: The role of computer investment. *Journal of Labor Economics*, 22(2), 397–430.

Gholami, R., Sang-Yong, T. L., & Heshmati, A. (2006). The causal relationship between information and communication technology and foreign direct investment. *The World Economy*, 29(1), 43–61.

Henderson, J., Dicken, P., Hess, M., Coe, N., & Yeung, H.W.-C. (2002). Global production networks and the analysis of economic development. *Review of International Political Economy*, 9(3), 436–464.

Hoskisson, R., Eden, L., Lau, C., & Wright, M. (2000). Strategy in emerging economies. *Academy of Management Journal*, 43(3), 249–267.

IMF (2001). *International Financial Statistics*. IMF, Washington, DC.

Indjikian, R. & Siegel, D. (2005). The impact of investments in it on economic performance: Implications for developing countries. *World Development*, 33(5), 681–700.

Jalava, J. & Pohjola, M. (2002). Economic growth in the new economy: Evidence from advanced economies. *Information Economics and Policy*, 14(2), 189–210.

Jorgenson, D. W. (2001). *Information technology and the US economy. American Economic Review*, 91(1), 1–32.

Jorgenson, D. W. & Stiroh, K. J. (2000). US economic growth in the new millennium. *Brooking Papers on Economic Activity*, 1, 125–211.

Kessides, C. (2004). Reforming infrastructure: Privatization, regulation and competition. World Bank Policy Report No. 28985.

Khouja, M. (1995). The use of data envelopment analysis for technology selection. *Computers and Industrial Engineering*, 28(1), 123–132.

Kraemer, K. L. & Dedrick, J. (2001). Information technology and economic development: Results and policy implications of cross-country studies. In *Information Technology, Productivity, and Economic Growth*, ed. M. Pohjola, Oxford University Press, Oxford.

Lam, P.-L. & Lam, T. (2005). Total factor productivity measures for Hong Kong telephone. *Telecommunications Policy*, 29(1), 53–69.

Madden, G. & Savage, S. J. (1999). Telecommunications productivity, catch-up and innovation. *Telecommunications Policy*, 23(1), 65–81.

Ngwenyama, O. & Morawczynski, O. (2009). Factors affecting ICT expansion in emerging economies: An analysis of ICT infrastructure expansion in five Latin American countries. *Information Technology for Development*, 15(4), 237–258.

OECD (2004). *DAC Network on Poverty Reduction: ICTs and Economic Growth in Developing Countries*. OECD, Paris.

OECD (2005a). Good practice paper on ICTs for economic growth and poverty reduction. *The DAC Journal*, 6(3), 1–69.

OECD (2005b). Background paper: The contribution of ICTs to pro-poor growth: No. 384. *OECD Papers*, 5(2), 15–52.

OECD (2005c). The contribution of ICTs to pro-poor growth: No. 379. *OECD Papers*, 5(1), 59–72.

Oliner, S. & Sichel, D. (2002). Information technology and productivity: Where are we now and where are we going? *Economic Review*, 3(3), 15–41.

Oliner, S. D. & Sichel, D. E. (2000). The resurgence of growth in the late 1990s: Is information technology the story? *Journal of Economic Perspectives*, 14(4), 3–22.

Piatkowski, M. (2002). The 'New Economy' and Economic Growth in Transition Economies. WIDER Discussion Paper No. 2002/63, Helsinki: WIDER.

Piatkowski, M. (2003). Does ICT Investment Matter for Output Growth and Labor Productivity in Transition Economies? TIGER Working Paper Series, No. 47. December. Warsaw. Available at http://www.tiger.edu.pl

Pohjola, M. (2002). New Economy in Growth and Development. WIDER Discussion Paper 2002/67, United Nations University World Institute for Development Economics Research (UNU/WIDER). Finland: Helsinki.

Pohjola, M. (ed.) (2001). *Information Technology, Productivity and Economic Growth: International Evidence and Implications for Economic Development. WIDER Studies in Development Economics*. Oxford University Press, United Kingdom.

Roztocki, N. & Weistroffer, H. (2008). Editorial preface: Information technology in transition economies. *Journal of Global Information Technology Management*, 11(4), 2–9.

Roztocki, N. & Weistroffer, H. (2011). From the special issue editors: Information technology in transition economies. *Information Systems Management*, 28(3), 188–191.

Sala-i-Martin, X. (1996). The classical approach to convergence analysis. *Economic Journal*, 106(4), 1019–1036.

Samoilenko, S. (2008). Contributing factors to information technology investment utilization in transition economies: An empirical investigation. *Information Technology for Development*, 14(1), 52–75.

Samoilenko, S. & Osei-Bryson, K. M. (2007). Increasing the discriminatory power of DEA in the presence of the sample heterogeneity with cluster analysis and decision trees. *Expert Systems with Applications*, 34(2), 1568–1581.

Samoilenko, S. & Osei-Bryson, K. M. (2008a). Strategies for telecoms to improve efficiency in the production of revenues: An empirical investigation in the context of transition economies. *Journal of Global Information Technology Management*, 11(4), 56–75.

Samoilenko, S. & Osei-Bryson, K. M. (2008b). An exploration of the effects of the interaction between ICT and labor force on economic growth in transitional economies. *International Journal of Production Economics*, 115(2), 471–481.

Samoilenko, S. & Osei-Bryson, K. M. (2010). Linking investments in telecoms and productivity growth in the context of transition economies within the framework of neoclassical growth accounting: Solving endogeneity problem with structural equation modeling. In *Proceedings of 18th European Conference on Information Systems*, Pretoria, South Africa, June 6th–9th, 2010.

Samoilenko, S. & Weistroffer, H. R. (2010a). Spillover Effect of Telecom Investments on Technological Advancement and Efficiency Improvement in Transition Economies. In *Proceedings of the SIG GlobDev 3rd Annual Workshop ICT in Global Development*; Saint Louis, Missouri, USA, December 12, 2010.

Samoilenko, S. & Weistroffer, H. R. (2010b). Improving the relative efficiency of revenue generation from ICT in transition economies: A product life cycle approach. *Journal of Information Technology for Development*, 16(4), 279–303.

Schreyer, P. (2000). The Contribution of Information and Communication Technology to Output Growth a Study of the G7 Countries. OECD Science, Technology and Industry Working Papers 2000/2, OECD, Directorate for Science, Technology and Industry.

Shao, B. & Lin, W. (2001). Measuring the value of information technology in technical efficiency with stochastic production frontiers. *Information and Software Technology*, 43(7), 447–456.

Siegel, D. (1997). The impact of computers on manufacturing productivity growth: A multiple-indicators, multiple-causes approach. *The Review of Economics and Statistics*, 79(1), 68–78.

The Economist (2004). Reaping the benefits of ICT: Europe's productivity challenge. A report from the Economist Intelligence Unit. Available online at http://graphics.eiu.com/files/ad_pdfs/MICROSOFT_FINAL.pdf

UN ICT Task Force Report (2005). Innovation and Investment: Information and Communication Technologies and the Millennium Development Goals. Report Prepared for the United Nations ICT Task Force in Support of the Science, Technologies and Innovation Task Force of the United Nations Millennium Project. Available online at http://www.unmillenniumproject.org/documents/Innovation_InvestmentMaster.pdf

WT/ICT Development Report (2006). Measuring ICT for social and economic development. *International Telecommunication Union's World Telecommunication/ICT Development Report*, 8th edition. Available online at http://www.itu.int/dms_pub/itu.../D-IND-WTDR-2006-SUM-PDF-E.pdf

Yearbook of Statistics (2009). Telecommunication Services Chronological Time Series. ITU Telecommunication Development Bureau (BDT), International Telecommunication Union.

Appendix

Table 15A.1 DEA Results, Scores of the Relative Efficiencies, 1993–2002 and 2003–2008

Period	DEA Orientation	DEA Model, Return to Scale	The Leaders	The Followers	Difference, %	t-test, p-value	Test of H03
1993–2002	Input oriented	Constant RS	0.62	0.54	15.03%	0.5065	H03 accepted
		Variable RS	0.73	0.73	0.21%		
		Non-increasing RS	0.62	0.54	15.07%		
	Output oriented	Constant RS	1.94	2.54	23.67%	0.0048	H03 rejected
		Variable RS	1.89	2.25	16.01%		
		Non-increasing RS	1.89	2.41	21.67%		
2003–2008	Input oriented	Constant RS	0.89	0.77	15.37%	0.0409	H03 rejected
		Variable RS	0.91	0.87	5.42%		
		Non-increasing RS	0.91	0.78	16.66%		
	Output oriented	Constant RS	1.15	1.51	31.50%	0.0000	H03 rejected
		Variable RS	1.11	1.48	33.82%		
		Non-increasing RS	1.11	1.50	35.09%		

Chapter 16

Understanding the Human Capital Dimension of Information and Communication Technology and Economic Growth in Transition Economies

Introduction

An important stream of research in the economics of information systems (IS) is the investigation of the relationship between investments in information and communication technologies (ICTs) and macroeconomic outcomes (Colecchia & Schreyer 2002; Pohjola 2002; Kraemer & Dedrick 2001). Many of these studies provide compelling evidence that ICT expansion and use have led to robust returns and economic growth in developed countries (OECD 2005; Oliner & Sichel 2002; Jalava & Pohjola 2002). In other contexts, research is scarce, and similar studies of emerging, developing, and transition economies (TEs) reveal significantly lower levels of returns on investments in ICT (Dewan & Kraemer 2000; Pohjola 2001;

Piatkowski 2003). It is important to note that while TEs share important economic characteristics, they differ from developed countries in terms of their levels of intensity of using ICT, competition and regulatory environments, relative costs of deployment of technology, quality of human capital, and flexibility of the business environment (OECD 2004). Consequently, insights obtained by analyzing the existing studies of developed economies cannot be applied easily to TEs, or to emerging and developing economies. As a result, researchers have suggested that more studies on global IT management in developing, emerging, and transition economies (Palvia 2006) are needed to identify specific factors inhibiting these economies from sharing similar levels of returns on ICT investments as developed countries (OECD 2004).

Despite the calls for additional studies, research on these issues remains sparse and scattered (Palvia 2006). One of the commonly noted reasons for the dearth of studies is the limited availability of reliable time series data (Hoskisson et al. 2000). Another challenge is the absence of a common taxonomy that clearly differentiates emerging, developing, and transition economies (Samoilenko 2008). While emerging, developing and transition economies are clearly distinct from developed economies, these countries do not represent a homogenous group (Roztocki & Weistroffer 2008; Hoskisson et al. 2000). Consequently, the widely cited in-depth case studies about ICT successes in selected developed countries are not generalizable to emerging, developing, and transition economies. Some researchers argue that to obtain a deeper understanding of the impact of ICT investments on macroeconomic outcomes, longitudinal studies are needed (Palvia 2006; Hoskisson et al. 2000); others suggest various levels of group analysis (Bagchi & Kirs 2009). However, even with reliable time series data from trusted public sources, researchers face two main challenges. The first challenge is associated with constructing a relatively homogenous sample from a heterogeneous population of emerging, developing, and transition economies, and the second challenge is represented by the general scarcity of theory-driven approaches and methodologies for analyzing samples of convenience and identification of relevant economy-specific antecedents.

The primary objective of this study is to propose a theory-based approach for formulating effective and efficient policies directed at increasing the macroeconomic impact of investments in ICT. A second objective is to conduct a longitudinal panel data study in the context of TEs to test the proposed approach and identify factors that may affect the impact of investments in ICT. Our study uses time series panel data on eight transition economies obtained from reliable and trusted public sources to construct a relatively homogenous sample of economies. Our analysis is guided by an established theoretical framework, supported by widely used, reliable methods. TEs are commonly defined as economies that are in transition from a centralized planning system to a free market economy (Roztocki & Weistroffer 2008); a specific description of the context can be found in "The Sample and Panel Data" section of the chapter. Our choice of these TEs is intentional. Due to the heterogeneity of emerging, developing, and transition economies, it is often difficult to generalize from single studies. However, as Samoilenko (2008) notes, TEs are advantageous as they share economic characteristics with both developed and less developed economic regions (see also OECD 2004). Therefore, TEs present a good vantage point from which a more general understanding of the relationship between the investments in ICT and economic growth can be developed.

Background

Recent research has found that increases in investment in ICT infrastructure alone do not result in proportional increases in revenues; multiple internal and external factors mediate this relationship (Dewan & Kraemer 2000; OECD 2004). Researchers in other sectors have long argued that complementary investments in human capital are essential to the realization of economic outcomes from technology investments (Parente & Prescott 1994; Benhabib & Spiegel 1994). Many ICT for development (IT4D) researchers now agree that insufficient attention to complementary investments in human capital is inhibiting the potential impact of ICT investments on the macroeconomic goals of developing and transition economies (Kraemer & Dedrick 2001; Pohjola 2002). The scarcity of highly skilled technology workers in many of these countries is seen as hindering the rate of ICT-related innovation and productivity improvements (Bollou 2006; OECD 2004). As ICT expansion increases, significant technological changes result. This leads to increased demand for technically specialized labor (Siegel et al. 1997; Bresnahan et al. 2002). The challenge for policy makers now is how to define requirements for and stimulate an appropriate human capital development to complement ICT technology changes (Indjikian & Siegel 2005).

Some researchers argue that the relatively homogenous environment of developed economies allows for easier sharing of best practices (Arcelus & Arocena 2000; Barro & Sala-i-Martin 1995; Sala-i-Martin 1996), which could help leverage the infrastructure and human capital to achieve spectacular returns (Baliamoune-Lutz 2003; Parente & Prescott 1994). However, TEs do not represent a homogenous group; they have different levels of infrastructure and human capital development, which makes it impossible to share best practices (Hoskisson et al. 2000). Consequently, TEs wishing to improve macroeconomic outcomes from ICT investments must face the challenges of creating and implementing unique strategies instead of copying from their more successful counterparts (Samoilenko & Osei-Bryson 2008). In this article, we propose an approach to formulating human capital strategies for national ICT policy. Our approach is general as it uses the neoclassical growth accounting framework and proposes a two-phase analytical process utilizing multivariate regression and data envelopment analysis (DEA).

The Research Problem

Policy makers tasked with the responsibility of increasing the impact of IT4D programs face two distinct problems: (1) determining an appropriate ICT investment strategy and (2) determining a complementary ICT human capital development and management strategy. To effectively address these problems, theoretically grounded approaches are needed for analyzing empirical situations in order to develop appropriate policies and strategies. While research has been directed at identifying specific ICT investment strategies in the context of TEs (Samoilenko & Osei-Bryson 2008), the question of how to determine a complementary human capital strategy remains open. The broad research question of this investigation concerns determining strategies for the requirements of high-technology human capital for IT4D initiatives in TEs. More specifically, we are interested in developing an approach that policy makers can adopt for the analysis and formulation of an ICT human capital development strategy. The key challenges for policy makers are as follows: (1) Given the specific empirical situation, what options exist when trying to achieve improvements in macroeconomic outcomes? (2) How can these options be analyzed and their implications understood so that appropriate strategies can be formulated?

Prior research on the productive efficiency of economies (where the economy can be an enterprise, an industrial sector, or a nation) has shown that improvements in outcomes can come from two main sources: increases in inputs or increases in the efficiency of converting inputs into outputs (also called productivity improvements) (Malmquist 1953; Caves et al. 1982; Färe et al. 1994). Consequently, the policy maker promoting a strategy to improve macroeconomic outcomes from investments in ICT has two distinct policy options that he/she can consider (see Table 16.1). He/she can increase the size of the workforce or increase the level of

Table 16.1 Policy Options for Improving ICT Macroeconomic Outcomes

Policy Options	Implications for Human Capital Strategy	Possible Human Capital Implementation Strategies
Increasing the level of investments in ICT staff	Increasing the size of the ICT workforce	1. Increase in the size of full-time ICT staff
		2. Increase in the number of contractors
Increasing the level of efficiency of conversion of investments into revenues	Increasing the productivity of the existing level of the ICT workforce	1. Increase in the quality/know-how of full-time staff
		2. Increase in the quality of technology utilized by full-time staff

productivity of the existing workforce. While these two strategies are not mutually exclusive, they are not likely to be implemented simultaneously. To increase the workforce, the manager can hire more full-time ICT staff, or conversely, hire contractors (i.e., any non-full-time employees). Similarly, the strategy of increasing the productivity of the workforce could be based on two factors. The first factor refers to improvements in the quality of the workforce itself (e.g., education, technical skills, know-how), and the second factor refers to improvements in the quality of the technology utilized by the workforce (e.g., better infrastructure, faster networks, more powerful workstations, etc.). The approach we outline here is designed to assist policy makers in analyzing these issues.

Determining Appropriate Policy Options

What should the policy maker do when faced with the problem of deciding between the two policy options? The policy maker could try: (1) increasing the size of the ICT staff or (2) increasing the level of productivity of ICT staff. Further, each policy option can be achieved by one of two possible implementation strategies (see Table 16.1). To determine the appropriate policy option and strategy, the policy maker will need to investigate the empirical situation. We suggest a two-step process for analyzing the empirical situation to determine which of the policy options and implementation strategies to adopt. In step 1, the policy maker needs to know if (a) increasing the level of the full-time ICT staff will have an impact on the macro outcome, and (b) given a positive answer to (a), what form the workforce expansion should take. The key question for this part of the analysis could be formulated as follows: In the context of TEs, would complementary investments in ICT staff improve the macroeconomic outcomes of the ICT sector? Given an affirmative answer to this question, the policy maker will then need to determine whether to expand the ICT workforce by hiring full-time ICT staff or by hiring contractors. From a theoretical perspective, the question concerns the complementarity (discussed in Chapter 2) of investments in ICT infrastructure and ICT staff. To interrogate this question rigorously, we propose the following null hypothesis:

> H01: In the context of TEs, the level of a full-time ICT staff is not complementary to the level of investments in ICT.

In simple terms, this hypothesis aims to test whether *there is evidence that additional investments in ICT combined with an increase in the number of full-time employees have a positive synergetic effect on the macroeconomic bottom line.* If we are able to reject H01, then there is empirical evidence to suggest that an increase in the level of investments in ICT should be accompanied by an increase in the level of full-time ICT staff, because, such an increase in full-time ICT staff is likely to improve the macro outcome. Alternatively, if H01 is not rejected, then it is empirically justified to suggest that any increase in investments in ICT should not be complemented with an increase in the size of the full-time ICT workforce. Consequently, the policy analysis should consider hiring contractors for any expansion of the ICT workforce.

In the second step, the policy maker is concerned with determining a complementary human capital strategy that could improve the productivity of full-time ICT staff (i.e., the efficiency of conversion of investments into revenues). The reader will recall that instead of hiring more ICT staff, the policy maker could implement a strategy aimed at increasing the productivity of the current full-time workforce, or increasing the quality of the technology utilized by the workforce. We can formulate a general question for investigation as follows: in order to achieve improvements in macroeconomic outcomes from ICT investments, should the policy maker follow a strategy of increased investments in workforce development or by investments in technology? Again, the policy maker would need to interrogate the empirical situation to determine the appropriate strategy that would lead to improvements in the macro outcome. To understand the empirical situation, he/she could empirically test the following null hypothesis:

> H02: In the context of TEs, productivity improvements (efficiency change—EC) resulting from investments in human capital development (ICT staff skills) are greater than productivity improvements resulting from investments in technology change (technological change—TC).

This hypothesis in concerned with testing whether the upgrading of skills and knowledge of the full-time ICT staff provides a greater contribution to the improvement in productivity than investments in better technologies. For clarification, we would like to point out that the change in productivity results from two sources, the quality of the available technology and skillful utilization of that technology. Effective utilization of the technology is dependent on the skills of users, and the skills of users can only be utilized to the extent of the technology that is available. In the case where TC > EC, productivity improvements are technology driven. Conversely, in the case where TC < EC, change in productivity results from changes in efficiency and is skill driven. Keeping this in mind, testing H02 allows us to determine whether the change in productivity is skill driven or technology driven. Now that we have articulated the central problems for analysis, we will outline the theoretical framework that is the basis of policy analysis approach and then describe the context of our study in more detail.

The theoretical framework of our research is represented by the *neoclassical growth accounting model* and the *theory of complementarity*. To operationalize our empirical analysis, we use DEA and multivariate regression.

Proposed Approach to Policy Analysis

As stated earlier, the analysis of policy options requires the analyst to follow a two-step process. Our methodology implements the analysis using the analytical techniques of multivariate regression (MR) and DEA. While each of the methods can provide valuable information to a decision maker, it is a synergy of the two that allows us to gain insight into the interplay between investments in ICT and the quantity and the quality of the workforce. Table 16.2 outlines the two-step approach of the analysis; we now outline the details of each step.

Step 1: Using MR to Determine the Presence of Complementarity

The purpose of the first step is to determine the presence of complementarity between the level of investments in ICT and the size of the full-time ICT staff. In order to do so, we utilize the formulation (1) of the translog function and investigate complementarity of investments in ICT and the size of ICT staff by testing for the presence of the interaction by evaluating the following hypothesis:

H0: β_5 is not statistically discernible from 0 at the given level of α

If the interaction term between investments in ICT and full-time ICT staff is significant (i.e., we reject the null hypothesis of $\beta_5 = 0$), then we have a reason to assume that such investments are complementary. Furthermore, if the direction of the interaction effect is positive (i.e., $+\beta_5$), then we have a reason to suggest that complementary investments in ICT and full-time ICT staff have a combined positive effect on the macroeconomic outcome. This implies that the policy maker should consider a complementary human capital development and ICT investment strategy and this strategy should be directed at the increasing the size of the full-time ICT staff. If, however, the direction of the interaction effect is negative (i.e., $-\beta_5$), then increasing the size of full-time ICT staff would be detrimental to the overall macroeconomic impact resulting from any increase in the level of investments in ICT. Table 16.3 summarizes the possible outcomes from this analysis.

Step 2: Using DEA to Determine the Sources of Relative Inefficiency

The second step of the analysis determines the appropriate course of action for improving the level of efficiency of the process of conversion of investments into revenues from ICT. Let us recall that according to Table 16.1, there are two routes by which the efficiency of the conversion of inputs into outputs could be improved: by means of

Table 16.2 A Summary of the Proposed Two-Step Approach to the Analysis

Step 1	Outcomes	Implications for Human Capital	Step 2	Outcomes	Implication for Human Capital
Determining the existence of complementarity between the levels of full-time ICT staff and annual investments in ICT	No interaction effect	Hire contractors	Determining the relative contributions of changes in technology and changes in efficiency to the process of conversion of investments into revenues	TC > EC	Increase the quality of the current workforce
	Interaction effect positive	Hire full-time workers		EC > TC	Increase the quality of technology
	Interaction effect negative	Improve efficiency of full-time workers			

Table 16.3 Summary of Possible Results and Implications from Step 1 of the Analysis

Results of the First Phase	Implication for Complementarity	Implications for Human Capital Strategy
Null hypothesis accepted	No complementarity between investments in ICT and full-time ICT staff	There is no empirical justification for hiring full-time ICT staff; an increase in the level of current workforce should come from hiring contractors
Null hypothesis rejected; interaction effect is positive	Combined effect of investments in ICT and full-time ICT staff is positive	There is an empirical justification for hiring full-time ICT staff; human capital strategy should be directed at the increase in the level of full-time ICT staff
Null hypothesis rejected; interaction effect is negative	Combined effect of investments in ICT and full-time ICT staff is negative	There is no empirical justification for hiring additional full-time ICT staff; human capital strategy should be directed at the increase in the level of efficiency of the current full-time ICT staff

utilizing better technology (which implies required additional investments in technology and infrastructure) or by means of improving the quality of the workforce (which implies required additional investments in the workforce development programs). Under the typical TE condition of limited resources, a policy maker will clearly benefit from identifying the appropriate and most economic route where additional investments are utilized to the greatest advantage with regard to the production of revenues.

In order to aid a policy maker with his/her decision, we suggest performing DEA and calculating the values of Malmquist index (*MI*) for the decision making units (DMUs) in our sample. Keeping in mind that MI can be decomposed into two components, *TC* (change associated with *changes in technology*) and *EC* (change associated with *changes in efficiency*), we can determine which component contributes more to the overall change in efficiency of the DMUs.

Accordingly, we can restate our second null hypothesis as follows:

$H0_2$: EC > TC

We summarize the possible outcomes and implications of the first phase of our methodology in Table 16.4.

Table 16.4 Summary of Possible Results and Implications from Step 2 of Analysis

Results of the Second Phase	Interpretation of the Results	Implications for Human Capital Strategy
TC > EC	Greater part of the change in efficiency comes from the changes associated with technology; efficiency of the full-time staff is lagging behind	Human capital strategy should consider introducing development programs for the full-time staff
EC > TC	Greater part of the change in efficiency comes from the changes associated with efficiency of the full-time staff; technology-associated change is lagging behind	Human capital strategy should consider increasing the quality of the technology used by full-time staff

The Sample and Panel Data

We demonstrate our approach using a panel data set of eight transition economies over two consecutive time periods: 1993–1997 and 1998–2002. The panel data are drawn from the *World Development Indicators* database* and the 2004 *Yearbook of Statistics* of the International Telecommunication Union (http://www.itu.int/ITU-D/ict/publications). A commonly accepted definition of TE is one that is in transition from central planning to a free market system (Roztocki & Weistroffer 2008). The TEs selected for this study are the Czech Republic, Estonia, Hungary, Latvia, Lithuania, Poland, Slovakia, and Slovenia. They were the most homogeneous of the group of 18 former socialist countries that have been labeled *transition economies in Europe and the former Soviet Union* (Bagchi & Kirs 2009). In an earlier study of these 18 TEs, Samoilenko and Osei-Bryson (2008) found that the eight listed here differed significantly from the rest in their level of ICT investments and revenues in the 10-year period from 1993 to 2002. It was also established that these eight TEs had higher levels of investments, revenues, and efficiency than the other ten (Samoilenko & Osei-Bryson 2008). These eight TEs exhibit a level of homogeneity and shared cultural characteristics that make them ideal for this kind of group-level analysis (Dimitrova & Beilock 2005; Deichmann et al. 2006; Palvia et al. 2002). They also offer a data set useful for the analyzing global ICT management strategy as well as the macroeconomic impact of investments in telecoms.

---

* http://web.worldbank.org/WBSITE/EXTERNAL/DATASTATISTICS.

Data Analysis

Analysis Procedure

Step 1: Analysis of H01

We used SAS *Enterprise Miner* to perform the data analysis of the first step of our methodology, the results of which are presented in Table 16.5. The following three variables (see Samoilenko & Osei-Bryson 2008 for a detailed overview and justification of the variables of this translog model) were used in the multiple regression analysis:

1. Gross domestic product (GDP, in current US$)
2. Annual telecom investment (% of GDP)
3. Full-time telecommunication staff (% of total labor force)

Table 16.5 presents a summary of the results of the multivariate regression. The interaction term is given in the form that it appeared in the actual model (Table 16.5, column 2). The column "β estimate" provides a value of the parameter estimate for the interaction term. The column labeled "P value" provides a two-tailed P value used in testing the null hypothesis that β = 0. We test the hypothesis at a 95% confidence level, i.e., a level of α = .05. As a result, a coefficient having a P value of .05 or less would be considered statistically significant, which would allow us to reject the null hypothesis.

Interpretation of Analysis

Based on regression analysis, we accept H01 for both periods: 1993–1997, and 1998–2002. Results suggest that for the period 1993–1997, for the right TEs, the level of full-time ICT staff is a complementary factor to investments in ICT. However, the direction of the effect is negative. The straightforward interpretation of this finding is that an increase in the level of investments in ICT and a

Table 16.5 Results of the Multivariate Regression

Period	Term in the Model	β Estimate	P Value	Adjusted R^2
1993–1997	ANNU_EL1*FULL_2KW Log (Annual Telecom Investment)*Log(Full-time ICT staff)	−18.066	.0416	.9927
1998–2002	ANNU_EL1*FULL_2KW Log (Annual Telecom Investment)*Log(Full-time ICT staff)	53.6891	<.0001	.9184

corresponding increase in the level of full-time ICT staff have a negative impact on the macroeconomic bottom line. However, it has been noted that an insufficient level of investments in ICT as well as insufficiently developed ICT infrastructure are among the main reasons why TEs have a limited impact from investments in ICT as compared to developed economies. Thus, increasing the level of investments in ICT would appear a prerequisite for TEs to develop levels of returns similar to the levels of developed economies. Consequently, based on the results of our analysis, the increase in the level of investments in ICT, while required for the increase in the macroeconomic outcomes of investments in ICT, must take place without a corresponding increase in the level of full-time ICT staff.

This suggests that the eight TEs during the period 1993–1997 had levels of the full-time ICT staff that were too high; during that period, these TEs would have benefited from reducing the quantity and increasing the quality of the remaining full-time ICT staff. Existing evidence that, in the context of TEs, macroeconomic impact of investments in ICT is associated with the smaller size and higher efficiency of ICT workforce (Samoilenko & Osei-Bryson 2010) supports this conjecture. Furthermore, a simulation study of Samoilenko (2009) demonstrated that reducing the level and increasing the productivity of the ICT workforce allows for establishing the link between investments in ICT and a macroeconomic growth in productivity even for underperforming TEs. Thus, it is reasonable to hypothesize that policies directed at the increase in efficiency of the reduced levels of the full-time ICT staff could reverse the negative direction of the interaction effect; however, we must test this hypothesis prior to making any definitive claims.

We also found a statistically significant interaction effect between the level of full-time ICT staff and investments in ICT in the eight TEs during the period from 1998 to 2002. The direction of the interaction effect is positive; thus, we can state that complementarity exists between investments in ICT and ICT staff. The results suggest that the increase in the level of investments in ICT should have been accompanied by the increase in the level of full-time ICT staff. What is the significance of the change in the direction of the interaction effect? It is reasonable to suggest that there is a certain "optimal level" of full-time ICT staff that is required for a given level of investments in ICT to contribute maximally to the increase in GDP. If this is the case, then there exists a "golden middle" in the investment-to-staff ratio. For example, it could be that up to the certain level of investments in ICT, the number of full-time staff must grow rapidly, while after that point, only an increase in productivity of the full-time ICT staff could provide increased contribution of the investments in ICT to macroeconomic growth.

Step 2: Analysis of H02

To perform DEA, we used the software application OnFront, version 2.02, produced by Lund Corporation (http://www.emq.com). The variables constituting the DEA model are listed in Table 16.6.

Table 16.6 List of Variables for DEA Models

Input Variables	Output Variables
GDP per capita (in current US$), Full-time telecommunication staff (% of total labor force) Annual telecom investment per telecom worker, Annual telecom investment (% of GDP in current US$) Annual telecom investment per capita Annual telecom investment per worker	Total telecom services revenue per telecom worker Total telecom services revenue (% of GDP in current US$), Total telecom services revenue per worker Total telecom services revenue per capita

Our choice of the variables for the DEA model was guided by the theoretical framework of our study. We included variables reflecting the levels of investments, revenues, as well as labor. All of the chosen variables were expressed as ratios in order to obtain a more objective assessment of the state of each TE relative to others in the sample. The variables in the DEA model represent the levels of investments and revenues relative to the whole population, the labor force, and the telecom industry. We believe that such relative representation provides a more objective depiction of not only the levels of investments and revenues, but also the economic and demographic environment within which the investments take place and the revenues are produced. However, an important limitation of our model is that it does not contain any variables reflecting social, economic, and institutional factors that potentially impact the levels of revenues and overall economic development of TEs. The results of DEA are summarized in Table 16.7.

Keeping in mind that our second null hypothesis is stated H02: EC > TC, the results of step 2 provide us with sufficient evidence to accept H02 for the period from 1993 to 1997, and to reject H02 for the period from 1998 to 2002.

Table 16.7 Values of Malmquist Index and EC and TC Components

Time Period	Criterion for Comparison	Annual Change in Productivity
First 5 years (1993–1997)	Malmquist index (MI)	1.18
	MI, TC component	1.08
	MI, EC component	1.14
Second 5 years (1998–2002)	MI	1.159
	MI, TC component	1.106
	MI, EC component	1.058

The results of DEA demonstrate that for the period 1993–1997, the greater part of the changes in productivity in the eight TEs came from changes in efficiency (EC). This means that the contribution to the macroeconomic bottom line from the efficiency of the workforce was greater than the contribution from the technology itself; in other words, at that period of time, this group of TEs was able to do more with less. It is reasonable to suggest that at that period of time, the eight TEs could have benefited the most from investments in better technology utilized by the workforce, for at that time, the skills of their full-time ICT labor were "ahead" of their tools, and the impact of investments in ICT was skill driven. Consequently, investments in the workforce development programs would have only increased the gap, providing workers with more skills than the existing technology allowed them to utilize. However, for the period 1998–2002, the situation reversed, and the greater part of changes in productivity came from changes in technology (TC) than from changes in efficiency (EC) of utilization of technology. The interpretation is fairly straightforward: the growth in productivity is technology driven, and any additional investments in better technology will probably make the existing gap between technology and the ability to utilize it only larger. This may suggest that the telecom workforce of the eight TEs could benefit from investing in the full-time workforce development programs.

Discussion and Conclusion

The global focus on ICT as an engine of development is presenting many challenges for policy makers in transition and developing economies. While it is argued that ICT expansion, utilization, and innovation in transition and developing economies will lead to economic expansion, such expansion comes with significant demands for development human capital. However, not all transition and developing economies have the necessary human capital capacity to achieve the level of economic returns of more advanced economies. Consequently, an important problem that these policy makers must face is how to shape complementary human capital and ICT development strategies to achieve the best possible economic outcomes. In this article, we presented an empirical-based approach that could assist policy makers in analyzing relevant data to determine appropriate strategies that could help to achieve a higher impact for investments in ICT. This approach offers a framework and a tool to assist policy makers in formulating theoretically sound and empirically justified policies for improving the quality of the ICT workforce. The approach focuses on determining the complementarity (the interaction effect) between investments in ICT and ICT staff. It allows the policy maker to determine the source of productivity and invest accordingly. We can see from the previous analysis (case 1) that when the interaction effect between levels of investments and labor is negative, efficiency change has a greater impact on growth in productivity than technology change. In such a situation, productivity growth is skill based, and it would be wise for the policy maker to follow a strategy of higher investment in improving the skills of workers and less on technology. On the other hand (case 2), when the interaction effect between investments in ICT and ICT staff is positive, technology change has a greater impact on productivity growth than efficiency change. In this situation, the policy maker would be wise to invest more in technology and less on expanding staff. However, in both situations, it is a question of balancing the investments for synergy.

Several researchers have pointed to the importance of the development of technical expertise in ICT expansion (Udo et al. 2008; Papazafeiropoulou 2004; Harindranath 2008). The challenge that policy makers have faced is how to determine investment strategies that take into account human capital requirements (Bob et al. 2007; Ifinedo 2006). The approach presented here is a contribution to the existing body of knowledge on global IT management. The approach is flexible and allows policy makers to analyze historical data to interrogate sources of relative inefficiencies and to formulate human capital strategies that explicitly take into consideration the level of investments in ICT. It offers both IT researchers and practitioners deeper insights and a wider perspective on how to analyze empirical situations and develop synergetic ICT policies that would deliver a higher macroeconomic impact from investments in ICT. Our approach is grounded in the widely accepted theoretical framework of neoclassical growth accounting and implemented using well-established data analytic methods of multivariate regression and DEA. Finally, the approach also allows for forecasting of the demand for a full-time workforce.

While our approach provides policy makers a framework and systematic procedures for analyzing and formulating complementary human capital strategies for IT4D, it is not without limitations. For example, it does not allow for consideration of unusual events, such as a sudden spike or a fall in the level of investments. Second, being based on historical data, our approach assumes an available workforce and a discernable trend in the levels of investments and workforce. Nevertheless, this research makes an important contribution to an important problem noted by researchers (Dewan & Kraemer 2000; OECD 2004), that complementary human capital policies are necessary, as increases in investment in ICT infrastructure alone do not result in proportional increases in economic growth. To date, no approaches can be found in the literature, and recently, some IT4D researchers pointed out that insufficient attention to complementary human capital policies is inhibiting IT4D macroeconomic goals of developing and transition economies (Kraemer & Dedrick 2001; Pohjola 2002). If we are to make progress in these countries, we need frameworks and approaches for addressing their human capital development strategies (Bollou 2006; OECD 2004; Indjikian & Siegel 2005).

Acknowledgment

Material in this chapter previously appeared in "Understanding the human capital dimension of ICT and economic growth: An approach to analyzing different

ICT workforce and technology investments policies," *Journal of Global Information Technology Management* 14:1, 59–79.

References

Aiken, L., & West, S. (1991). *Multiple Regression: Testing and Interpreting Interactions*. Sage Publications, Newbury Park, CA.

Arcelus, F. J. & Arocena P. (2000). Convergence and productive efficiency in fourteen OECD countries: A non-parametric frontier approach. *International Journal of Production Economics*, 66, 105–117.

Bagchi, K. & Kirs, P. (2009). Group analysis at regional levels can be meaningful in global IS research. *Journal of Global Information Technology Management*, 12(4), 1–5.

Baliamoune-Lutz, M. (2003). An analysis of the determinants and effects of ICT diffusion in developing countries. *Information Technology for Development*, 10(3), 151–169.

Barro, R. & Sala-i-Martin, X. (1995). *Economic Growth*. McGraw Hill, Boston.

Benhabib, J. & Spiegel, M. (1994). The Role of Human Capital nin Economic Development: Evidence from Aggregate Cross-Country Data. *Journal of Monetary Economics*, 34, 143–173.

Bob, T., Borislav, J., Kajan, E., Vidas-Bubanja, M., & Vuksanovic, E. (2007). E-Commerce in Serbia: Where roads cross electrons will flow. *Journal of Global Information Technology Management*, 10(2), 34–56.

Bollou, F. (2006). ICT Infrastructure Expansion in Sub-Saharan Africa: An analysis Of Six West African Countries from 1995–2002. *Electronic Journal on Information Systems in Developing Countries*, 26(5), 1–16.

Braumoeller, B. (2004). Hypothesis testing and multiplicative interaction terms. *International Organization*, 58(4), 807–820.

Bresnahan, T. F., Brynjolfsson, E., & Hitt, L. M. (2002). Information technology, workplace organization, and the demand for skilled labor: Firm level evidence. *Quarterly Journal of Economics*, 117, 339–376.

Caves, D. W., Christensen, L. R., & Diewert, W. E. (1982). The economic theory of index numbers and the measurement of input, output, and productivity. *Econometrica*, 50, 1393–1414.

Colecchia, A. & Schreyer, P. (2002). ICT investment and economic growth in the 1990s: Is the United States a unique case? A comparative study of nine OECD countries. *Review of Economic Dynamics*, 5, 408–442.

Deichmann, J., Eshghi, B., Haughton, D., Masnghetti, M., Sayek, S., & Topi, L. (2006). Exploring break-points and interaction effects among predictors of the international digital divide. *Journal of Global Information Technology Management*, 9(4), 47–71.

Dewan, S. & Kraemer, K. (2000). Information technology and productivity: Evidence from country level data. *Management Science (Special Issue on the Information Industries)*, 46(4), 548–562.

Dimitrova, D. & Beilock, R. (2005). Where freedom matters: Internet adoption among the former socialist countries. *The International Journal for Communication Studies*, 67(2), 173–187.

Färe, R., Grosskopf, S., Norris, M., & Zhang, Z. (1994). Productivity growth, technological progress, and efficiency in industrialized countries. *American Economic Review*, 84, 374–380.

Harindranath, G. (2008). ICT in a transition economy: The case of Hungary. *Journal of Global Information Technology Management*, 11(4), 33–55.

Hoskisson, R., Eden, L., Lau, C., & Wright, M. (2000). Strategy in emerging economies. *Academy of Management Journal*, 43(3), 249–267.

Ifinedo, P. (2006). Key information systems management issues in Estonia for the 2000s and a comparative analysis. *Journal of Global Information Technology Management*, 9(2), 22–44.

Indjikian, R. & Siegel, D. (2005). The impact of investments in IT on economic performance: Implications for developing countries. *World Development*, 33(5), 681–700.

Jalava, J. & Pohjola, M. (2002). Economic growth in the new economy: Evidence from advanced economies. *Information Economics and Policy*, 14(2), 189–210.

Kraemer, K. L. & Dedrick, J. (2001). Information technology and economic development: Results and policy implications of cross-country studies. In *Information Technology, Productivity, and Economic Growth*, ed. M. Pohjola, Oxford University Press, Oxford.

Malmquist, S. (1953). Index numbers and indifference surfaces. *Trabajos de Estatistica*, 4, 209–242.

OECD (2004). ICTs and economic growth in developing countries. *The DAC Journal*, 5(4), 1–30.

OECD (2005). Good practice paper on ICTs for economic growth and poverty reduction. *The DAC Journal*, 6(3), 1–69.

Oliner, S. & Sichel, D. (2002). Information technology and productivity: Where are we now and where are we going? *Economic Review*, 3(3), 15–41.

Palvia, C., Palvia, S., & Whitworth, J. (2002). Global information technology: A meta analysis of key issues. *Information and Management*, 39, 403–414.

Palvia, P. (2006). Key IS management issues: Need for an international research program. *Journal of Global Information Technology Management*, 9(2), 1–4.

Papazafeiropoulou, A. (2004). Inter-country analysis of electronic commerce adoption in south Eastern Europe: Policy recommendations for the region. *Journal of Global Information Technology Management*, 7(2), 54–69.

Parente, S. & Prescott, E. (1994). Barriers to Technology Adoption and Development. *Journal of Political Economy*, 102, 298–321.

Piatkowski, M. (2003). The Contribution of ICT Investment to Economic Growth and Labor Productivity in Poland 1995–2000. TIGER Working Paper Series No. 43, Warsaw, July. Available online at http://www.tiger.edu.pl

Pohjola, M. (2002). New Economy in Growth and Development. WIDER Discussion Paper 2002/67, United Nations University World Institute for Development Economics Research (UNU/WIDER). Helsinki, Finland.

Pohjola, M. (ed.) (2001). Information technology, productivity and economic growth: International evidence and implications for economic development. *WIDER Studies in Development Economics*, Oxford, Oxford University Press, United Kingdom.

Rao, M., Earls, T., & Sanchez, G. (2007). International collaboration in transorganizational systems development: The challenges of global insourcing. *Journal of Global Information Technology Management*, 10(3), 52–69.

Roztocki, N. & Weistroffer, H. (2008). Editorial preface: Information technology in transition economies. *Journal of Global Information Technology Management*, 11(4), 2–9.

Sala-i-Martin, X. (1996). The classical approach to convergence analysis. *Economic Journal*, 106(4), 1019–1036.

Samoilenko, S. (2008). Contributing factors to information technology investment utilization in transition economies: An empirical investigation. *Information Technology for Development*, 14(1), 52–75.

Samoilenko, S. & Osei-Bryson, K.-M. (2008). Determining strategies for ICT to improve efficiency in the production of revenues: An empirical investigation in the context of transition economies using DEA, decision trees, and neural nets. *Journal of Global Information Technology Management*, 11(4), 56–75.

Samoilenko, S. (2009). *Impact of the Investments in ICT on Total Factor Productivity: Empirical Investigation of Macroeconomic Outcomes of Investments in Telecoms in the Context of Economies in Transition*. VDM Verlag, Germany.

Samoilenko, S. & Osei-Bryson, K. M. (2010). Linking investments in telecoms and productivity growth in the context of transition economies within the framework of neoclassical growth accounting: Solving endogeneity problem with structural equation modeling. *Proceedings of 18th European Conference on Information Systems*, Pretoria, South Africa, June 6th–9th, 2010.

Siegel, D. S., Waldman, D., & Youngdahl, W. E. (1997). The adoption of advanced manufacturing technologies: Human resource management implications. *IEEE Transactions on Engineering Management*, 44(3), 288–298.

Solow, R. (1957). Technical change and the aggregate production function. *Review of Economics and Statistics*, 39(3), 312–320.

Stafford, T., Turan, A., & Khasawneh, A. (2006). MIDDLE-EAST.COM: Diffusion of the Internet and online shopping in Jordan and Turkey. *Journal of Global Information Technology Management*, 9(3), 43–61.

Udo, G., Bagchi, K., & Kirs, P. (2008). Diffusion of ICT in developing countries: A qualitative differential analysis of four nations. *Journal of Global Information Technology Management*, 11(1), 6–27.

Chapter 17

An Exploration of the Effects of the Interaction between Information and Communication Technology and Labor Force on Economic Growth in Transition Economies

Introduction

It is by now a commonly accepted fact that investments in information and communication technologies (ICTs) can facilitate macroeconomic growth in developed countries (OECD 2005a). Numerous studies have shown that such investments resulted in generous payoffs for the United States (Jorgenson 2001; Jorgenson & Stiroh 2000; Oliner & Sichel 2002; Stiroh 2002) and a number of economies of the European Union (Colecchia & Schreyer 2001; Van Ark et al. 2002; Daveri 2002; Jalava & Pohjola 2002). There appears to be no immediate reason why other types of countries, namely, developing, least developed, or transition economies (TEs),* would not profit from investments in ICT on the macroeconomic level. However, at this point, any conclusion regarding this matter is premature, for the existing empirical evidence is limited.

It would also appear that the nature of the relationship between investments in ICT and a macroeconomic outcome is not straightforward. There is a consensus among the research and development communities that absence or an insufficient level of complementary investments serves as one of the reasons why the impact of ICT investments does not manifest itself at the macroeconomic level (Kraemer & Dedrick 2001; Pohjola 2002). Indeed, when ICT-driven economic growth took place in the context of the developed countries, ICT deployment was accompanied by complementary investments. Research conducted by the Organisation for Economic Co-operation and Development (OECD 2004) demonstrated that a number of complementary factors could influence the extent of the impact of investments in ICT. One of these factors refers to the amount and quality of available human capital, where the better-skilled and better-equipped workforce is more likely to achieve higher rates of ICT-related innovation and increased productivity (OECD 2004).

Complementarity of investments has been investigated in the context of research and development portfolios by Lambertini (2003), Lin and Saggi (2002), and Rosenkranz (2003) and in the context of process and product innovation by Athey and Schmutzler (1995). In a more relevant context to this research, Giuri et al. (2005) explored the complementarity between skills, organizational change, and investments in information and ICT; Bugamelli and Pagano (2004) studied the complementarity between investment in ICT and the related investment in human and organizational capital. Gera and Wulong (2004) examined complementarity of the investment in ICT and organizational changes and worker skills, and Loukis and Sapounas (2004) inquired into the complementarity between Information Systems' investment and a set of IS management factors. Varis et al. (2004) conducted an inquiry into how complementarity of resources affects partnering needs in the context of the companies operating in technology-intensive fields.

At this point, we would like to outline the research problem of this study. The goal of our inquiry is to explore the existence of a complementary relationship between ICT (in the form of annual *investments in telecoms*) and human capital (in the form of *full-time telecom staff*), and then to determine the effect of this interaction on gross domestic product (*GDP*) in the context of TEs. We would like to provide a brief justification of the importance of this research problem.

1. First, overall, our inquiry fits into the established stream of research investigating the relationship between investments in ICT and their macroeconomic outcomes (Colecchia & Schreyer 2001, 2002; OECD 2003a,b,c, 2004, 2005a,b,c; IMF 2001; Lee & Khatri 2003; Piatkowski 2003a,b, 2004; Pohjola

---

* There exists a common agreement regarding the designation of TEs as countries in the process of transitioning from a centrally planned economy to a market-oriented economy.

2000, 2001, 2002; Avgerou 1998; Morales-Gomez & Melesse 1998; Kraemer & Dedrick 2001). Despite the ample evidence of the positive impact of ICT investments on the economies of the developed countries, the research concerning the effects of ICT on developing and transition economies is scarce. It was noted that "substantive research is urgently required if investment commitments are to be made—by the private sector or development agencies—with any real understanding of likely outcomes" (OECD 2004, p. 4).
2. Second, this investigation focuses on the important facet of the multidimensional construct, namely, the presence of complementary investments and their effect on the relationship between investments in ICT and economic growth. Telecommunications are expected to facilitate productivity and economic growth (Andonova & Diaz-Serrano 2007); however, not much at this point is known about complementary factors that determine the successful translation of ICT investments into GDP, especially in the context of TEs.
3. Third, TEs present a particularly interesting case for inquiry not only because research conducted in this context is scarce but, additionally, because TEs share economic characteristics with both developed and less developed economic regions (OECD 2004). Consequently, TEs present a good vantage point from which a complementarity of investments in ICT can be investigated and the results of the study generalized to a broader context.
4. Fourth, the subject of our study is a complementarity of investments in telecoms because not only do investments in telecoms represent a subset of investments in ICT, but they are also a common type of investment regularly made by almost any economy of the world.
5. Finally, we investigate complementarity of investments in telecoms and full-time telecom staff because the amount and quality of available human capital was identified as one of the complementary factors to investments in ICT in the context of developed countries (OECD 2004), but no study to our knowledge has corroborated this finding in the context of TEs.

We present our inquiry in a sequence of five parts. First, we offer a description of the data used in this study and provide a brief overview of the results of our previous investigations, which constitute a background of the current inquiry. Second, we discuss the theoretical framework that supports our inquiry. Third, we provide an overview of the data analytic method used in this study. Fourth, we present the results of our inquiry. The final part is dedicated to the discussion of the results, followed by a brief conclusion.

Overview of the Data and Background of this Study

The data for our inquiry were collected from the *World Development Indicators* database (http://web.worldbank.org/WBSITE/EXTERNAL/DATASTATISTICS), the *World Bank*'s (http://web.worldbank.org) comprehensive database containing development data, and the *Yearbook of Statistics* (2004, http://www.itu.int/ITU-D /ict/publications), an annual publication of the *International Telecommunication Union* (ITU; http://www.itu.int). To minimize the heterogeneity of our sample, we wanted to use TEs that belong to the same group and started the transition at about the same time. We have chosen 25 countries classified as *transition economies in Europe and the former Soviet Union* by IMF (2000).

Unfortunately, data on seven TEs that we wanted to include in our analysis, namely, Croatia, former Yugoslav Republic of Macedonia (FYR Macedonia), Georgia, Russia, Tajikistan, Turkmenistan, and Uzbekistan, were not available, or contained too many missing data points to be useful in the analysis. Consequently, we concentrated on the following 18 TEs: Albania, Armenia, Azerbaijan, Belarus, Bulgaria, Czech Republic, Estonia, Hungary, Kazakhstan, Kyrgyzstan, Latvia, Lithuania, Moldova, Poland, Romania, Slovakia, Slovenia, and Ukraine. Overall, for our 18 TEs, we were able to construct a data set covering the period from 1993 to 2002.

While compiling our data set, we had to deal with the issue of how to find a scale that minimizes the bias arising from comparing and contrasting countries that are different in size, population, and level of wealth. We have decided to use percentages rather than actual number values. Thus, for example, while the variable *full-time telecom staff* was represented in the *Yearbook of Statistics* as a number value, we have transformed it into the percentage of the variable *labor force, total*, provided by the World Development Indicators (WDI) data set.

In a similar fashion, we have transformed numerical dollar figures of investments into percentages of GDP by using the same approach. The reasoning was that percentages would allow for minimizing the bias associated with the size and level of economic development of a given TE, thus representing a structure of the economy and the labor force of each TE more objectively.

Consequently, we ended up with three variables that we use in the current study:

1. GDP (in current US$)
2. Annual telecom investment (% of GDP)
3. Full-time telecommunication staff (% of total labor force)

For the purposes of our study, we adapt the definition of investment in telecoms provided by the *Yearbook of Statistics* as an investment that

> …refers to expenditure associated with acquiring the ownership of telecommunication equipment infrastructure (including supporting land and buildings and intellectual and non-tangible property such as computer software). These include expenditure on initial installations and on additions to existing installations.

In our previous inquiries (Samoilenko 2008; Samoilenko & Osei-Bryson 2008a, 2008b), we researched the relationship between *investments in telecoms* (a subset of investments in ICT) and *revenues from telecoms* in the context listed for the aforementioned 18 TEs.

First, by using *cluster analysis* (CA), we determined, that these 18 TEs are not homogenous in terms of levels of investments in telecoms and revenues from telecoms; results of CA yielded a two-cluster solution (Samoilenko 2008). Further, by using *data envelopment analysis* (DEA), we determined that these two clusters of TEs differ in terms of the relative efficiency of the utilization of resources and relative efficiency of the production of revenues; we labeled the group with higher averaged efficiency scores the *leaders* and the group with lower scores the *followers* (Samoilenko & Osei-Bryson, 2008a). We also established, by using *decision tree* (DT) analysis, that the leaders differ from the followers in some important areas, such as militarization of the economy, quality of human resources, and level of sociotechnical development (Samoilenko & Osei-Bryson, 2008b). By using *neural network* (NN) simulation, we demonstrated that an increase in the level of investments would not reduce the level of relative inefficiency of the followers, for the latter is a result of inefficient processes of conversion of investments into revenues (self-reference, under review). Further, by using *Malmquist index* (MI) to estimate the yearly changes in total factor productivity (*TFP*), we obtained evidence of conversion of the *leaders* and *followers* subsets of 18 TEs in terms of the efficiency of utilization of inputs and efficiency of the production of revenues (Samoilenko & Osei-Bryson, 2010). One of the purposes of the current inquiry is to find out whether some of the reasons for disparity between the leaders and followers could be, in part, due to the complementarity between investments in telecoms and full-time telecom staff.

Theoretical Framework

To approach our research problem, we rely on a neoclassical framework of growth accounting.

In the case of this study, we use the following variables constituting the neoclassical production function:

Y = GDP
A = TFP
K = investments in telecoms
L = full-time telecom staff

For every point in time (i.e., every year from 1993 to 2002), neoclassical production function allows us to relate investments in telecoms, full-time telecom staff, and GDP in the following fashion:

$$\text{GDP} = f(\text{TFP, investments in telecoms, full-time telecom staff})$$

As a result, for each TE in the study, for the period of 10 years, we have 10 sets of variables constituting the function. This approach of determining the relationship between variables at a *point in time*, rather than *over a period of time*, has advantages. First, we do not need to account for depreciation of the telecom infrastructure, assuming the same rate of depreciation for all TEs in the study. Second, because the numbers for full-time telecom staff and labor force, total, are reported annually, it allows us to treat the number of full-time telecom staff as being constant over the period of any given year. Third, because we are dealing not with the flow of investments, but with investments at the point in time, such an approach also allows us to treat annual investment in telecoms as an annual level of added capital stock. The limitation of this approach, however, is that we cannot inquire into the relationship between changes in the flow of investments and labor force and corresponding changes in GDP.

Rarely should any type of investment made on the macroeconomic level be perceived and considered in isolation. In this chapter, we use the *theory of complementarity* as a theoretical framework that supports our search for the complementarity between investments in telecoms and full-time telecom staff. A brief overview of the theory of complementarity is offered next.

Theory of Complementarity and Translog Production Function

Initially introduced in economics by Edgeworth (1881), the concept of *complementarity* refers to the notion that the increase in one factor could result in the increased benefit received from its complementary factors. We apply the theory of complementarity to our research problem to argue that if benefits of the investments in telecoms are to be reaped successfully at the macroeconomic level, then such investments should not be made without considering the state of full-time telecom staff. Thus, if the two factors are more effective when taken jointly, rather than separately, we consider such factors complementary. However, even if the complementarity of the investments exists within a given production function, it could not be identified through the formulation offered by the Cobb–Douglas production function. Complementarity of the investments could only be discerned if the formulation allows for the presence of the interaction term between the specified investments. Thus, we turn our attention to the transcendental logarithmic (*translog*) production function that offers an opportunity for exploring interactions. This function is defined as follows:

$$\log Y = \beta_0 + \beta_1 * \log K + \beta_2 * \log L + \beta_3 * \log K^2 \\ + \beta_4 * \log L^2 + \beta_5 * \log K * \log L + e$$

It is easy to see that the Cobb–Douglas function is "nested" in the translog function, and a test of the hypothesis that both functions describe production process equally well would entail testing the following null hypothesis:

H0: $\beta_3 = \beta_4 = \beta_5 = 0$

The translog production function is more flexible than the Cobb–Douglas function in the sense that it allows testing for the presence of the interactions between the variables, where the test for the presence of the interaction would involve testing of the following hypothesis:

H0: β_5 is not statistically discernible from 0 at the given level of α.

For example, let us say that we are interested in investigating the following production function:

$$Y = f(A, K_{TC}, L_{TC}),$$

where Y = output (GDP), A = TFP, K_{TC} = annual investments in telecoms, and L_{TC} = full-time telecom staff.

Then we can conduct a test for the presence of the statistically discernible interaction between the two variables using the translog function as follows:

$$\log Y = \beta_0 + \beta_1 * \log K_{TC} + \beta_2 * \log L_{TC} + \beta_3 * \log K_{TC}^2 \\ + \beta_4 * \log L_{TC}^2 + \beta_5 * \log K_{TC} * \log L_{TC} + \varepsilon$$

Consequently, the test for the presence of the interaction would involve testing of the following hypothesis:

H0: β_5 is not statistically discernible from 0 at the given level of α.

If the interaction term between investments in telecoms and full-time telecom staff is significant (i.e., we reject the null hypothesis of $\beta_5 = 0$), then we have a reason to assume that such investments are complementary. Next, we offer a brief overview of multiple regression (MR), the data analytic tool that we use in our study to determine the presence of the interaction effect, followed by the formal definition of the research problem of our inquiry.

Formal Definition of the Research Problem

The purpose of the MR (Pearson 1908) is to model the relationship between a single dependent variable and the multiple independent variables. Generally, MR procedure estimates a linear relationship within the following model:

$$Y = a + b_1 * X_1 + b_2 * X_2 + b_3 * X_3 + \ldots + b_n * X_n + e$$

where
  Y = dependent variable
  a = intercept
  b_k = slope coefficient
  X_k = independent variable
  e = error term

Unlike the *general linear model* (GLM), the purpose of which is to test the relationship between the dependent and a single independent variable of the model

$$Y = a + a + b_1 * X_1 + e,$$

the interpretation of the MR is not as straightforward. While GLM models the dependent variable as a function of the intercept and the product of a slope and the value of the independent variable, MR relates each independent variable in a partial fashion to a dependent variable, meaning that a coefficient of each independent variable represents a partial contribution to the dependent variable, while controlling for the rest of the independent variables in the MP equation. As a result, unlike the GLM, which directly correlates the independent and dependent variables, MR correlates the dependent and independent variables in a partial fashion, which is often referred to as a *partial correlation* (Yule 1907).

In the case of this investigation, while using MR, we are interested in the interaction effect of the independent variables on the dependent variable. Then the general model of MR takes form of

$$Y = a + b_1 * X_1 + b_2 * X_2 + b_3 * X_1 X_2 + \ldots + b_n * X_n + b_k * X_k + b_m * X_n * X_k + e$$

The test for interaction amounts to testing the null hypothesis

$$H0: b_3 = 0$$

And in the case of $b_3 \neq 0$, we are able to reject the null hypothesis of no interaction between X_1 and X_2.

The interpretation of the interaction term in MR, however, is not as straightforward as the interpretation of the slope coefficient of an independent variable (Aiken and West, 1991). For example, b_3 in the previous equation reflects the relationship between Y and X_1 and X_2 when X_1 and X_2 increase *jointly*. Furthermore, b_3 in the equation above reflects a conditional relationship between Y and X_1 and X_2, for the impact of X_1 on Y would depend on the level of X_2 and vice versa.

For the purposes of this research, we are only interested in testing the null hypothesis of no interaction between the investments in telecoms and full-time telecom staff. As a result, we are not going to inquire into issues potentially affecting the interaction term such as the presence of thresholds, level-dependent dynamic of the interacting variables, and so on. At this point, we formulate the research question that we are addressing in this study.

The objective of this investigation is to establish whether there exists a statistically significant interaction effect between investments in telecoms and full-time telecom staff, the presence of which affects the relationship between investments in telecoms and GDP. Thus, we state the null hypothesis corresponding to the research question as follows:

> H01: For a given sample of 18 TEs, there exists no statistically significant interaction effect between investments in telecoms and full-time telecom staff presence that produces a statistically discernible change in relationship between investments in telecoms and GDP.

Furthermore, keeping in mind the differences, discovered in the course of our previous investigations, between the *leaders* and *followers* subsets of our sample, we also put forward the following H02 and H03:

> H02: For the *leaders* subset of the sample of 18 TEs, there exists no statistically significant interaction effect between investments in telecoms and full-time telecom staff the presence of which produces a statistically discernible change in the relationship between investments in telecoms and GDP.
> H03: For the *followers* subset of the sample of 18 TEs, there exists no statistically significant interaction effect between investments in telecoms and full-time telecom staff the presence of which produces a statistically discernible change in the relationship between investments in telecoms and GDP.

Resultantly, the test for the presence of a statistically significant interaction effect will involve testing the null hypotheses in the three different contexts, namely, the full set, the *leaders* subset, and the *followers* subset of 18 TEs.

In the case of our study, we can test for the presence of the statistically discernible interaction between the two variables by using the translog function as presented here:

$$\log Y = \beta_0 + \beta_1 * \log K_{TC} + \beta_2 * \log L_{TC} + \beta_3 * \log K_{TC}^2$$
$$+ \beta_4 * \log L_{TC}^2 + \beta_5 * \log K_{TC} * \log L_{TC} + \varepsilon$$

Thus, a test for the presence of the interaction would involve testing of the following hypothesis:

H0: β_5 is not statistically discernible from 0 at the given level of α.

If the interaction term turns out to be statistically significant, allowing us to reject H0, we would be able to state that we determined the presence of complementarity between investments in telecoms and full-time telecom staff.

Results of the Data Analysis

In this section, we present the results of our inquiry. We used SAS *Enterprise Miner* to perform the data analysis. The general approach that we have followed is depicted in Figure 17.1. Next, we are going to describe in detail all the relevant steps that have been taken at each node of the diagram.

In the first step of our analysis, we used the *data source* node primarily for two purposes. First, we selected the data set that we were going to use for each case (i.e., full set, leaders, and followers), and second, we defined the GDP variable as a "target" of the data analysis. However, the original values contained in the data set could not be used in the translog function, since translog requires logs, squared logs, and log interaction terms. The transformation of the data was performed in the next step of our analysis.

The second step involved the use of two *transform variables* nodes. In the first *transform variables* node, we used log transformation to produce logs of the original

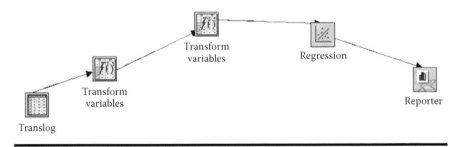

Figure 17.1 Process flow of the data analysis.

data values in the data set defined in the *data source* node. Thus, at the first *transform variables* node, we ended up with the transformed original variables. However, we still needed squared logs and interaction terms.

In the second *transform variables* node, we used the option "create new variable" to create all of the required squared log terms. Consequently, we ended up with the logs of the original variables, plus with the squared logs of the original variables. At this point, we still needed to define the interaction terms prior the data analysis.

The third step involved the use of the *regression* node. First, we used "interaction builder" to create and add to the model the required interaction term. Finally, we chose to "suppress interrupt" with "general linear model" as options of the regression model and then run the data analysis.

The *reporter* node was used to obtain the results of the data analysis in a better-formatted way than was offered by the output of the *regression* node. Results of the data analysis are presented in Table 17.1. The interaction term is given in the form that it appeared in the actual model (column "term in the model"). The column "estimate" provides a value of the parameter estimate for β, an interaction term. The column labeled "*P* value" provides a two-tailed *P* value used in testing our null hypothesis that $\beta = 0$. We tested the hypothesis at 95% confidence level, or, as it is commonly expressed, at a level of $\alpha = .05$. As a result, a coefficient having a *P* value of .05 or less would be considered statistically significant, which would allow us to reject the null hypothesis of $\beta = 0$ at the level of $\alpha = .05$.

The result of the data analysis of the complete data set consisting of 18 TEs suggests that the interaction term of the variables annual investment in telecoms and full-time telecom staff is not statistically significant; we accepted the null hypothesis of no interaction. It is, however, a counterintuitive finding; *we would expect that the full-time telecom staff serve as caretakers of investments in telecoms.* Pinjala et al. (2006), for example, noted the importance of highly skilled professional staff to the capability of the companies to complete in the business market. It makes sense to think that full-time telecom staff serves as one of the major components that is involved in the process of transformation of investments into revenues, and ultimately, into GDP.

Hoskisson et al. (2000) brings attention to the possibility that even within the same geographical region, economies of the same type could significantly differ in terms of their regime, starting points of and paths to transition, and the degree of the achieved progress. Subsequent separate data analyses of the *leaders* and *followers* subsets of the original sample produced markedly different results; in both cases, we were able to reject null hypotheses of no interaction. However, despite the similarity with regard to the significance of the interaction term, these two groups of 18 TEs differ with regard to the direction of the interaction effect. We can see that complementarity is positive for the leaders and is negative for the followers. This change in the direction of the effect suggests the presence of the threshold.

One of the possible interpretations of this finding is the existence of an optimal ratio of full-time telecom staff that is required for annual investments in telecoms

Table 17.1 Results of the Data Analysis

Cluster	Term in the Model	Estimate (Value of β)	P Value	Adjusted R^2	Result
Complete set (180 data points)	ANNU_EL1 Log (Annual Telecom Investment)	42.8020	.8613	.8655	H01 accepted
	FULL_2KW Log(Full-time Telecom Staff)	−43.9723	<.0001		
	ANNU_EL1*FULL_2KW Log (Annual Telecom Investment)*Log(Full-time Telecom Staff)	57.4954	.8156		
The leaders (66 data points)	ANNU_EL1 Log (Annual Telecom Investment)	42.8020	<.0001	.9167	H02 rejected
	FULL_2KW Log(Full-time Telecom Staff)	−43.9723	<.0001		
	ANNU_EL1*FULL_2KW Log (Annual Telecom Investment)*Log(Full-time Telecom Staff)	57.4954	<.0001		

(Continued)

Table 17.1 (Continued) Results of the Data Analysis

Cluster	Term in the Model	Estimate (Value of β)	P Value	Adjusted R^2	Result
The followers (114 data points)	ANNU_EL1 Log (Annual Telecom Investment)	−2.6459	.0008	.9874	H03 rejected
	FULL_2KW Log(Full-time Telecom Staff)	−50.6696	<.0001		
	ANNU_EL1*FULL_2KW Log (Annual Telecom Investment)*Log(Full-time Telecom Staff)	−2.1280	.0087		

to contribute optimally to the increase in GDP. If this is the case, then there exists a "golden middle" of the investment-to-staff ratio. For example, it is possible that up to a certain level of the investment in telecoms, the number of the staff must grow rapidly, while after that point, only an increase in productivity of the full-time telecom staff could provide increased contribution of investments in telecoms to the macroeconomic growth. Moreover, once the certain level of productivity is achieved, the number of full-time telecom staff must increase, and so on in a spiral fashion. This suggestion seems to be quite reasonable, for Tohidi and Tarokh (2006) state that "beyond some value of team size, the marginal cost of an additional team member exceeds the marginal value of team's production" (p. 614).

While, at this point, we cannot test the validity of our interpretation and determine whether the full-time telecom staff of the leaders is more productive than that of the followers, we can obtain some insights regarding this matter. In order to do so, we calculated the averaged ratios of investments in telecoms to full-time telecom staff for each group. This ratio could be thought of as the amount of investments in telecoms handled by each full-time telecom staff. We also obtained the averaged values of annual revenues from telecoms per full-time telecom worker for each group. The results presented in Table 17.2 suggest, that the leaders invest more in telecoms per telecom worker than the followers (third column). We also can see the marked difference between the averaged values of annual revenues from telecoms per full-time Telecom worker (last column). We also decided to take into consideration a possible effect of the political institutions on the diffusion of the telecommunication technologies (Andonova & Diaz-Serrano 2007) and see if the leaders differ from followers in terms of level of political risks. It is reasonable to suspect that countries with a lower level of political risk should do better in terms of attracting investments and converting them into revenues. To compare the leaders and followers in this regard, we used averaged values of the Political Constraint (POLCON) index that was reported and used in the investigation of Andonova and Diaz-Serrano (2007). This variable, which ranges from 0 to 1, was chosen because it is "an objective and conservative measure among the available indices of political risks," which could be interpreted as an "objective measure of the degree to which investors' interests are protected by a given polity" (Andonova & Diaz-Serrano 2007). The results of the comparison are presented in the last column of Table 17.2. Again, we can see that the leaders differ favorably from the followers in this regard also, possibly by offering a greater degree of political commitment for investor protection (Andonova & Diaz-Serrano 2007).

We have also included the graphical representation (Figures 17A.1–17A.5) of the comparison in the Appendix of the chapter. One of the possible explanations for such disparity is that the workforce of the followers is less efficient and effective than that of the leaders. This is a reasonable suggestion; after all, Wacker et al. (2006) have confirmed the effect of human resources on productivity, and Martınez-Lorente et al. (2004) noted the relationship between the quality and operational performance of the firm and qualified human resources.

Table 17.2 Comparison of the Leaders and Followers in Terms of Investments per Telecom Worker

Data Set 1993–2002	Full-Time Telecom Staff (% of Labor Force), Averaged	Annual Telecom Investment (% of GDP), Averaged	Ratio of Annual Telecom Investment to Full-Time Telecom Staff, Averaged	Annual Revenue from Telecoms per Full-Time Telecom Worker (in Current US$), Averaged	Value of POLKON Variable, Averaged
Full Set	0.4778	0.8178	1.7117	33,506.51	0.31
Followers	0.4497	0.5861	1.3031	9,240.25	0.25
Leaders	0.5228	1.1905	2.2770	75,420.98	0.42

Summary and Conclusion

In this chapter, we described our inquiry into the complementarity of investments in telecoms and full-time telecom staff in the context of 18 TEs. We determined the presence of the statistically significant interaction effect that can influence positively or adversely the macroeconomic outcome of investments in telecoms. This study makes several contributions to the existing body of knowledge.

1. First, our research was driven by accepted and well-established theoretical frameworks, which have previously been successfully applied in the context of developed countries. It is true that in some areas of inquiry, such as research on strategies in emerging market economies, "theories promulgated for developed market economies may not be appropriate for emerging economies" (Hoskisson et al. 2000). However, we were able to demonstrate that the same theoretical frameworks that drive inquiries in the context of developed countries can drive research on the complementarity of investments in the context of TEs.
2. Second, we corroborate the findings of earlier studies regarding the complementary factors affecting the economic outcomes of investments in ICT. Namely, we demonstrate that the level of investments in telecoms and the level of the full-time telecom staff are complementary factors.
3. Third, in addition to corroborating the results of the previous studies, our research obtained some new empirical findings. Namely, we were able to demonstrate that the direction of the interaction effect on the macroeconomic outcome is not constant, but varies, possibly depending on the levels of the interacting variables.

The results of the study could be of benefit to the community of practitioners as well. Any policy maker or investor shall benefit from taking into consideration the importance of the human resource factor affecting the economic outcomes of investments in telecoms, acknowledging, therefore, that no significant increase in the level of such investments shall directly result in a proportional increase in the bottom line.

In the course of our investigation, we identified that leaders differ from followers in many important aspects, which are summarized in Table 17.2. At this point, the impact of the differentiating variables on complementarity between investments in telecoms and full-time telecom staff is not clear; this could offer an interesting direction for the future inquiries in this area.

Despite the contributions that this study makes, our research is not without its limitations. The major limitation of this study is associated with the quantity of the data. Clearly, our research might have offered richer insights if more variables were available from the data sources that we used. This limitation, however, is not unique to our study, but is characteristic of the research in this area in general

(Hoskisson et al. 2000). In addition, it may be argued that the time series data covering a 10-year period could be insufficient to inquire into the nature of the events taking place on a macroeconomic level. Nevertheless, in the area of research where it "would appear to be a major need for longitudinal studies" (Hoskisson et al. 2000), we feel that the contributions provided by our study outweigh its limitations.

Acknowledgment

Material in this chapter previously appeared in "An exploration of the effects of the interaction between ICT and labor force on economic growth in transitional economies," *International Journal of Production Economics* 115, 471–481.

References

Aiken, L. & West, S. (1991). *Multiple Regression: Testing and Interpreting Interactions*. Sage Publications, Newbury Park, CA.

Andonova, V. & Diaz-Serrano, L. (2007). Political Institutions and the Development of Telecommunications (January 2007). IZA Discussion Paper No. 2569. Available online at http://ssrn.com/abstract=961371

Athey, S. & Schmutzler, A. (1995). Product and process flexibility in an innovative environment. *RAND Journal of Economics*, 26(4), 557–574.

Avgerou, C. (1998). How can IT enable economic growth in developing countries? *Information Technology for Development*, 8, 15–28.

Braumoeller, B. (2004). Hypothesis testing and multiplicative interaction terms. *International Organization*, 58(4), 807–820.

Bugamelli, M. & Pagano, P. (2004). Barriers to investment in ICT. *Applied Economics*, 36(20), 2275–2286.

Colecchia, A. & Schreyer, P. (2001). The Impact of Information Communications Technology on Output Growth. STI Working Paper 2001/7, OECD, Paris.

Colecchia, A. & Schreyer, P. (2002). ICT investment and economic growth in the 1990s: Is the United States a unique case? A comparative study of nine OECD countries. *Review of Economic Dynamics*, 5, 408–442.

Daveri, F. (2002). The new economy in Europe, 1992–2001. *Oxford Review of Economic Policy*, 18(3), 345–362.

Gera, S. & Wulong, G. (2004). The effect of organizational innovation and information and communications technology on firm performance. *International Productivity Monitor, Centre for the Study of Living Standards*, 9, 37–51.

Giuri, P., Torrisi S., & Zinovyeva, N. (2005). ICT, Skills and Organizational Change: Evidence from a Panel of Italian Manufacturing Firms. LEM Papers Series 2005/11, Laboratory of Economics and Management (LEM), Sant'Anna School of Advanced Studies, Pisa, Italy.

Hoskisson, R., Eden, L., Lau, C., & Wright, M. (2000). Strategy in emerging economies. *Academy of Management Journal*, 43(3), 249–267.

IMF (2001). *International Financial Statistics*. IMF, Washington, DC.
Jalava, J. & Pohjola, M. (2002). Economic growth in the new economy: Evidence from advanced economies. *Information Economics and Policy*, 14(2), 189–210.
Jorgenson, D. W. (2001). Information technology and the US economy. *American Economic Review*, 91(March), 1–32.
Jorgenson, D. W. & Stiroh, K. J. (2000). Raising the speed limit: US economic growth in the information age. *Brookings Papers on Economic Activity*, 2(1), 125–211.
Kraemer, K. L. & Dedrick, J. (2001). Information technology and economic development: Results and policy implications of cross-country studies. In *Information Technology, Productivity, and Economic Growth*, ed. M. Pohjola, Oxford University Press, Oxford.
Lambertini, L. (2003). The monopolist's optimal R&D portfolio. *Oxford Economic Papers*, 55, 561–78.
Lee, I.-H. & Khatri, Y. (2001). *The Role of the New Economy in East Asia*. Unpublished. International Monetary Fund, Asia and Pacific Department, Washington. Available online at http://www.er.cna.it/biblioteca/files/20050606162556200505140103 5103 WEO_2001_09_11__01_09_2610.pdf
Lin, P. & Saggi, K. (2002). Product differentiation, process R&D, and the nature of market competition. *European Economic Review*, 46, 201–211.
Loukis, E. & Sapounas, I. (2004). The impact of information systems investment and management on business performance in Greece. ECIS 2005 Conference Paper.
Madden, G. & Savage, S. (1998). CEE telecommunications investment and economic growth. *Information Economics and Policy*, (10), 173–195.
Martinez-Lorente, A., Sanchez-Rodriguez, C., & Dewhurst, F. (2004). The effect of information technologies on TQM: An initial analysis. *International Journal of Production Economics*, 89, 77–93.
Morales-Gomez, D. & Melesse, M. (1998). Utilizing information and communication technologies for development: The social dimensions. *Information Technology for Development*, 8, 3–13.
OECD (2003a). *The Sources of Economic Growth in OECD Countries*. OECD, Paris.
OECD (2003b). *ICT and Economic Growth: Evidence from OECD Countries, Industries and Firms*. OECD, Paris.
OECD (2003c). *Integrating Information and Communication Technologies in Development Programmes*. Policy Brief, OECD, Paris.
OECD (2004). *DAC Network on Poverty Reduction: ICTs and Economic Growth in Developing Countries*. OECD, Paris. Available online at http://www.oecd.org/dataoecd/15/54/34663175.pdf
OECD (2005a). Good practice paper on ICTs for economic growth and poverty reduction. *The DAC Journal*, 6(3), 1–69.
OECD (2005b). Background paper: The contribution of ICTs to pro-poor growth: No. 384. *OECD Papers*, 5(2), July, 15–52.
OECD (2005c). The contribution of ICTs to pro-poor growth: No. 379. *OECD Papers*, 5(1), March, 59–72.
Oliner, S. & Sichel, D. (2002). Information technology and productivity: Where are we now and where are we going? *Economic Review*, 3(3), 15–41.
Pearson, K. (1908). On the generalized probable error in multiple normal correlation. *Biometrika*, 6, 59–68.

Piatkowski, M. (2003a). The contribution of ICT investment to economic growth and labor productivity in Poland 1995–2000. *TIGER Working Paper Series*, No. 43. July. Warsaw. Available online at http://www.tiger.edu.pl

Piatkowski, M. (2003b). Does ICT investment matter for output growth and labor productivity in transition economies? *TIGER Working Paper Series*, No. 47. December. Warsaw. Available online at http://www.tiger.edu.pl

Piatkowski, M. (2004). The Impact of ICT on Growth in Transition Economies. *TIGER Working Paper Series*, No. 59. July. Warsaw. Available online at http://www.tiger.edu.pl/publikacje/TWPNo59.pdf

Pinjala, K., Pintelon, L., & Vereecke, A. (2006). An empirical investigation on the relationship between business and maintenance strategies. *International Journal of Production Economics*, 104, 214–229.

Pohjola, M. (2000). Information Technology and Economic Growth: A Cross-Country Analysis. United Nations University World Institute for Development Economics Research (UNU/WIDER), Helsinki, Finland.

Pohjola, M. (2002). New Economy in Growth and Development. WIDER Discussion Paper 2002/67, United Nations University World Institute for Development Economics Research (UNU/WIDER). Helsinki, Finland.

Pohjola, M. (ed.) (2001). Information Technology, Productivity and Economic Growth: International Evidence and Implications for Economic Development. WIDER Studies in Development Economics, Oxford University Press, Oxford, UK.

Rosenkranz, S. (2003). Simultaneous choice of process and product innovation when consumers have a preference for product variety. *Journal of Economic Behavior & Organization*, 50(2), 183–201.

Schreyer, P. (2000). The Contribution of Information and Communication Technology to Output Growth: A Study of the G7 Countries. *STI Working Papers* 2000/2, OECD, Paris.

Samoilenko, S. (2008). Contributing Factors to Information Technology Investment Utilization in Transition Economies: An Empirical Investigation, *Journal of Information Technology for Development*, 14(1), 52–75.

Samoilenko, S. and Green, L. "Convergence and Productive Efficiency in the Context of 18 Transition Economies: Empirical Investigation Using DEA" in *Proceedings of the Southern Association for Information Systems Conference*, Richmond, VA, USA March 13–15, 2008.

Samoilenko, S. and Osei-Bryson, K. M. (2008a). Strategies for Telecoms to Improve Efficiency in the Production of Revenues: An Empirical Investigation in the Context of Transition Economies, *Journal of Global Information Technology Management*, 11(4), 56–75.

Samoilenko, S. and Osei-Bryson, K. M. (2008b). An Exploration of the Effects of the Interaction between ICT and Labor Force on Economic Growth in Transitional Economies, *International Journal of Production Economics*, 115, 471–481.

Samoilenko, S. and Osei-Bryson, K. M. (2010). Determining Sources of Relative Inefficiency in Heterogeneous Samples: Methodology Using Cluster Analysis, DEA and Neural Networks, *European Journal of Operational Research*, 206, 479–487.

Stiroh, K. (2002). Information technology and the U.S. productivity revival: What do the industry data say? *American Economic Review*, 92(5), 1559–1576.

Tohidi, H. & Tarokh, M. (2006). Productivity outcomes of teamwork as an effect of information technology and team size. *International Journal of Production Economics*, 103, 610–615.

Van Ark, B., Melka, J., Mulder, N., Timmer, M., & Ypma, G. (2002). ICT Investments and Growth Accounts for the European Union, 1980–2000. Research Memorandum GD-56, Groningen Growth and Development Centre, Groningen. Available online at http://www.eco.rug.nl/ggdc/homeggdc.html

Varis, J., Virolainen, V.-M., & Puumalainen, K. (2004). In search for complementarities—Partnering of technology-intensive small firms. *International Journal of Production Economics*, 90, 117–125.

Wacker, J., Yang, C.-L., & Sheu, C. (2006). Productivity of production labor, non-production labor, and capital: An international study. *International Journal of Production Economics*, 103, 863–872.

Whelan, K. (1999). Tax incentives, material inputs, and the supply curve for capital equipment. Federal Reserve Board. Finance and Economics Discussion Series Paper, April 21, 1999. Available online at http://www.federalreserve.gov/pubs/feds/1999/199921/199921pap.pdf

Yule, G. U. (1907). On the theory of correlation for any number of variables, treated by a new system of notations. *Proceedings of Royal Society*, 79A, 182–193.

Appendix

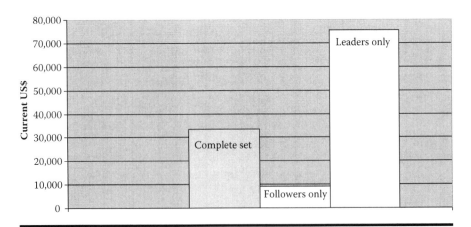

Figure 17A.1 Annual revenues from telecoms, per full-time telecom worker.

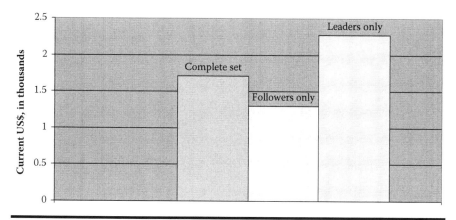

Figure 17A.2 Annual investments in telecoms per full-time telecom worker.

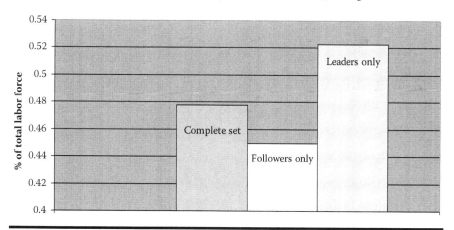

Figure 17A.3 Full-time telecom staff.

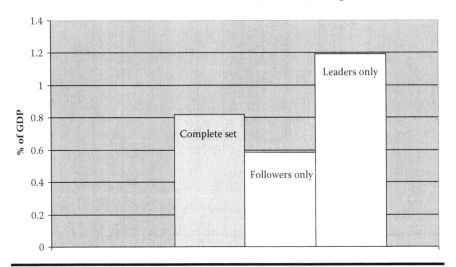

Figure 17A.4   Annual telecom investment.

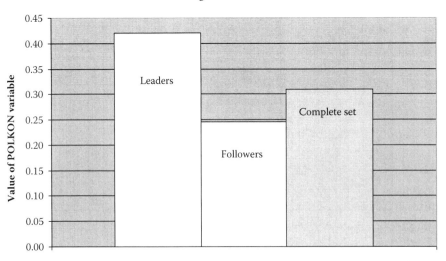

Figure 17A.5   Averaged values of POLKON indicator.

Chapter 18

Contributing Factors to Information Technology Investment Utilization in Transition Economies: An Empirical Investigation

Introduction

What are the transition economies (TEs) and where, in the economic map of the world, do they fit? It is not an easy task to categorize the economies of the world according to the single taxonomy, for no single commonly agreed upon taxonomy exists. Perhaps, multiple perspectives would do a better job of outlining the context of this study.

The World Bank classifies all economies of the world based on the *gross national product (GNP) per capita* of a given country in 1999. If the value was less than $755, the country is considered a *low-income* country. If the value was above $9,266, then the country is labeled as a *high-income* country, and if the value falls somewhere in between, then it is a *middle-income* country (World Bank 2002). Middle-income countries can be further subdivided into two subgroups of *lower-middle-income* countries, where GNP per capita is between $756 and $2,995, and *upper-middle-income* countries, where GNP per capita is between $2,996 and $9,265.

The United Nations designates those low-income economies that are characterized by weak human assets (as measured through a composite *Human Assets Index*) and economic vulnerability as *least-developed* countries (UNCTAD 2004), while conceding that no agreed upon criteria exist for categorizing either *developed* or *developing* countries (UNCD 1999). According to the perspective of the World Bank, however, developed countries are the high-income economies that have a high standard of living (often represented in terms of *Human Development Index*; UNDP 1998), while developing countries are the low- and middle-income economies that have a low to moderate standard of living.

The term *transition economies* (TEs) is used to refer to countries in the process of transitioning from a government or a state-controlled centrally planned economy (Ollman 1997; Myers 2004) to a market-oriented economy, where the market, rather than government or state, plays the "invisible hand" (Smith 1776). It does not mean, however, that TEs constitute a homogenous group in terms of the level of economic development. The World Bank, for example, may group some of them with the developed, and some with the developing countries, depending on the level of industrialization.

Hoskisson et al. (2000) combined two groups of "51 high-growth developing countries in Asia, Latin America, and Africa/Middle East, and 13 transition economies in the former Soviet Union" into the category of *emerging economies*. They define "emerging economy" as a country that "satisfies two criteria: a rapid pace of economic development, and government policies favoring economic liberalization and the adoption of a free market system." They warn, "at present, there is no standard list of countries agreed to be emerging economies," nor that there exists a common agreement on the meaning of the term; consequently, the "term 'emerging market economy' may also mean different things to different researchers."

At this point, there is no single accepted taxonomy to classify economies of the world. There are differences between the countries that defy a single framework; nevertheless, there are common traits among them as well. One such trait is that the great majority of the countries in the world invest in *information and communication technologies* (*ICTs*).

Revenue generation serves as a major means by which investments in ICT contribute to macroeconomic growth (UN ICT Task Force Report 2005; WT/ICT Development Report 2006). Consequently, improving the effectiveness and efficiency of revenue production is a possible route to increase the macroeconomic impact of investments in ICT. And while these investments have been consistently contributing to economic growth by producing significant and reliable streams of revenues in developed countries, the result of such investments is not as clear-cut in the context of the TEs. In order to make investments in ICT attractive to domestic and international investors, however, TEs must be able to demonstrate their ability to produce revenues from such investments in a reliable and efficient manner.

There is little doubt that investments in ICT could and do produce robust returns and contribute to the overall economic growth in the context of developed economies (OECD 2005; Jorgenson 2001; Jorgenson & Stiroh 2000; Oliner & Sichel 2002; Stiroh 2002; Colecchia & Schreyer 2001; Van Ark et al. 2002;

Daveri 2002; Jalava & Pohjola 2002). In the contexts of developing countries and economies in transition, however, the levels of the returns on investments differ significantly.

The TEs present a particularly interesting case for the research because they share economic characteristics with both developed and less developed economic regions (OECD 2004). For example, domestic markets of TEs "include *both* substantial populations with significant disposable income *and* large numbers of people without" (OECD 2004, p. 12). As a result, TEs present a good vantage point from which the relationship between the investments in ICT and economic growth can be investigated.

Multiple research studies conducted in this area have identified a group of factors that affect the return on investments in ICT. It has been suggested that the differences in capital stock and infrastructure (Dewan & Kraemer 2000; Piatkowski 2002), a level of investments in ICT (Murakami 1997; Piatkowski 2002), as well as the amount and quality of the available human capital (OECD 2004), are some of the variables that impact the level of returns on investments in ICT.

We point the attention of the reader to another, rather obvious, but overlooked, determinant of effective production of revenue, namely, the efficiency of utilization of investments. As mentioned above, there is a consensus in the research community that capital stock and infrastructure, human capital, and level of investments in ICT are factors that affect the economic outcome of investments. However, do these factors also affect the efficiency of utilization of investments? It would appear to be the case in the context of developed countries, but, to our knowledge, there have been no studies undertaken to answer this question in the context of TEs.

Furthermore, as economies gradually acquire better infrastructures, invest more in ICT, and improve the level of human capital, do these economies obtain higher levels of returns on investments because of the corresponding gradual increases in efficiency as well? We could not find any reported evidence in the published literature regarding this question either.

Let us elaborate on the importance of this research problem a little further. It is reasonable to propose that investments in ICT must be gauged in accordance with the level of efficiency with which these investments can be utilized. The simple supporting reason for this statement is that efficiency is a relative term. It is relative to the external context (i.e., how efficient is country A relative to country B?), as well as to the internal context, and it is the internal context that deals with the level of the available resources. If country A can efficiently utilize X level of the investments in ICT, in the absence of perfect scalability, the efficiency would likely change if the level of investments in ICT drastically changes.

However, the issue of the relationship and the interplay between the level of investments in ICT and the efficiency of utilization of investments is simply too complex to be tackled in this study, for there are smaller questions that we must answer first. One such question concerns the factors that affect the level of efficiency of utilization of investments in ICT.

In line with related research in this area, this study too is ultimately concerned with the economic outcome of investments in ICT. Unlike the other studies, however, this research inquires specifically into the efficiency of the process by which investments in ICT are utilized, as well as into factors that possibly affect the level of efficiency.

More formally, in this research we look at a subset of investments in ICT, specifically *investment in telecoms*, and investigate how efficiently TEs utilize these investments to produce revenues, and what factors contribute to the efficiency of investment utilization. For the purposes of our study, we adapt the definition of investment in telecoms provided by the Yearbook of Statistics (2004) (see more about this source in the next section), namely, as an investment that

> ...refers to expenditure associated with acquiring the ownership of telecommunication equipment infrastructure (including supporting land and buildings and intellectual and non-tangible property such as computer software). These include expenditure on initial installations and on additions to existing installations.

In order to address this question we employ a three-phase methodology utilizing *data envelopment analysis* (Charnes et al. 1994; Cooper et al. 2004), *cluster analysis* (CA; Aldenderfer & Blashfield 1984), and *decision trees* (Breiman et al. 1984). To approach our research problem, we rely on a neoclassical framework of growth accounting (Solow 1957) and theory of complementarity (Edgeworth 1881).

Theoretical Framework

Growth Accounting

A neoclassical production function relates output and inputs in the following manner:

$$Y = f(A, K, L)$$

where, in the case of this study:
 Y = revenues from telecoms
 A = TFP
 K = investments in telecoms
 L = quantity of full-time telecom employees

Based on the function provided above, growth accounting uses a Cobb–Douglas production function:

$$Y = A * K^\alpha * L^\beta$$

where α and β are constants determined by technology.

By taking the logarithm, the following formulation is obtained:

$$\log Y = \log A + \alpha \log K + \beta \log L$$

Extension to the above formulation of the Cobb–Douglas production function, called the transcendental logarithmic (*translog*) production function, is provided below:

$$\log Y = \beta_0 + \beta_1 \cdot \log K + \beta_2 \cdot \log L + \beta_3 \cdot \log K^2 + \beta_4 \cdot \log L^2 + \beta_5 \cdot \log K * \log L + e$$

It is easy to see that the Cobb–Douglas function is "nested" in the translog function, and testing whether both functions describe the production process equally well would entail testing the following null hypothesis:

$$H0: \beta_3 = \beta_4 = \beta_5 = 0$$

The translog production function is more flexible than Cobb–Douglas function in the sense that it allows testing for the presence of the interactions between the variables, where the test for the presence of the interaction would involve testing of the following hypothesis:

$$H0: \beta_5 \text{ is not statistically discernible from 0 at the given level of } \alpha$$

Complementarity of investments has been investigated in the context of research and development (R&D) portfolios by Lambertini (2003), Lin and Saggi (2002), and Rosenkranz (2003) and in the context of process and product innovation by Athey and Schmutzler (1995). In more relevant context to this research, Giuri et al. (2005) explored the complementarity between skills, organizational change, and investments in ICT. Bugamelli and Pagano (2004) studied the complementarity between investment in ICT and the related investment in human and organizational capital. Gera and Wulong (2004) examined complementarity of the investment in ICT and organizational changes and worker skills, and Loukis and Sapounas (2004) inquired into the complementarity between IS investment and the set of IS management factors.

Overview of the Data

In our choice of data analytic tools, we were restricted by our selection of the data, which represents a sample of convenience. Use of any parametric method would have required an assumption of data normality, which a sample of convenience may not satisfy. Thus, we decided to use data envelopment analysis, CA, and decision trees, all well-known and dependable nonparametric methods.

The data for this study were obtained from the *World Development Indicators* database (web.worldbank.org/WBSITE/EXTERNAL/DATASTATISTICS), which is the *World Bank*'s (web.worldbank.org) comprehensive database on development data, and the *Yearbook of Statistics* (2004) (http://www.itu.int/ITU-D/ict/publications), which is published yearly by the *International Telecommunication Union* (ITU; www.itu.int). In our choice of variables, we were greatly restricted by the availability of the data. For example, while the development data of the World Bank's database cover more than 600 indicators for 208 economies, data on many of the indicators relevant to our research were not available, or were available only for a few countries, or contained too few data points to be useful in data analysis.

In our choice of TEs to include in the study, we tried to identify and select a group of countries that started the transition process at approximately the same time. Thus, we decided on the following 18 transitional economies: Albania, Armenia, Azerbaijan, Belarus, Bulgaria, Czech Republic, Estonia, Hungary, Kazakhstan, Kyrgyzstan, Latvia, Lithuania, Moldova, Poland, Romania, Slovakia, Slovenia, and Ukraine.

In terms of the length of the time series, we were restricted to the period from 1992 to 2004, for which data were provided by the Yearbook of Statistics of ITU. We decided to begin our analysis with year 1993 because we believe that year provides a common starting point for the transitional economies. Our reasoning here is that it took a year from the dissolution of the Soviet block in 1991 for the transition process to begin, and using year 1992 as a starting point may favor "early starters."

Methodology: Searching for the Determinants of the Efficiency of Utilization of Investments in Telecoms

In this part of the paper, we describe, in a step-by-step fashion, the methodology used in this study.

Phase 1: DEA

The cornerstone of our approach is DEA, which we utilize to obtain the relative efficiency score for each TE in our sample. Our data set spans 10-year period; consequently, we perform DEA 10 times, one for each year for the period from 1993 to 2002. As a result, we obtain 10 scores of the relative efficiency, one for every year, for each TE in our sample. These scores refer to the relative efficiency of transforming investments in telecoms into revenues from telecoms. From the set of available data, we select a subset representing inputs and a subset representing outputs. These are the variables that are used in the specification of the DEA model.

Data Used to Perform DEA

For the DEA part of the methodology, we have identified a model consisting of the six input and four output variables. We present the description of the model first, and then follow with the justification of the variables that comprise our model.

Input variables of the DEA model:

1. GDP per capita (in current US$)
2. Full-time telecommunication staff (% of total labor force)
3. Annual telecom investment per telecom worker
4. Annual telecom investment (% of GDP in current US$)
5. Annual telecom investment per capita
6. Annual telecom investment per worker

Output variables of the DEA model:

1. Total telecom services revenue per telecom worker
2. Total telecom services revenue (% of GDP in current US$)
3. Total telecom services revenue per worker
4. Total telecom services revenue per capita

The main goal that we pursue in performing DEA is to find out how efficient the 18 TEs are in converting investment inputs into the revenue outputs. Therefore, we did not include any other types of inputs or outputs such as those related to infrastructure, capabilities, utilization, etc. It should be mentioned that the purpose of our DEA model is not to reflect the path by which the investments are transformed into the revenues over the course of 1 year; rather, the intent of our model is to depict a "fiscal efficiency" of the TEs regarding their investments in telecoms.

Upon the close inspection of the chosen variables, one can see that all of them are expressed as ratios. We intentionally present the levels of investments and revenues not in absolute dollar terms but in relative units. The intent in doing so is to lessen the impact of the differences between TEs in terms of their size, population, and level of wealth, while representing the investments and revenues more broadly (i.e., relative to the whole population, labor force of a country, and the telecom industry). We argue that such relative representation provides a more objective depiction of not only the investments and revenues themselves but also the economic and demographic environment within which the investments take place and the revenues produced.

There are no objective criteria according to which the "best" DEA model can be constructed; instead, the decision about including input and output variables is usually delegated to the purview of the investigator. We would like to, however, provide some justification regarding our choice of the input and output variables in our DEA model.

We include the input variable *GDP per capita (in current US$)* in order to take into consideration the differences between the levels of the economic development of 18 TEs in the study. Let us recall that according to the World Bank, economies can be classified as low income, lower middle income, upper middle income, and high income. Consequently, inclusion of the input variable *GDP per capita (in current US$)* allows us to account for the possible differences in the level of industrialization of these countries.

The reason for the inclusion of the input variable *full-time telecommunication staff (% of total labor force)* is intuitive; according to the assumption of the study, investments in telecoms are converted into revenues by full-time telecom employees, who represent one of the essential input components of the revenue-generating process (i.e., without employees, investments cannot be converted into revenues). The inclusion of the rest of the input and output variables of the DEA model is based on the theoretical framework used in our study, neoclassical growth accounting. Again, the reason for representing the variables *annual telecom investments* and *total telecom services revenues* relative to GDP, total population, total labor force, and total telecom employees was to counter the differences between TEs in terms of their size, population, and level of wealth. At the same time, such an approach allows us to obtain some sort of representation of the structure of the economies within which investments convert to revenues.

Phase 2: CA

In the second phase of our inquiry, we use CA to determine whether the TEs in our sample are similar in terms of their relevant characteristics, as represented by the input and output variables of the DEA model. Thus, more formally, in this part of the study, we test the null hypothesis that there are no discernable clusters of TEs with respect to their level of investments in and revenues from ICT. This hypothesis can be stated as follows:

> **H0:** *The sample of 18 TEs is homogenous in terms of the levels of annual telecom investments and total telecom services revenues.*

Consequently, we reject the null hypothesis if CA results in more than one cluster and, given a set of data points representing a transitional economy over the 10-year period of time, every cluster contains a complete set of data points representing a given economy.

If the results of CA reveal the presence of multiple subgroups in the sample, then we calculate the relative efficiency scores averaged over the 10-year period (identified in phase 1) for each group. We can expect that if heterogeneous subgroups are identified in our sample, then these subgroups have different average relative efficiencies.

Data Used to Perform CA

To perform CA, we reduce the data set used to conduct DEA by removing the *GDP per capita* and *full-time telecommunication staff* variables. While these two variables are important as the inputs of the DEA model, in CA we aim to test homogeneity of the sample in terms of the investments and revenues only; thus, we decided against using these variables in CA. The complete list of the variables used to perform CA is provided below.

Variables used to perform CA:

1. Annual telecom investment per telecom worker (current US$)
2. Annual telecom investment (% of GDP)
3. Annual telecom investment per capita (current US$)
4. Annual telecom investment per worker (current US$)
5. Total telecom services revenue per telecom worker (current US$)
6. Total telecom services revenue (current US$)
7. Total telecom services revenue per worker (current US$)
8. Total telecom services revenue per capita (current US$)

Phase 3: DT

In the third phase of our study, we use DT analysis to identify the most important dimensions that differentiate the heterogeneous subgroups in our sample. The goal of this phase is to identify some of the variables that may be responsible for the differences in average relative efficiencies across the subgroups. In order to do so, we create a new categorical target variable with its domain of values equal to the number of subgroups identified in phase 2. Once the first, most important split is made, we record the name of the variable and the value at which the split was made. After that, we remove that variable from further analysis and repeat the procedure.

Data Used to Perform DT

Before conducting the DT analysis, we identified the largest set of the data available to us; in our analysis, we were able to use 34 variables, which are listed below. Let us recall that, in this study, we are interested in finding a set of factors that may contribute to the efficiency of the utilization of the investments in telecoms by full-time telecom employees. Consequently, the aim of this part of the data analysis is to identify a pool of factors that are possibly complementary to the quantity of full-time telecom employees.

Variable used for DT analysis:

1. Exports of computer, communications, and other services (% of commercial service exports)
2. High-technology exports (% of manufactured exports)

3. Imports of computer, communications, and other services (% of commercial service imports)
4. Military expenditure (% of GDP)
5. Military personnel (% of total labor force)
6. Fixed line and mobile phone subscribers (per 1,000 people)
7. International telecom, outgoing traffic (minutes per subscriber)
8. Internet users (per 1,000 people)
9. Mobile phones (per 1,000 people)
10. Telephone mainlines (per 1,000 people)
11. Telephone mainlines per employee
12. Health expenditure per capita (current US$)
13. Health expenditure, private (% of GDP)
14. Health expenditure, public (% of GDP)
15. Health expenditure, total (% of GDP)
16. Immunization, DPT (% of children age 12–23 months)
17. Immunization, measles (% of children age 12–23 months)
18. Pupil/teacher ratio, primary
19. School enrollment, secondary (% gross)
20. School enrollment, tertiary (% gross)
21. R&D expenditure (% of GDP)
22. Researchers in R&D (% of total labor force)
23. Technicians in R&D (% of total labor force)
24. Roads, paved (% of total roads)
25. Roads, total network (km)
26. Full-time telecommunication staff (% of total labor force)
27. Annual telecom investment (% of GDP in current US$)
28. Urban population (% of total)
29. Urban population growth (annual %)
30. Population growth (annual %)
31. Foreign direct investment, net inflows (% of GDP)
32. GDP growth (annual %)
33. GDP per capita (constant US$2,000)
34. GDP per capita growth (annual %)

Results

Results: DEA

In this section, we describe the results of the DEA. To perform DEA, we used the software application "OnFront," version 2.02, produced by Lund Corporation (http://www.emq.com).

In using "OnFront" to obtain the efficiency scores, we have chosen to use Farrel Input-Saving Measure of Efficiency as a direct efficiency measure for the three types of models: constant return to scale (CRS), variable return to scale (VRS), and non-increasing return to scale (NIRS). In Table 18.1, we provide the scores of the relative efficiency of the 18 TEs averaged over 10 years. The complete results of DEA, with the scores for each year, are provided in Table 18A.1 through 18A.3 in The Appendix.

The results of DEA show that a number of transitional economies in some years obtained a rating of 100% relative efficiency. This, sometimes overly generous assignment of efficiency scores, is a common characteristic of the most DEA models (Lins et al. 2003). Another common characteristic of DEA models is that they tend to evaluate as efficient those DMUs that have the smallest input values, or the DMUs with the largest outputs (Ali 1994).

Table 18.1 Averaged Scores of Relative Efficiency, per Country, per DEA Model

Country	CRS	VRS	NIRS
Albania	0.98	1.00	0.98
Armenia	0.89	0.91	0.89
Azerbaijan	0.73	0.97	0.73
Belarus	0.51	0.68	0.51
Bulgaria	0.92	0.94	0.94
Czech Republic	0.85	0.90	0.86
Estonia	0.96	0.97	0.96
Hungary	1.00	1.00	1.00
Kazakhstan	0.64	0.77	0.67
Kyrgyzstan	0.93	1.00	0.93
Latvia	0.76	0.89	0.76
Lithuania	0.66	0.83	0.67
Moldova	0.92	1.00	0.92
Poland	0.93	0.96	0.93
Romania	0.58	0.70	0.58
Slovakia	0.81	0.84	0.81
Slovenia	0.95	0.98	0.95
Ukraine	0.91	0.93	0.93

Consequently, based only on the results of the DEA analysis, we cannot determine the true nature of the relative efficiency of the TEs in our sample. It is possible that the relatively efficient TEs obtained their status because they are indeed more efficient in the utilization of the inputs than the relatively inefficient ones. It is also possible, however, that the status of being relatively efficient was awarded to the TEs with the lowest levels of investments in telecoms.

Results: CA

Let us recall that we conducted CA with the purpose of identifying the presence of possible differences between the TEs in terms of the levels of investments and the revenues.

The variables subjected to CA are not measured on the same scale, so, prior to CA, the data had to be standardized. We used SAS Enterprise Miner (EM) to perform CA. We started our inquiry by choosing the "Automatic" setting, which did not require any input from the investigator regarding the desired number of clusters.

The "Automatic" setting of EM uses *Standard Least Squares* clustering criterion (which minimizes the sum of squared distances of data points from the cluster means), *Ward's Minimum Variance** as a clustering method, and limits the minimum number of clusters to 2 and the maximum to 40. By beginning with this setting, which resulted in a five-cluster solution, we were able to determine the starting point in our analysis. By requesting fewer and fewer number of clusters, we then gradually derived four-, three-, and two-cluster solutions.

By using CA, we were able to come up with a solution that partitions our data set into two clusters. The membership of each cluster is provided in Table 18.2. One of the clusters contains the data points completely representing Poland, Czech Republic, Hungary, and Slovenia over the 10-year period, while the second cluster contains the data points completely representing Albania, Armenia, Azerbaijan, Belarus, Kazakhstan, Kyrgyzstan, Moldova, Romania, and Ukraine. Thus, these results suggest that we are able to reject the null hypothesis regarding the homogeneity of 18 TEs in terms of investments and revenues from telecoms.

Once the results of CA were obtained, we separated our data set into the two subgroups and calculated the scores of the averaged relative efficiency for each cluster. According to these calculations, one of the clusters, members of which include Czech Republic, Hungary, Poland, and Slovenia, has higher averaged relative efficiency scores than the cluster containing Albania, Armenia, Azerbaijan, Belarus, Kazakhstan, Kyrgyzstan, Moldova, Romania, and Ukraine. Subsequently, we call the first group the "Leaders" and the second group the "Majority."

---

* *SAS System Documentation* offers the following description: "Ward's method tends to join clusters with a small number of observations, and it is strongly biased toward producing clusters with roughly the same number of observations. It is also very sensitive to outliers" (Milligan, 1980).

Table 18.2 Membership of the Two-Cluster Solution

Majority	Leaders
Albania (1993–2002) **Armenia (1993–2002)** **Azerbaijan (1993–2002)** **Belarus (1993–2002)** Bulgaria (1993–2001) Estonia (1993) **Kazakhstan (1993–2002)** **Kyrgyzstan (1993–2002)** Latvia (1993, 1996) Lithuania (1993–1998) **Moldova (1993–2002)** **Romania (1993–2002)** Slovakia (1993, 1994, 1999) **Ukraine (1993–2002)**	Bulgaria (2002) **Czech Rep (1993–2002)** Estonia (1994–2002) **Hungary (1993–2002)** Latvia (1994, 1995, 1997–2002) Lithuania (1999–2002) **Poland (1993–2002)** **Slovenia (1993–2002)** Slovakia (1995–1998, 2000–2002)

These findings are consistent with the results reported by Piatkowski (2003), who studied eight TEs of Europe (Bulgaria, Czech Republic, Hungary, Poland, Romania, Russia, Slovakia, and Slovenia) and concluded that in the period between 1995 and 2000, ICT capital has most potently contributed to output growth in the Czech Republic, Hungary, Poland, and Slovenia.

Let us elaborate some of the importance of the results of CA. Hoskisson et al. (2000) state that even within the same geographic region, emerging market economies are not homogenous, and the differences between the countries make comparisons in small samples problematic. The results of CA, first, confirm that our set of 18 TEs is, indeed, not homogenous. Second, the results demonstrate that heterogeneity of the sample established in this study is very specific to the telecommunication industry. Meaning, it may be possible to produce entirely different groupings in the case of comparison of 18 TEs in terms of the investments in and revenues from, let us say, international tourism, or R&D. It may even be that the 18 TEs are homogenous in some regard, such as, let us suppose, percentage of paved roads. However, we aimed to test the null hypothesis regarding the homogeneity of 18 TEs in terms of the investments in and revenues from the investments in telecoms, and we rejected it based on the results of CA. For the convenience of the reader, we provide summarized results in Table 18.3.

Results: DT

In this part of our analysis, we use DT to identify the characteristics of those TEs, which are the most efficient in utilizing their investments in telecoms. First, we identified the largest (in terms of the number of variables) set of the data that

Table 18.3 Comparison of the Clusters Based on DEA

Criterion for Comparison	"Leaders" Cluster	"Majority" Cluster	Difference	Difference %
Average efficiency score, CRS	0.89	0.79	0.10	12.54%
Average efficiency score, VRS	0.95	0.88	0.07	7.48%
Average efficiency score, NIRS	0.89	0.80	0.09	11.63%

Yearbook of Statistics and *WDI Database* could yield (the complete list of the variables is provided in section "Data Used to Perform DT").

Second, we created a binary dummy variable, which was set as a "target" of the DT analysis. We assigned the value of "1" of the target variable to the countries comprising the "Leaders" cluster, and we assigned the value of "0" to the members of the "Majority" cluster.

The third step consisted of iterative generation of multiple DT models, where every iteration involved the following three steps. The first step consisted of generating a DT model. The second step involved identifying the variable that was used for the top split, as well as recording the split value for that variable. During the third step, the top split variable of a current iteration was taken out from the data set (this was obtained by setting its status to "don't use" in the Decision Tree node of EM). Then the process was repeated.

In our evaluation of the resulting models, we were looking for those variables, splits along which resulted in the cleanest possible separation of the data set according to the value of the target variable. Consequently, after analysis of the generated DT models, we ended up with 15 variables that vividly differentiate the "Leaders" from the "Majority." Moreover, we have calculated the average value of the split variable, as well as the (approximate) percentile within which the value of the split falls.

The results of the DT analysis suggest that on average, "Leaders" have higher

- GDP per capita
- Level of annual telecom investment
- Level of international telecom traffic
- Number of mobile phones

- Number of telephone mainlines
- Percentage of the Internet users among the population
- Number of teachers per pupil in the system of primary education
- Percentage of the total labor force employed as R&D technicians
- Level of spending on health care

At the same time, "Leaders" have lower

- Level of military expenditure
- Percentage of the labor force serving in military

than the "Majority."

All the compiled information allows us to conclude that the "Leaders" appear to be wealthier, in general, than the "Majority," having better infrastructure and smaller armies. In addition to providing some new insights, our findings are congruent with the results of the previous research conducted in the context of TEs.

Cornia and Popov (2001) state that "socialist economies differed considerably among each other in terms of the military sector" and suggest that the resulting "structural distortions" affect the ability of TEs to sustain output during the transition. Results of our inquiry confirm that higher levels of militarization are associated with lower levels of macroeconomic output. It was also noted that the excessive military expenditures are associated with the reduction of investments in social sectors. Our findings demonstrate that higher levels of military expenditure of the "Majority" are associated with their lower levels of spending on health care. Campos and Coricelli (2002) state that a "quality of infrastructure fundamental for the functioning of a market economy, proxied by telephone lines" and human capital play an important role in the macroeconomic growth of TEs. Again, our investigation found evidence that TEs with better infrastructure and higher quality of human capital tend to have higher levels of macroeconomic output.

Based on the results of the DT analysis, we summarize a list of the contributing factors affecting the efficiency of utilization of investments in telecoms, as well as a per factor distribution of the "Majority" and the "Leaders," in Table 18.4.

We also provide the averages for each group and the levels of split in terms of the absolute values for each variable in Table 18.5. The information that was obtained from the analysis of each of the DT models is presented as a graph in Figure 18.1.

The list of contributing factors affecting the efficiency of utilization of investments in telecoms yielded by DT analysis can be reduced to a smaller number of general factors. One of the possible groupings is presented in Table 18.6.

Table 18.4 List of Contributing Factors and Corresponding Distribution per Group of TEs

Variable/Contributing Factor	Distribution Majority	Distribution Leaders
GDP per capita (constant US$2,000)	97% are in bottom 60%	90% are in top 40%
Annual telecom investment (% of GDP)	62% are in bottom 43%	98% are in top 57%
Fixed line and mobile phone subscribers	94% are in bottom 75%	63% are in top 25%
International telecom, outgoing traffic, minutes per subscriber (per 1,000 people)	88% are in bottom 55%	90% are in top 45%
Telephone mainlines (per 1,000 people)	75% are in bottom 43%	100% are in top 57%
Telephone mainlines per employee	65% are in bottom 37%	100% are in top 63%
Health expenditure, public (% of GDP)	100% are in bottom 75%	67% are in top 25%
Health expenditure per capita (current US$)	88% are in bottom 50%	100% are in top 50%
Military personnel (% of labor force)	72.5% are in top 40%	100% are in bottom 60%
Military expenditure (% of GDP)	63% are in top 36%	100% are in bottom 64%
Pupil/teacher ratio (primary)	100% are in bottom 70%	63% are in top 30%
Internet users (per 1,000 people)	75% are in bottom 45%	93% are in top 55%
Mobile phones (per 1,000 people)	60% are in bottom 34%	100% are in top 66%
Technicians in R&D (% of labor force)	50% are in bottom 27%	100% are in top 73%
R&D expenditure (% of GDP)	67% are in bottom 45%	83% are in top 55%

Table 18.5 List of Contributing Factors, Averages in Absolute Values, and Levels of Split

Variable/Contributing Factor	Averages (in Absolute values)		Level of Split
	Majority	Leaders	
GDP per capita (constant US$2,000)	1034.49	4,636.11	2,720.51
Annual telecom investment (% of GDP)	0.60	1.11	1.11
Fixed line and mobile phone subscribers (per 1000 people)	158.78	485.24	351.27
International telecom, outgoing traffic, minutes per subscriber	95.13	98.86	117.08
Telephone mainlines (per 1000 people)	160.46	296.63	249.96
Telephone mainlines per employee	71.06	154.89	122.90
Health expenditure, public (% of GDP)	3.15	5.05	5.00
Health expenditure per capita (current US$)	77.05	383.27	151.00
Military personnel (% of labor force)	1.12	1.95	1.64
Military expenditure (% of GDP)	2.27	1.95	2.09
Pupil/teacher ratio (primary)	18.57	14.22	15.71
Internet users (per 1,000 people)	114.55	27.72	22.86
Mobile phones (per 1,000 people)	329.70	73.37	81.03
Technicians in R&D (% of labor force)	0.06	0.11	0.048
R&D expenditure (% of GDP)	0.52	0.87	0.57

Contribution of the Study

This study makes several contributions to the existing body of knowledge. First, our research was driven by accepted and well-established theoretical frameworks, which have previously been applied successfully in the context of developed countries. It is true that in some areas of inquiry, such as research on strategies in emerging market economies, "theories promulgated for developed market economies may not be appropriate for emerging economies" (Hoskisson et al. 2000). However, we demonstrated that the same theoretical frameworks that drive inquiries in the context of developed countries can drive research on the efficiency of the utilization of resources, effectiveness of the production of revenue, and complementarity of the investments, in the context of TEs.

304 ■ *Theoretical Research Frameworks Using Multiple Methods*

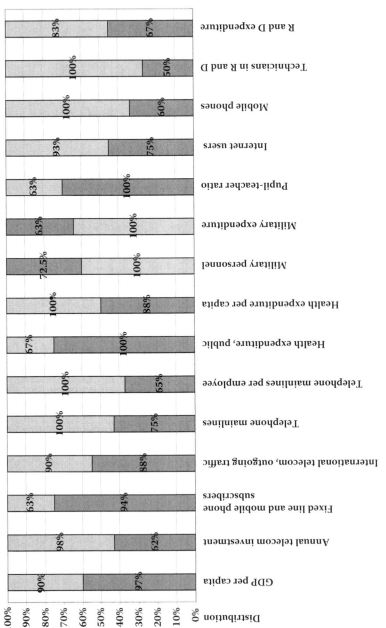

Figure 18.1 "Majority" (dark gray) vs. "Leaders" (light gray): comparison in terms of 15 criteria.

Legend: "GDP per capita –90% of the "Leaders" are in the top 40% while 97% of the "Majority" are in the bottom 60%

Table 18.6 General Factors Contributing to Efficiency of Utilization of ICT Investment in TEs

General Factors	Contributing Factor
Level of economic development	GDP per capita (constant US$2000)
Level of investments in telecoms	Annual telecom investment (% of GDP)
Level of accumulated ICT capital	Telephone mainlines (per 1,000 people) Telephone mainlines per employee Mobile phones (per 1,000 people)
Level of utilization of accumulated ICT capital	Fixed line and mobile phone subscribers International telecom, outgoing traffic, minutes per subscriber Internet users (per 1000 people)
Level of sociotechnical development	Health expenditure, public (% of GDP) Health expenditure per capita (current US$) Pupil/teacher ratio (primary) Technicians in R&D (% of labor force) R&D expenditure (% of GDP)
Level of militarization	Military personnel (% of labor force) Military expenditure (% of GDP)

Second, our research suggests a theory-driven methodology that can be used for the purpose of identification of the pool of possibly complementary factors. We also outlined how our research can be extended (i.e., via translog function) to include the actual test for interaction between the factors in the study.

Third, we corroborate the findings of earlier studies regarding the factors affecting the economic outcomes of investments in ICT in the context of TEs. Namely, we demonstrate that the efficiency of utilization of investments in telecoms, one of the determinants of the economic outcomes of investments, is affected by the factors reflecting the levels of investments in telecoms, level of accumulated telecom capital, and level and quality of the available human capital.

Fourth, in addition to corroborating the results of the previous studies, our research obtained some new empirical findings in the form of a set of factors affecting the level of the economic outcome of investments in telecoms. Namely, we were able to demonstrate that the overall level of economic development, level of utilization of the accumulated telecom capital, and level of militarization of the economy are among the factors that affect the efficiency of the utilization of investments in telecoms.

Fifth, from the theoretical standpoint, this research provides a contribution to the existing body of knowledge by suggesting the additional set of variables that should be included in the model describing the relationship between investments in ICT and the economic outcomes of such investments. This study also sheds some new light regarding the type of the complementary investments that must take place in parallel with investments in ICT.

The results of the study could be of benefit to the community of practitioners as well. Any policy maker or investor should be well served by taking into consideration the multiple factors affecting the economic outcomes of investments in telecoms, acknowledging, therefore, that no significant increase in the level of investments in telecoms should directly result in the similar increase in revenues from such investments.

Moreover, the results of the study also suggest that increase in efficiency of the utilization of the resources should go hand-in-hand with the increase in investments, and vice versa. Meaning, any significant increase in investments in telecoms should be accompanied by the increase in the efficiency of utilization of these investments, while it is unlikely that any significant increase in efficiency of utilization of resources can take place without any additional investments.

This bears an important implication on strategy of investments in telecoms in the context of TEs. The highly industrialized and efficient (in terms of the utilization of resources) TEs can handle, effectively and efficiently, one-time large investments. However, the investments in the context of the less industrialized TEs, characterized by inefficient utilization of resources, should be made gradually, in step-by-step fashion. Consequently, the results of each investment step must be evaluated against the criteria of increasing efficiency of utilization of investments before any additional investments are made.

Summary and Conclusion

In this study, we searched for some of the factors affecting the efficiency of utilization of investments in telecoms in the context of 18 TEs. Use of DEA allowed us to determine the relative efficiency of the utilization of investments by each TE in the sample. By using CA, we were able to demonstrate the presence of the two heterogeneous subgroups within our sample of 18 TEs. By incorporating the results of CA with the results of DEA, we determined the presence of a significant difference between the two groups of TEs in terms of the relative efficiency of the utilization of investments in telecoms. We named the subgroup with the higher averaged relative efficiency the "Leaders" and the subgroup with the lower averaged relative efficiency the "Majority."

By using DT, we were able to demonstrate that the "Leaders" differ from the "Majority" in terms of 15 factors that reflect the differences in the level of economic development, level of investments in telecoms, level of accumulated telecom capital, as well as level of utilization of telecom capital, level of sociotechnical development, and level of militarization of TE. The results of our study strongly suggest that these factors affect the level of the efficiency of the utilization of investments in telecoms in the context of TEs.

Based on the results of our study, we hypothesize that the "Majority" have a lower level of relative efficiency of utilization of investments in telecoms because of the following three reasons: first, the "Majority," in comparison with the "Leaders," simply do not invest enough in telecoms to be concerned with the issue of efficiency. Second, even if the "Majority" do invest at a relatively sufficient level, a large part of the investments is directed not toward obtaining revenue but toward building the required supporting infrastructure. Because only a part of the overall investments is involved in revenue production, it is reflected as an overall low-efficiency score. Third, as the level of the telecom infrastructure increases, it brings about an increased complexity associated with the utilization of the infrastructure. Consequently, the process by which investments are transformed into revenues becomes more complicated, and the "Majority" lacks the necessary sociotechnical "know-how" to manage that increased complexity.

Testing of these hypotheses represents the outline of some of the possible directions for future research in this area. Intuitively, it would make sense to expect a gradual increase in investments as being accompanied by an associated gradual increase in the level of learning regarding the utilization of the investments, in the form of the sociotechnical "know-how." However, at this point, it is not clear whether the increase in efficiency brings about the increase in investments, or whether the additional investments allow for obtaining the higher efficiency of the utilization of the investments.

Despite the contributions that this study makes, our research is not without its limitations. First, the use of the different clustering criteria may produce a clustering solution that is different from the one obtained in this research. Second, during DEA, we used a model where all the input variables and the output variables were weighted equally. However, some of the variables may be more important than other variables and, therefore, such variables perhaps should be weighted more heavily than other variables. Moreover, while we provided rationalization for the DEA model used in our study, ultimately, there are no objective criteria according to which such model can be constructed. Hence, a different DEA model could result in different scores of relative efficiency.

Another limitation of this study is associated with the quantity of the data. Clearly, our research may have offered richer insights if more variables were available from the data sources that we used. This limitation, however, is not unique to our study, but is characteristic to the research in this area in general (Hoskisson et al. 2000). In addition, it may be argued that the time series data covering a 10-year period could be insufficient to inquire into the nature of the events taking place on the macroeconomic level. Nevertheless, in the area of research where it "would appear to be a major need for longitudinal studies" (Hoskisson et al. 2000), we feel that the contributions provided by our study outweigh its limitations.

Acknowledgment

Material in this chapter previously appeared in "Contributing factors to information technology investment utilization in transition economies: An empirical investigation," *Journal of Information Technology for Development* 14:1, 52–75.

References

Aldenderfer, M. S. & Blashfield, R. K. (1984). *Cluster Analysis*. Sage, Newbury Park, CA.

Ali, A. I. (1994). Computational aspects of DEA. In *Data Envelopment Analysis: Theory, Methodology and Applications*, eds. A. Charnes, W. W. Cooper, A. Lewin, & L. M. Seiford, Kluwer Academic Publishers, Boston, pp. 63–88.

Athey, S. & Schmutzler, A. (1995). Product and process flexibility in an innovative environment. *RAND Journal of Economics, The RAND Corporation*, 26(4), 557–574.

Breiman, L., Friedman, J. H., Olshen, R. A., & Stone, C. J. (1984). *Classification and Regression Trees*. Chapman & Hall, New York.

Bugamelli, M. & Pagano, P. (2004). Barriers to investment in ICT. *Applied Economics, Taylor and Francis Journals*, 36(20), 2275–2286.

Campos, N. & Coricelli, F. (2002). Growth in transition: What we know, what we don't and what we should. *Journal of Economic Literature, American Economic Association*, 40(3), 793–836.

Charnes, A., Cooper, W. W., Lewin, A. Y., & Seiford, L. M. (eds.). (1994). *Data Envelopment Analysis: Theory, Methodology, and Applications*. Boston, Kluwer.

Colecchia, A., & Schreyer, P. (2001). The Impact of Information Communications Technology on Output Growth. STI Working Paper 2001/7, OECD, Paris.

Cooper, W. W., Seiford, L. M., & Zhu, J. (2004). Data envelopment analysis: History, models and interpretations. In *Handbook on Data Envelopment Analysis*, eds. W. W. Cooper, L. M. Seiford, & J. Zhu, Kluwer Academic Publishers, Boston, pp. 1–39.

Cornia, G. A. & Popov, V. (2001). Structural and institutional factors in the transition to the market economy: An overview. In *Transition and Institutions: The Experience of Late Reformers*, eds. G. A. Cornia & V. Popov, Oxford University Press, Oxford, UK, pp. 3–29.

Daveri, F. (2002). The new economy in Europe, 1992–2001. *Oxford Review of Economic Policy*, 18(3), 345–362.

Dewan, S. & Kraemer, K. (2000). Information technology and productivity: Evidence from country level data. *Management Science (Special Issue on the Information Industries)*, 46(4), 548–562.

Edgeworth, F. Y. (1881). *Mathematical Psychics*. Kegan Paul, London.

Gera, S. & Wulong, Gu. (2004). The effect of organizational innovation and information and communications technology on firm performance. *International Productivity Monitor, Centre for the Study of Living Standards*, 9, 37–51.

Giuri, P., Torrisi S., & Zinovyeva, N. (2005). ICT, Skills and Organizational Change: Evidence from a Panel of Italian Manufacturing Firms. LEM Papers Series 2005/11, Laboratory of Economics and Management (LEM), Sant'Anna School of Advanced Studies, Pisa, Italy.

Hoskisson, R., Eden, L., Lau, C., & Wright, M. (2000). Strategy in emerging economies. *Academy of Management Journal*, 43(3), 249–267.

Jalava, J. & Pohjola, M. (2002). Economic growth in the new economy: Evidence from advanced economies. *Information Economics and Policy*, 14(2), 189–210.

Jorgenson, D. W. (2001). Information technology and the US economy. *American Economic Review*, 91(March), 1–32.

Jorgenson, D. W. and Stiroh, K. J. (2000). Raising the speed limit: US economic growth in the information age. *Brookings Papers on Economic Activity*, 2(1), 125–211.

Lambertini, L. (2003). The monopolist's optimal R&D portfolio. *Oxford Economic Papers*, 55, 561–78.

Lin, P. & Saggi, K. (2002). Product differentiation, process R&D, and the nature of market competition. *European Economic Review*, 46, 201–211.

Lins, M. P. E., Gomes, E. G., de Mello, J. C. C. B. S., & de Mello, A. J. R. S. (2003). Olympic ranking based on a zero sum gains DEA model. *European Journal of Operational Research*, 148(2), 312–322.

Loukis, E. and Sapounas, I. (2004). The impact of information systems investment and management on business performance in Greece. ECIS 2005 Conference paper.

Milligan, G. W. (1980). An examination of the effects of six types of error perturbation on fifteen clustering algorithms. *Psychometrika*, 45, 325–342.

Murakami, T. (1997). The impact of ICT on economic growth and the productivity paradox. Available online at http://www.tcf.or.jp/data/19971011_Takeshi_Murakami_2.pdf

Myers, D. (2004). *Construction Economics*. Spon Press, UK.

OECD (2004). DAC Network on Poverty Reduction: ICTs and economic growth in developing countries. OECD, Paris: Available on line at http://www.oecd.org/dataoecd/15/54/34663175.pdf

OECD (2005). Good practice paper on ICTs for economic growth and poverty reduction. *The DAC Journal*, 6(3).

Oliner, S. D. & Sichel, D. E. (2002). Information technology and productivity: Where are we now and where are we going? *Economic Review*, 3(3), 15–41.

Ollman, B. (1997). *Market Socialism: The Debate among Socialists*. Routledge, UK.

Piatkowski, M. (2002). The "New Economy" and Economic Growth in Transition Economies. WIDER Discussion Paper No. 2002/63, Helsinki: WIDER.

Piatkowski, M. (2003). Does ICT Investment Matter for Output Growth and Labor Productivity in Transition Economies? *TIGER Working Paper Series*, No. 47. Available online at http://www.tiger.edu.pl

Rosenkranz, S. (2003). Simultaneous choice of process and product innovation when consumers have a preference for product variety. *Journal of Economic Behavior & Organization*, 50(2), 183–201.

Solow, R. (1957). Technical change and the aggregate production function. *Review of Economics and Statistics*, 39(3), 312–20.

Smith, A. (1776). The Wealth of Nations. Available online at http://www.marxists.org/reference/archive/smith-adam/works/wealth-of-nations/

Stiroh, K. (2002). Information technology and the U.S. productivity revival: What do the industry data say? *American Economic Review*, 92(5), 1559–1576.

United Nations Conference and Development (UNCTAD) (2004). The Least Developed Countries Report, United Nations Publications.

United Nations Development Program (UNDP) (1998). *Human Development Report*. Oxford University Press, New York.

UN ICT Task Force Report (2005). Innovation and Investment: Information and Communication Technologies and the Millennium Development Goals. Report Prepared for the United Nations ICT Task Force in Support of the Science, Technologies and Innovation Task Force of the United Nations Millennium Project. Available online at http://www.unicttaskforce.org/

United Nations Statistics Division (UNCD) (1999). Standard country or area codes for statistical use. Statistical Papers, Series M, No. 49 Rev. 4. United Nations, New York.

Van Ark, B., Melka, J., Mulder, N., Timmer, M., & Ypma, G. (2002). ICT Investments and Growth Accounts for the European Union, 1980–2000. Research Memorandum GD-56, Groningen Growth and Development Centre, Groningen. Available online at http://www.eco.rug.nl/ggdc/homeggdc.html

World Bank (2002). Global Economic Prospects and the Developing Countries. World Bank. Available online at http://web.worldbank.org/WBSITE/EXTERNAL/EXTDEC/EXTDECPROSPECTS/GEPEXT/EXTGEP2002

WT/ICT Development Report (2006). Measuring ICT for social and economic development. International Telecommunication Union's World Telecommunication/ICT Development Report, 8th edition. Available online at http://www.itu.int/ITU-D/ict/publications/wtdr_06/index.html

Yearbook of Statistics (2004). Telecommunication Services Chronological Time Series 1993–2002. ITU Telecommunication Development Bureau (BDT), International Telecommunication Union. Available online at http://www.itu.int/ITU-D/ict/publications

Appendix

Table 18A.1 Farrel Input-Saving Measure of Efficiency, CRS

Country	1993	1994	1995	1996	1997	1998	1999	2000	2001	2002
Albania	1	1	1	1	1	1	1	0.80	1	1
Armenia	0.79	1	1	1	1	0.89	0.81	0.73	0.86	0.80
Azerbaijan	0.49	0.60	1	1	1	0.88	0.51	0.84	0.49	0.48
Belarus	0.56	0.31	0.68	0.64	0.56	0.47	0.57	0.44	0.39	0.52
Bulgaria	1	1	1	0.88	0.86	0.83	0.92	1	0.72	1
Czech Republic	1	1	0.78	0.60	0.81	0.97	0.98	0.81	0.68	0.90
Estonia	0.81	0.79	1	1	1	1	1	1	1	1
Hungary	1	1	1	1	1	1	1	1	1	1

(Continued)

Table 18A.1 (Continued) Farrel Input-Saving Measure of Efficiency, CRS

Country	1993	1994	1995	1996	1997	1998	1999	2000	2001	2002
Kazakhstan	0.80	0.40	0.60	0.67	0.75	0.65	0.51	0.54	0.51	1
Kyrgyzstan	1	1	1	0.72	0.56	1	1	1	1	1
Latvia	0.41	0.99	0.94	0.94	1	0.85	0.69	0.67	0.55	0.52
Lithuania	0.82	0.82	0.74	0.85	0.45	0.46	0.50	0.46	0.55	0.96
Moldova	0.22	1	1	1	1	1	1	1	1	1
Poland	0.93	1	0.90	0.76	0.85	1	0.84	1	1	1
Romania	0.55	0.52	0.62	0.59	0.48	0.46	0.55	0.52	0.63	0.90
Slovakia	0.82	0.80	0.79	0.87	0.55	0.51	0.71	1	1	1
Slovenia	1	1	1	1	1	1	1	0.71	0.81	1
Ukraine	1	0.70	0.70	1	0.92	1	1	1	0.85	0.88

Factors to Information Technology Investment Utilization in TEs ■ 313

Table 18A.2 Farrel Input-Saving Measure of Efficiency, VRS

Country	1993	1994	1995	1996	1997	1998	1999	2000	2001	2002
Albania	1	1	1	1	1	1	1	0.95	1	1
Armenia	1	1	1	1	1	0.89	0.83	0.75	0.87	0.80
Azerbaijan	0.82	0.92	1	1	1	0.95	1	1	1	1
Belarus	0.75	0.59	0.77	0.72	0.64	0.60	0.64	0.67	0.66	0.77
Bulgaria	1	1	1	0.93	0.95	0.83	1	1	0.72	1
Czech Republic	1	1	0.80	0.75	0.87	0.97	1	0.95	0.74	0.94
Estonia	0.86	0.83	1	1	1	1	1	1	1	1
Hungary	1	1	1	1	1	1	1	1	1	1
Kazakhstan	1	0.48	0.62	0.70	1	0.65	0.70	0.78	0.75	1
Kyrgyzstan	1	1	1	1	1	1	1	1	1	1
Latvia	0.57	1	1	0.95	1	0.93	0.90	0.94	0.83	0.76
Lithuania	0.87	0.93	0.82	0.87	0.60	0.63	0.81	0.91	0.86	1
Moldova	1	1	1	1	1	1	1	1	1	1
Poland	0.93	1	0.91	0.85	0.93	1	0.96	1	1	1
Romania	0.61	0.63	0.67	0.66	0.63	0.65	0.73	0.75	0.72	0.91
Slovakia	0.84	0.82	0.80	0.87	0.68	0.66	0.74	1	1	1
Slovenia	1	1	1	1	1	1	1	0.96	0.87	1
Ukraine	1	0.70	0.75	1	1	1	1	1	0.86	1

Table 18A.3 Farrel Input-Saving Measure of Efficiency, NIRS

Country	1993	1994	1995	1996	1997	1998	1999	2000	2001	2002
Albania	1	1	1	1	1	1	1	0.80	1	1
Armenia	0.79	1	1	1	1	0.89	0.81	0.73	0.86	0.80
Azerbaijan	0.49	0.60	1	1	1	0.88	0.51	0.84	0.49	0.48
Belarus	0.56	0.31	0.68	0.64	0.56	0.47	0.57	0.44	0.39	0.52
Bulgaria	1	1	1	0.93	0.95	0.83	1	1	0.72	1
Czech Republic	1	1	0.78	0.60	0.81	0.97	1	0.81	0.68	0.90
Estonia	0.81	0.79	1	1	1	1	1	1	1	1
Hungary	1	1	1	1	1	1	1	1	1	1
Kazakhstan	0.80	0.40	0.60	0.67	1	0.65	0.51	0.54	0.51	1
Kyrgyzstan	1	1	1	0.72	0.56	1	1	1	1	1
Latvia	0.41	1	1	0.94	1	0.85	0.69	0.67	0.55	0.52
Lithuania	0.82	0.93	0.74	0.87	0.45	0.46	0.50	0.46	0.55	0.96
Moldova	0.22	1	1	1	1	1	1	1	1	1
Poland	0.93	1	0.90	0.76	0.85	1	0.84	1	1	1
Romania	0.55	0.52	0.62	0.59	0.48	0.46	0.55	0.52	0.63	0.90
Slovakia	0.82	0.80	0.79	0.87	0.55	0.51	0.74	1	1	1
Slovenia	1	1	1	1	1	1	1	0.71	0.81	1
Ukraine	1	0.70	0.70	1	1	1	1	1	0.85	1

Chapter 19

Socioeconomic Impact of Information and Communication Technology Capabilities in Sub-Saharan Economies: Using Association Rules to Describe the Structure of Complex Systems

Overview

In this chapter, we demonstrate an application of concepts of complex systems theory (CST) to investigating the socioeconomic impact of ICT in the context of three groups of sub-Saharan economies. The three groups differ in terms of their levels of income—they are classified by the International Monetary Fund as being low-income, low-middle-income, and upper-middle-income economies of the region.

We utilize the framework of the Networked Readiness Index (NRI), allowing for representation of the internal structure of each of the group in terms of a combination of paths connecting *drivers* and *impacts*. The major premise of this investigation is that a group of economies has its own properties, and those properties are expressed by a specific structure that could be described in terms of a combination of paths between drivers and impacts. The minor premise is that the classification reflects a transition, a development, of a complex system through the stages of development—in this case, economic development. CST informs us that every state of a complex system is supported by a specific structure, and as a system goes through steps in its development, the structure also changes. Consequently, the conclusion is that each group of sub-Saharan economies would have its own distinct structure that could be described in terms of the unique set of paths between drivers and impacts.

Introduction

At this point, it is expected that the impact of information and communication technologies (ICTs) at the country level extends beyond purely economic gains (e.g., via growth in productivity) and into the sphere of social development (Eide 2015). While wealthier economies of the world may look toward optimization of the economic impact of ICT, poorer countries of sub-Saharan Africa (SSA) should be in a position of reaping a transformational level of socioeconomic benefits of ICT. It is hard to determine whether the transformational impact within the context of SSA is indeed taking place, but we could start the assessment by investigating the link between the state of ICT and its socioeconomic impacts. The premise is that for a sustained transformational impact to take place, an economy needs to obtain and maintain an efficient path of transforming ICT capabilities into socioeconomic outcomes.

Benchmarking is one of the tools by which improvements in efficiency could be obtained, and we suggest that this tool could be utilized by SSA economies to improve the performance of their ICT capabilities. However, it is a bridge too far for SSA economies to benchmark developed countries outright, for the disparity in the levels of accumulated and developed ICT infrastructure and annual investments is too great to disregard. Consequently, we suggest that as a first step, we investigate the efficiency of socioeconomic impact of ICT capabilities within a group of SSA. Such investigation would entail identification of the economies that are more efficient in obtaining the socioeconomic benefits of ICT, and then proceeding with identifying the characteristics of such economies vis-à-vis characteristics of the less efficient economies, all within the context of SSA.

We feel that such inquiry is well justified because "…the complex relationships between ICTs and socioeconomic performance are not fully understood and their causality not fully established" (Di Battista et al. 2015, p. 4). We formulate the overall goal of this investigation as follows:

> The purpose of the inquiry is to investigate the presence of complex relationships between the state of ICT and the socioeconomic impacts of ICT in the context of sub-Saharan economies.

We aim to achieve the overall goal by accomplishing two research objectives. The first objective of this investigation is as follows:

> To develop a methodology allowing for uncovering context-specific complex associations between combinations of factors reflecting the state of ICT and socioeconomic impacts of ICT

By achieving the first goal, we equip ourselves with a tool allowing for investigating the existence of nonobvious combinations of factors within various contexts. Once the methodology is developed, we apply it in action to achieve our second objective, which is as follows:

> To conduct an empirical investigation in the context of sub-Saharan economies with the purpose of uncovering some of the complex associations between the factors reflecting the state of ICT and its socioeconomic impact.

Resultantly, by conducting this investigation, we aim to contribute to the existing body of knowledge in the area of information and communication technologies for development (ICT4D) in more than one way. First, we develop a methodology allowing for identifying a combination of characteristics describing various groups of SSA. While decision tree analysis could be performed to identify the factors specific to various groups of economies (Samoilenko & Osei-Bryson 2014), this technique is not well suited for identifying *combinations* of factors. In this study, we demonstrate how association rule mining (ARM) could be used to identify a set of attributes differentiating various groups of SSA economies. Second, while the previous inquiries concentrated either on economic or on social impacts of ICT capabilities, our study aims to be more comprehensive in this regard—we investigate both the social and economic impact of ICT capabilities within the same sample of SSA economies. Finally, the results of empirical analysis should provide valuable information to policy and decision makers working in the area of ICT4D within the context of SSA.

We conduct our investigation within the context of 27 economies of SSA, using the data set for the period of 2011–2014. The analysis of the data is supported by a three-phase methodology utilizing data envelopment analysis (DEA) and ARM.

Research Framework

In our investigation, we rely on the framework of networked readiness (Dutta et al. 2015), the adapted version of which is depicted in Figure 19.1. The framework relies on 4 subindexes and their 10 subcategories (or *pillars*) to obtain the value of the NRI, which reflects the capacity of economies to benefit from ICT. An increase in the value of NRI for a given economy is indicative of the increase of the impact of ICT on innovation and productivity (Dutta & Jain 2003). Interestingly, the original framework does not explicitly connect the *environment*, *readiness*, and *usage* subindexes (referred to as *drivers* within the framework) with the *impact* subindex (referred to as *impact*), despite relying on a principle that "…the environment, readiness, and use—interact, co-evolve, and reinforce each other to create greater impact" (Di Battista et al. 2015, p. 4).

We scope our inquiry by only considering relationships between drivers (environment, readiness, and use) and impact (socioeconomic impact of drivers)—as indicated by arrows in Figure 19.1. All possible interactions within drivers (e.g., between environment, readiness, and use), we consider to be beyond the scope of our investigation. In our inquiry, we use the framework depicted in Figure 19.1 to investigate, via DEA, the efficiency of the process by which the environment, readiness, and usage subindexes impact the two subcategories of the Impact subindex—*economic* and *social* impacts. Once we identify the better and worse performers of the sample, we could use ARM, via *market basket analysis* (MBA), to attempt to identify a combination of subcategories describing the groups.

Data

We obtained the data from a publicly available source—the World Economic Forum Global Information Technology Report (GITR) 2015 (available online at http://reports.weforum.org/global-information-technology-report-2015/network-readiness-index/). In 2012, the representation of NRI has partially changed in terms of the number and representation of the pillars of three subindexes of NRI; it was also the year when the impact subindex was introduced. Given the changes that took place between 2011 and 2012, we decided to concentrate on the new version of NRI and collect the data provided in GITR 2012, 2013, 2014, and 2015. In some cases, the representation of SSA economies was inconsistent—for example, we could not include Angola, Seychelles, Liberia, Gabon, Sierra Leone, and Guinea in our sample because the data for some of the years was missing. We were able to compile the data set representing 27 economies of SSA (the classification of the International Monetary Fund as of October 2014). The sample consists of 14 low-income economies, 9 low-middle economies, and 4 upper-middle economies (the classification of the World Bank as of July 2014). Membership of each group of the sample is provided in Table 19.1.

Socioeconomic Impact of ICT in Sub-Saharan Economies ■ 319

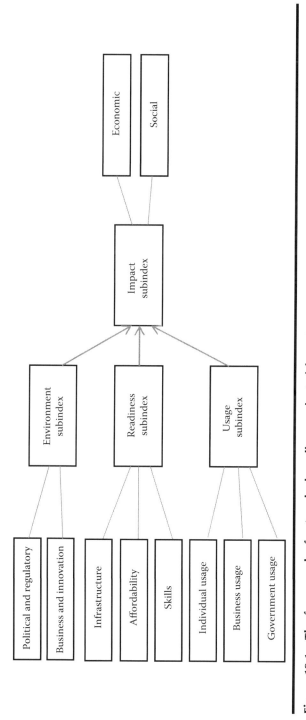

Figure 19.1 The framework of networked readiness, adapted format.

Table 19.1 Sample of Sub-Saharan Economies, by Income Level

Income Level	Sub-Saharan Economies
Low income	Burkina Faso, Burundi, Chad, Ethiopia, Gambia, Kenya, Madagascar, Malawi, Mali, Mozambique, Rwanda, Tanzania, Uganda, Zimbabwe
Low-middle income	Cameroon, Cape Verde, Côte d'Ivoire, Ghana, Lesotho, Nigeria, Senegal, Swaziland, Zambia
Upper-middle income	Botswana, Mauritius, Namibia, South Africa

Methodology

We employ a three-phase methodology that relies on DEA and ARM to support our investigation. We describe each phase next.

Phase 1: DEA

During the first phase, we rely on DEA to evaluate the relative efficiency of three *Drivers->Impact* paths. We will use the *variable return to scale* (VRS) DEA model to conduct the analysis, for it is reasonable to argue that SSA economies have not yet reached the point of developing a level of ICT infrastructure allowing accrual of the benefits yielded by capitalizing on economies of scale.

Given a 4-year time period, we will run DEA 12 times. Consequently, for each economy in the sample, we are going to have four scores of relative efficiency for each of the three models. At this point, we need to provide a justification for the inputs and outputs included in our models. With regard to outputs, the reasoning is intuitive—first, we would like to assess the efficiency of the overall impact, and then, each type of the impact separately. This is because an economy could be efficient in obtaining one type of impact (e.g., economic) and not efficient with regard to another impact (e.g., social).

With regard to the choice of the inputs of DEA, model justification our approach is methodological (Table 19.2). While we are free to use eight subcategories of drivers as inputs of a DEA model, the general rule of thumb is that for a reasonable level of discrimination, the number of economies (or *decision-making units* in DEA terms) must be at least twice the product of inputs and outputs (Dyson et al. 2001). In our case, we have a sufficient number of economies in our set, but if we use a DEA model with eight inputs and two outputs, then we would need to have at least 2*8*2 = 32 economies in the sample.

Furthermore, and more importantly, the greater the number of factors included in the DEA model, the lower the level of discrimination of the model (Dyson et al. 2001). However, we would like to use all the data available to us so we could inquire, for example, whether relatively efficient economies are

Table 19.2 DEA Models of the Study

DEA Model	Inputs of DEA Model	Outputs of DEA Model
Drivers->Overall_Impact (DOI)	Environment subindex Readiness subindex Usage subindex	Impact subindex
Drivers->Economic_Impact (DEI)	Environment subindex Readiness subindex Usage subindex	Economic impact subcategory
Drivers->Social_Impact (DSI)	Environment subindex Readiness subindex Usage subindex	Social impact subcategory

characterized by specific combinations of subcategories of drivers. We would use ARM to do so, but this would require transformation of the data that we perform in phase 2.

Phase 2: Data Transformation for ARM

In our data set, all of the values for NRI, subindexes, and subcategories are expressed in the form of continuous variables. To perform Association Rules (AR) analysis, we need to convert the data available to us into a set of transactions, where each transactions represents a list of items. There are various ways according to which such conversion could take place, and we leave the specifics of determining the conversion key to be under the purview of a domain knowledge expert.

Because one of the major goals of this study is to develop and test a methodology, we concentrate on illustrating the overall approach that we advocate and dedicate less attention to the justification of the conversion key. In our case, we decided to use a quartile-based conversion, for this approach presents the data in easy-to-understand categories (e.g., first quartile or first 25% = "low," third quartile or 50–75% = "midhigh"). We used Microsoft Excel 2013 to perform the conversion, and we offer an illustration of the results in Table 19.3.

Once the data is converted into a new format, we need to add another variable, "transactionID"—this variable comes from the analysis of the results of DEA. We use the scheme of "Economy_Income_Level_Relative_Efficiency" to create the variables for the new attribute. For example, a relatively efficient low-income economy is encoded as "LI1," and a relatively inefficient upper-middle-income economy receives a value of "UM0."

We also create a variable that reflects the income level of the country—this is with the purpose of discovering patterns of subindexes that differentiate the economies of SSA in the sample. In this case, the encoding is based on the classification

Table 19.3 Quartile-Based Data Conversion: Illustration of the Original versus Cnverted Data

Economy/Subindex	Environment	Readiness	Usage	Impact
Original Data (2015)				
Botswana	3.93	3.26	3.43	3.09
Burkina Faso	3.41	2.15	2.73	2.85
Burundi	2.86	2.61	2.08	2.11
Converted Data (2015)				
Botswana	Midhigh	Midhigh	High	Midhigh
Burkina Faso	Low	Low	Midlow	Midlow
Burundi	Low	Midlow	Low	Low

of the World Bank—"LI" for "low income," "LM" for "low middle," and "UM" for "upper middle."

Phase 3: Market Basket Analysis via ARM

The purpose of phase 3 is to find possible patterns, associations, or causal structures that may exist in our data. One of the main advantages of MBA is that it is suitable for undirected data mining; thus, we will aim to discover naturally occurring associations between the factors (subindexes of drivers and impact)—components of NRI. MBA could be classified as being either confirmatory or exploratory in nature. In the case of our investigation, we employ exploratory MBA, for we do not have any theoretical support for why certain relationships between the subindexes of NRI should exist. A very common approach to generating associations between the variables, or *itemsets*, via MBA involves using the *Apriori* algorithm (Agrawal & Ramakrishnan 1994)—we will rely on this approach in the current investigation.

Research Questions and Null Hypotheses of the Study

At this point, we can operationalize two objectives of this investigation in the form of the specific research questions and corresponding null hypotheses.

The first research question operationalizes the first objective as follows:

> Whether the developed methodology is capable of generating sets of association rules for a given set of criteria

One of the possible criteria is associated with the level of income of economies (e.g., low income vs. low middle vs. upper middle), while another criterion could be a relative level of efficiency (e.g., relatively efficient vs. relatively inefficient) of the *Drivers->Impact* path, and the third one could be a combination of the first two (e.g., relatively efficient low income vs. relatively inefficient low income).

We can answer this research question by testing the corresponding null hypothesis:

> H01: The methodology will fail to generate sets of association rules for a given set of criteria.

We will test H01 under the minimal conditions of *support* > 20%, *confidence* > 1.0, and *lift* > 1.0.

The second research question operationalizes the second objective as follows:

> Whether the choice of context-specific criteria impacts the combination of factors describing relationships between drivers and impact of ICT

Basically, we would like to find out if the different contexts (e.g., efficient low income, inefficient low middle income, upper middle income, etc.) could be characterized via different sets of criteria—this allows us to inquire into the specificity of a setting expressed as a combination of subcategories of NRI.

We will answer the second research question after testing the second null hypothesis of the study:

> H02: No combination of factors contained in generated association rules would be unique to a given context.

The simple side-by-side comparison of the generated association rules will serve as a sufficient criterion for testing H02.

Results of the Data Analysis

Phase: 1 DEA

We offer a summary of the results from our application of the DEA method. If a given economy has been determined to be relatively efficient at least three times over a period of 4 years, we have labeled such economy as "efficient" for the whole period of 4 years. Because our economies fall within three distinct groups—LI, LM, and UM—we also determined the relative efficiency of each economy over the 4 years within its group—we will use this information in phase 3 when we perform MBA.

Results demonstrated that seven economies out of the full sample are relatively efficient with regard to the impact of drivers on social, economic, and overall impact of ICT. Additionally, we identified relatively efficient economies for each of the income-level groups; in some cases (e.g., Burundi, Chad, Kenya, Mali, Rwanda, and Senegal), the economies that are relatively efficient within their group are also efficient overall. In other cases (e.g., Swaziland, Lesotho, Botswana, Mauritius, Namibia, and South African Republic), the economies that are relatively inefficient overall end up being efficient within their respective group (Table 19.4).

Phase 2 and Phase 3: Data Transformation and MBA

The results summarized in Table 19.5 allow us to test our null hypotheses.

First, the results allow us to reject H01, for the application of MBA did result in generating multiple association rules under the criteria of *support* > 20%, *confidence* > 1.0, and *lift* > 1.0.

Second, the results also allow us to reject H02, for the rules generated by MBA contain context-specific combinations of factors.

We also offer a summary of some of the impact-specific rules for low-income, low-middle-income, and upper-middle-income economies in Table 19.6.

Discussion of the Results

The results of the data analysis offered in the previous section offer evidence that we successfully addressed two research questions of this study; namely, we developed and tested a methodology allowing for investigating complex context-specific relationships between the factors reflecting the state and the impact of ICT capabilities. The discussion of the results is presented along the four important points that we considered noteworthy—these are presented in no particular order of significance.

First,

> It does appear that while the relationships between the subindexes and subcategories of NRI are complex and idiosyncratic, the common themes could be traced.

We point out that while a variety of association rules have been generated for a different set of criteria, a common line could also be glanced—some subcategories of NRI subindexes (e.g., related to skills, business, individual usage, political and regulatory environment, etc.) appear more frequently than other subcategories.

Second,

> Categorization of SSA economies based on income level has a counterpart in the categorization of SSA economies based on NRI level.

Table 19.4 Results of DEA

Income Level	Economy	DOI, Overall	DEI, Overall	DSI, Overall	DOI, within Its Income Group	DEI, within Its Income Group	DSI, within Its Income Group
LI	BFA	x	x	x	–	x	–
LI	BDI	x	x	x	x	x	x
LI	TCD	x	x	x	x	x	x
LI	ETH	x		x	–	–	x
LI	GMB	–	–	–	x	–	–
LI	KEN	x	x	x	x	x	x
LI	MDG	–	–	–	x	–	–
LI	MWI	–	–	–	–	–	–
LI	MLI	x	x	x	x	x	x
LI	MOZ	–	–	–	–		–
LI	RWA	x	x	x	x	x	x
LI	TZA	–	–	–	–	–	–
LI	UGA	–	–	–	–	–	–
LI	ZWE	–	–	–	–	–	–

(Continued)

Table 19.4 (Continued) Results of DEA

Income Level	Economy	DOI, Overall	DEI, Overall	DSI, Overall	DOI, within Its Income Group	DEI, within Its Income Group	DSI, within Its Income Group
LM	CMR	–	–	–	x	x	x
LM	CPV	–	–	x	–	–	x
LM	GHA	–	–	–	–	–	–
LM	NGA	–	x	–	x	x	–
LM	SEN	x	x	x	x	x	x
LM	SWZ	–	–	–	x	x	x
LM	ZMB	–	–	–	–	–	–
LM	CIV	–	–	–	x	x	–
LM	LSO	–	–	–	x	x	x
UM	BWA	–	–	–	x	x	x
UM	MUS	–	–	–	x	x	x
UM	NAM	–	–	–	x	x	x
UM	ZAF	–	x	–	x	x	x

Note: x, relatively efficient economy.

Socioeconomic Impact of ICT in Sub-Saharan Economies ■ 327

Table 19.5 Results of Market Basket Analysis

Condition	Generated Rules			Sup.	Conf.	Lift
Low income	low POL&REG_ENV, low ECON_IMP	=>	low SOCIO_IMP	25%	1.0	2.7
	low GOV_USE, low SOCIO_IMP	=>	low POL&REG_ENV	25%	1.0	3.5
	low INFR_READ, low BUS_USE	=>	low SKILL_READ	29%	1.0	2.1
Low middle	midlow ECON_IMP	=>	midhigh BUS_USE	21%	1.0	2.5
Upper middle	high INFR_READ, high SKILL_READ	=>	high IND_USE	63%	1.0	1.15
	high POL&REG_ENV, high SKILL_READ	=>	high IND_USE	75%	1.0	1.15
	high POL&REG_ENV, high INFR_READ, high SKILL_READ	=>	high IND_USE	63%	1.0	1.15
Low income, inefficient	low INFR_READ, low BUS_USE	=>	low SKILL_READ	36%	1.0	1.9
	low IND_USE, low BUS_USE	=>	low SKILL_READ	32%	1.0	1.9

(Continued)

Table 19.5 (Continued) Results of Market Basket Analysis

Condition	Generated Rules			Sup.	Conf.	Lift
Low income, efficient	low SOCIO_IMP, low ECON_IMP	=>	low GOV_USE	40%	1.0	2.33
	low SOCIO_IMP, low ECON_IMP	=>	low POL&REG_ENV	40%	1.0	2.33
	low GOV_USE, low SOCIO_IMP, low ECON_IMP	=>	low POL&REG_ENV	40%	1.0	2.33
	low POL&REG_ENV, low GOV_USE	=>	low SOCIO_IMP	43%	1.0	2.33
Low middle, inefficient	midlow INFR_READ, midhigh SKILL_READ	=>	midhigh ECON_IMP	50%	1.0	1.7
	midhigh POL&REG_ENV, midhigh SKILL_READ	=>	midhigh ECON_IMP	50%	1.0	1.7
	midhigh POL&REG_ENV, midlow INFR_READ, midhigh SKILL_READ	=>	midhigh ECON_IMP	42%	1.0	1.7

(Continued)

Table 19.5 (Continued) Results of Market Basket Analysis

Condition	Generated Rules			Sup.	Conf.	Lift
Low middle, efficient	low BUS_USE, low GOV_USE	=>	low SOCIO_IMP	25%	1.0	4.0
	low BUS_USE, low GOV_USE	=>	low ECON_IMP	25%	1.0	3.0
	low GOV_USE, low SOCIO_IMP, low ECON_IMP	=>	low BUS_USE	25%	1.0	3.0
	low BUS_USE, low GOV_USE, low ECON_IMP	=>	low SOCIO_IMP	25%	1.0	4.0

Table 19.6 Impact-Specific Rules for Low-Income and Low-Middle-Income Economies

Condition	Generated Rules			Sup.	Conf.	Lift
Low income	low POL&REG_ENV, low ECON_IMP	=>	low SOCIO_IMP	25%	1.0	2.7
	low POL&REG_ENV, low GOV_USE	=>	low ECON_IMP	25%	0.9	2.3
	low IND_USE, low BUS_USE	=>	low ECON_IMP	21%	0.7	1.9
	low SKILL_READ, low BUS_USE	=>	low ECON_IMP	21%	0.6	1.5
Low middle	midhigh BUS_USE	=>	midlow SOCIO_IMP	20%	0.5	2.5
	midlow INFR_READ	=>	midhigh SOCIO_IMP	25%	0.6	1.9
	midhigh POL&REG_ENV	=>	midhigh SOCIO_IMP	25%	0.6	1.7
	midhigh SKILL_READ	=>	midhigh SOCIO_IMP	25%	0.5	1.5
	midhigh INFR_READ	=>	high ECON_IMP	20%	0.6	2.9
	midlow GOV_USE	=>	midhigh ECON_IMP	20%	0.7	1.8

(Continued)

Table 19.6 (Continued) Impact-Specific Rules for Low-Income and Low-Middle-Income Economies

Condition	Generated Rules			Sup.	Conf.	Lift
Upper middle	high BUS&INNOV_ENV, high AFFORD_READ, high GOV_USE	=>	high SOCIO_IMP	32%	1.0	2.3
	high SKIL_READ, high AFFORD_READ, high GOV_USE	=>	high SOCIO_IMP	32%	1.0	2.3
	high POL&REG_ENV, high AFFORD_READ, high GOV_USE	=>	high SOCIO_IMP	31%	1.0	2.3
	high BUS&INNOV_ENV, high INFR_READ, high BUS_USE	=>	high ECON_IMP	31%	0.85	2.3
	high BUS&INNOV_ENV, high IND_USE, high BUS_USE	=>	high ECON_IMP	31%	0.85	2.2
	high POL&REG_ENV, high BUS&INNOV_ENV, high INFR_READ, high BUS_USE, high GOV_USE	=>	high ECON_IMP	31%	0.85	2.2

Results suggest that wealthier economies tend to have higher scores of NRI subindexes, and vice versa. The presence of a simple association between the level of income of an economy and the scores of NRI subindexes seems to be apparent.

Third,

> There are no miracles with regard to the impact of ICT capabilities.

We were not able to generate any association rules that would produce a relationship of the type *low Drivers->high Impact*. Hence, we suggest that one of the ways of improving the level of impact of *ICT drivers* is by improving the level of ICT drivers. We were not able to generate any rules that produced a result of the type *low Drivers->high Impact* or *high Drivers->low Impact*.

Fourth,

> Benchmarking should not be rejected as a tool for improving the impact of ICT capabilities.

Results summarized in Table 19.6 suggest the presence of a pattern, where higher scores of ICT drivers are associated with higher scores of ICT impacts. We also observe that poorer economies tend to have lower scores of drivers and impacts, and wealthier economies tend to have higher scores of drivers and impacts. While it is premature to draw a broad conclusion, it is reasonable to suggest that in order for poorer economies to obtain similar levels of impacts available to wealthier economies, they should engage in benchmarking and concentrate on raising their scores of drivers to levels comparable to their higher-income counterparts.

Conclusion

In this investigation, we developed and applied a methodology allowing for generating sets of association rules from the combination of factors describing relationships between drivers and impact of ICT. The results of the data analysis do confirm the notion that the relationships between the factors representing drivers and impact are indeed complex. However, the underlying complexity of the relationships could be made more transparent to researchers and practitioners by the methodology developed in this study.

While it is too early to answer the question of "Why do certain relationships exist?," at this point, we are well equipped to answer the question of "What relationships do exist?" As a result, we hope that this study contributed to the overall body of knowledge in the area of ICT4D by increasing the transparency of the process by which ICT capabilities impact socioeconomic outcomes in various contexts.

Acknowledgment

Material in this chapter previously appeared in "Disparity of Social and Economic Impact of ICT Capabilities in Sub-Saharan Economies: Empirical Investigation of Differentiating Factors" in Proceedings of the SIG GlobDev Pre-ECIS Workshop ICT in Global Development, Istanbul, Turkey, June 12, 2016.

References

Agrawal, R. & Ramakrishnan, S. (1994). Fast algorithms for mining association rules in large databases. In Proceedings of the 20th International Conference on Very Large Data Bases (VLDB '94), eds. Jorge B. Bocca, Matthias Jarke, and Carlo Zaniolo, Morgan Kaufmann Publishers Inc., San Francisco, pp. 487–499.

Di Battista, A., Dutta, S., Geiger, T., & Lanvin, B. (2015). The networked readiness index 2015: Taking the pulse of the ICT revolution. *Global Information Technology Report*, 3–28.

Dutta, S. & Jain, A. (2003). The networked readiness of nations. In *The Global Information Technology Report 2002–2003*, eds. S. Dutta, B. Lanvin, & F. Paua, Oxford University Press, New York, Oxford.

Dutta, S., Geiger, T., & Lanvin, B. (eds.) (2015). Global Information Technology Report 2015. ICTs for Inclusive Growth. Geneva: World Economic Forum and INSEAD. Available online at http://www3.weforum.org/docs/WEF_Global_IT_Report_2015.pdf, retrieved April 20, 2015.

Dyson, R. G., Allen R., Camanho A. S., Podinovski, V. V., Sarrico C. S., & Shale, E. A. (2001). Pitfalls and protocols in DEA. *European Journal of Operational Research*, 132, 245–259.

Eide, E. B. (2015). Preface. In *Global Information Technology Report*, p. v.

Samoilenko, S. & Osei-Bryson, K. M. (2014). Formulation of context-dependent and target-specific strategies of the impacts of ICT on development. In Proceedings of the 7th Annual SIG GlobDev Pre-ICIS Workshop ICT in Global Development; Auckland, New Zealand, December 14, 2014.

Chapter 20

Improving the Relative Efficiency of Revenue Generation from ICT in Transition Economies: A Product Life Cycle Approach

Introduction

Investments in information and communications technologies (ICTs) have been shown to have a significant macroeconomic impact in the context of developed countries (OECD 2001, 2005; Colecchia & Schreyer 2001, 2002; Jorgenson 2001; Jorgenson & Stiroh 2000; Oliner & Sichel 2002; Stiroh 2002; Van Ark et al. 2002; Daveri 2002; Jalava & Pohjola 2002). In transition economies (TEs), however, macroeconomic outcomes of investments in ICT have been mixed. TEs are economies that have recently changed or are in the process of changing from planned economies to free market economies, such as the countries of the former Soviet Bloc (Roztocki & Weistroffer 2008). Some of these countries, viz., Poland, Czech Republic, Hungary, and Slovenia, seem to be able to benefit from investments in ICT to a greater extent than other TEs that struggle to exhibit any significant results of such investments on a macroeconomic scale (Piatkowski 2003).

Revenue generation serves as one of the major means by which investments in ICT contribute to macroeconomic growth (UN ICT Task Force Report 2005; WT/ICT Development Report 2006). Consequently, improving the efficiency of revenue creation is a possible route by which the macroeconomic impact of investments in ICT may be increased. This route may require, first, identifying existing inefficiencies in the process of revenue production, second, determining where in the process of revenue creation intervention is required, and, third, choosing the appropriate course of action that will lead to improvements in efficiency. We suggest, therefore, that the process of increasing the level of efficient production of revenues can be broken down into three steps: *identification, isolation,* and *intervention*. During the first step, the presence of inefficiencies in the process of generation of revenues is identified. During the second step, the isolation of the target area takes place, the target area being the period in the process of revenue generation where an improvement in efficiency shows the most promise. Finally, during the intervention step, the most effective measures or actions for improving efficiency are determined.

The justification of this approach is intuitive, for even in the case of a one-time investment in ICT, i.e., an investment at a single point in time, it is likely that revenues from investments are generated over a period of time. Furthermore, we argue that the sector of ICT is very likely service- or product-driven, where a substantial portion of the revenues from investments in ICT is associated with the sales of a (possibly new) product; this area of revenue production is the subject of our inquiry. More specifically, we can limit the scope of this investigation to that part of the revenue production process that is associated with the sale of an ICT product. Consequently, one of the assumptions of our investigation is that the country-level data on investments and revenues from ICT are representative of investment/revenue dynamics associated with the sale of an ICT product. In this study, we define an *ICT product* as any *ICT-related product or service introduced for the purposes of satisfying customer needs*. Consequently, because the revenues are generated over the whole life cycle of the ICT-related product or service, the level of effectiveness of revenue generation may change depending on the stage within the life cycle of the product.

In accordance with the suggested three-step approach, we propose three research questions. The first question deals with finding the least efficient part of the process of revenue generation:

> *Which part of the revenue production process is characterized by minimum efficiency?*

The second question aims at the identification of the most effective area of intervention within the revenue production process with respect to the return on investment. The reasoning behind the second research question is that an intervention during the least efficient stage is not necessarily the most investment-efficient action. This is because the efficiency of utilization of resources also varies during the different stages of the revenue production process. Thus, we are searching for the stage in the revenue production process during which the allocated resources can be utilized with the greatest efficiency, i.e., produce "most bang for the buck." We state our second research question as follows:

> *Which part of the revenue production process exhibits maximum efficiency with respect to utilization of additional resources?*

Finally, the third research question deals with the issue of intervention. The reasoning behind the third research question is that the level of investments and the level of the efficiency of the revenue production process are two factors that positively impact the outcome of the process of the production of revenues. However, it is not clear which policy is the most effective, increasing the level of investments, or allocating additional resources to improve the efficiency of the process of conversion of investments into revenues. Consequently, we state our third research question as follows:

> *Given the current level of investments in ICT, should additional resources be allocated as additional investments in ICT, or should these resources be used to increase the efficiency of the current investments?*

To answer these research questions, we investigate investments in telecommunications technology (telecoms), as representative of investments in ICT, in the context of 18 TEs, using data envelopment analysis (DEA), cluster analysis (CA), and neural networks (NNs). We also incorporate into our investigation the product life cycle (PLC) model to determine where in the process of revenue production intervention holds the most promise. For the purposes of our study, an investment in telecoms

> …refers to expenditure associated with acquiring the ownership of telecommunication equipment infrastructure (including supporting land and buildings and intellectual and non-tangible property such as computer software). These include expenditure on initial installations and on additions to existing installations. (Yearbook of Statistics 2004)

We chose to use investments in telecoms because they represent an important subset of investments in ICT and because they constitute a common type of investments regularly made by almost any economy in the world.

The theoretical framework supporting our inquiry is based on the framework of neoclassical growth accounting and the PLC model.

Theoretical Framework

PLC Model

We adopt a definition of the PLC model as a "time dependent model of the volume of sales and earnings during different stages of the life of a certain product" (Bescherer 2005). The PLC model is commonly represented as consisting of either four or five stages. A typical rendering of a five-stage model is provided in Figure 20.1. A four-stage PLC model is similar to the five-stage model, with the difference that a four-stage model integrates the *Saturation* stage into the *Maturity* stage.

The actual *Sales* curve may vary from the generic shape of the general PLC curve as shown in Figure 20.1. Despite such variations, however, the distinctive S-shape of the PLC *Sales* curve is commonly observed. Indeed, Hauser et al. (2006), after review of Rogers (2003), Sultan et al. (1990), Van den Bulte and Stremersch (2004), and Mahajan et al. (1995), point out an emerging consensus among researchers that the *Sales* curve over PLC is usually S-shaped and that the "S-shaped curve seems to hold for successive generations of the product." Consequently, in this paper, we assume that the actual PLC *Sales* curve is S-shaped.

Let us briefly review the five stages of this model. During the *Introduction* stage, penetration of the market by the product takes place. This stage is characterized by a period of slow growth. In the context of TEs, however, this stage of the PLC may compare favorably to the general context (viz., developed economies), as TEs are less likely to develop a completely new product from scratch, but rather may introduce

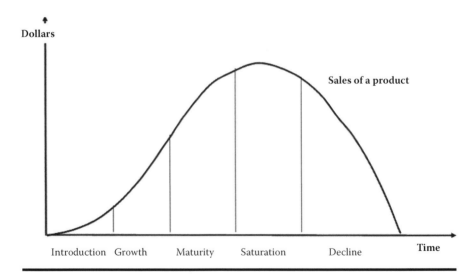

Figure 20.1 Five-stage product life cycle model. (Adapted from Meffert, H. (2000). Marketing—Grundlagen marktorientierter Unternehmensführung, Konzepte—Instrumente—Praxisbeispiele, 9th edition. Gabler-Verlag, Wiesebaden.)

a product that has already been developed and tested in other markets, possibly in developed countries (Gielens & Steenkamp 2004; Gruber & Verboven 2001a,b).

While it is unlikely that any telecom-related product generates a significant profit during the *Introduction* stage, the *Growth* stage represents a period of rapid growth of sales and, consequently, growth of revenues. It is during the *Growth* stage that economies of scale become the driver of the increase in output production; therefore, in this stage, additional investments are often made to increase market share and production capacities. This stage of the PLC in the context of TEs may compare unfavorably to the general model of developed economies, as TEs have a lower level of economic and technological maturity relative to developed countries, and it is the level of economic wealth that determines the pattern of growth (Stremersch & Tellis 2004; Talukdar et al. 2002). Consequently, a lack of the required production capacity, as well as difficulties in achieving economies of scale, may adversely affect the growth in sales and profits at this stage.

In the *Maturity* stage, profits continue to increase, albeit at a slower rate than in the *Growth* stage. During this stage, it is common to see investments associated with modifications of products and with improvements in the production process, as well as improvements in overall quality of the products. In the context of TEs, this stage may compare favorably with the general model, as it is likely that TEs make use of already established technologies, and the quality of the products is already high (Hall 1980; Anderson & Zeithaml 1984). On the other hand, it means that in TEs the peak level of sales will be determined not so much by the effort expanded during the *Maturity* stage, but by the production capacities developed in the *Growth* stage.

The peak in the level of sales of the product characterizes the *Saturation* phase. Once that peak is reached, a negative growth rate settles in. As stated earlier, in the four-stage PLC model, this phase is part of the *Maturity* stage.

During the *Decline* stage, sales of the product start falling, and so do the revenues from the sales of the product. In the context of TEs, this stage may compare unfavorably to the general model, as the global marketing strategies used to extend the life of the product in other settings (viz., developed countries) may not always account for the specific context of TEs, and thus be suboptimal (Stremersch & Tellis 2004).

Despite the popularity of the PLC model, it only represents a general pattern of a product from its growth to decline; it should not be taken as a *prescriptive* standard for making decisions. Nevertheless, Thorelli and Burnett (1981) suggest that the *normative* characteristics of the PLC model are of value to researchers and practitioners alike, while Magnan et al. (1999) identify it as a desirable planning framework.

Description of the Data and the Background of the Study

The data for this study were obtained from two sources: the *World Development Indicators* database, which is the World Bank's comprehensive database on development data (web.worldbank.org/wbsite/external/datastatistics), and the *Yearbook of Statistics*, which is the annual publication of *International Telecommunication Union* (ITU; www.itu.int/ITU-D/ict/publications).

In our choice of TEs to include in the study, we aimed to find a group of countries that started the transition process at approximately the same time, and decided on the following 18 TEs: Albania, Armenia, Azerbaijan, Belarus, Bulgaria, Czech Republic, Estonia, Hungary, Kazakhstan, Kyrgyzstan, Latvia, Lithuania, Moldova, Poland, Romania, Slovakia, Slovenia, and Ukraine.

Overall, we were able to construct 10-year time-series data spanning the period from 1993 to 2002. We decided to begin our analysis with the year 1993 because we believe that year provides a common starting point for the economic transition in the countries in our group. Our reasoning here is that it took a year from the dissolution of the Soviet Bloc in 1991 for the transition process to start, and using year 1992 as a starting point would favor "early starters."

In a previous study (Samoilenko & Osei-Bryson 2007), CA was used to determine that the sample of 18 TEs is not homogenous in terms of the investments and revenues from telecoms (the variables used for CA are shown in Table 20A.1 in the Appendix); the results yielded the two-cluster solution presented in Table 20.1. Using DEA, the authors were also able to determine that Cluster 2, which includes Czech Republic, Hungary, Poland, and Slovenia, has a higher average relative efficiency score than Cluster 1 (Samoilenko & Osei-Bryson 2007). Thus, cluster 2 was designated as the *Leaders* and cluster 1 as the *Followers*.

These results are in agreement with observations by Hoskisson et al. (2000) that even within the same geographic region, emerging market economies are not homogenous. Indeed, the results strongly suggest that the 18 TEs are not homogenous in terms of investments and revenues from telecoms, and corroborate the

Table 20.1 Results of CA: Two-Cluster Solution

Cluster 1 (the Followers)	*Cluster 2 (the Leaders)*
Albania (1993–2002), Armenia (1993–2002)	Bulgaria (2002), Czech Rep (1993–2002)
Azerbaijan (1993–2002), Belarus (1993–2002)	Estonia (1994–2002),
Bulgaria (1993–2001), Estonia (1993)	Hungary (1993–2002),
Kazakhstan (1993–2002), Kyrgyzstan (1993–2002)	Latvia (1994, 1995, 1997–2002)
Latvia (1993, 1996), Lithuania (1993–1998)	Lithuania (1999–2002)
Moldova (1993–2002), Romania (1993–2002)	Poland (1993–2002), Slovenia (1993–2002)
Slovakia (1993,1994, 1999), Ukraine (1993–2002)	Slovakia (1995–1998, 2000–2002)

Table 20.2 Variables for the DEA Model

Input Variables	Output Variables
GDP per capita (in current US$)	Total telecom services revenue per telecom worker
Full-time telecommunication staff (% of total labor force)	Total telecom services revenue (% of GDP)
Annual telecom investment per telecom worker	Total telecom services revenue per worker
Annual telecom investment (% of GDP)	Total telecom services revenue per capita
Annual telecom investment per capita	
Annual telecom investment per worker	

findings of Piatkowski (2003) that in the period "between 1995 and 2000 ICT capital has most potently contributed to output growth in the Czech Republic, Hungary, Poland, and Slovenia." The current investigation incorporates these previous results of CA and proceeds further with DEA and NN, for which we identified a model consisting of six input and four output variables, presented in Table 20.2. The main goal in performing DEA is to determine how efficient the 18 TEs are in converting investment inputs into revenue outputs. Therefore, we do not consider any other types of inputs or outputs, such as those related to infrastructure, capabilities, utilization, etc.

Overview of the Methods and Techniques Used in the Study

The component data analytic techniques that we use in our methodology are CA, DEA, and NNs. We summarize the general purposes of these techniques as well as their specific use in the current study in Table 20.3.

While these three nonparametric techniques that we employ in our investigation are well established and commonly utilized in IS research, they are rarely used in a stand-alone fashion. Instead, investigators often combine multiple techniques to inquire into research problems from different vantage points provided by different methods. Review of the research literature shows that not only is DEA a widely used method for the purpose of evaluating productivity and performance (e.g., Khouja 1995; Shao & Lin 2001; Samoilenko & Green 2008; Bollou & Ngwenyama 2008), but also that it is commonly combined with other techniques, such as CA (e.g., Shin & Sohn 2004; Hirschberg & Lye 2001; Lemos et al. 2005), NNs (e.g., Samoilenko & Osei-Bryson 2008a; Çelebi & Bayraktar 2008; Emrouznejad & Shale 2009; Mostafa 2009; Wu 2009), decision trees (e.g., Samoilenko & Osei-Bryson 2007;

Table 20.3 Summary of Techniques Used in the Study

Method	General Purpose	Outcome	Use in This Study
CA	To identify naturally occurring subgroups (clusters) within a data sample	Subgroups of the sample that differ with respect to the variables representing the entities in the sample	To identify the presence of subgroups of TEs, which differ in terms of the levels of investments and revenues from telecoms (to represent ICT)
DEA	To evaluate the relative efficiency of the conversion of inputs into outputs by each entity (DMU) in a sample	Scores for the relative efficiency for each entity (DMU) in the sample	To determine scores for the relative efficiency of the process of converting ICT investments into revenues by each of the subgroups identified earlier by CA
NNs	To model the relationships between inputs and outputs	A "black-box" model of the process by which inputs are transformed into outputs	To construct a model for the transformation of ICT investments into revenues for each of the subgroups (as identified by CA) with different levels of relative efficiency (as calculated by DEA)

Samoilenko 2008; Wu 2009), and regression analysis (e.g., Cooper & Tone 1997; Bollou & Ngwenyama 2008; Parthasarathy & Anbazhagan 2008; Samoilenko & Osei-Bryson 2008b). Thus, the use of a multitechnique type of methodology is not novel to our inquiry. Furthermore, because such nonparametric methods as DEA, CA, and NN are not bound by the strict assumptions typical of parametric methods, and do not alter, but simply present different perspectives on the data of the sample, combining these methods in a sequential fashion does not present a threat of introducing potential problems with multiplicative confounding errors.

Methodological Approach

In order to answer the research questions of this study, we employ the five-phase approach depicted in Figure 20.2.

The combination and sequence in which we use CA, DEA, and NN in this study is important. We use CA (phase 1) first to identify distinct clusters within our sample of TEs. We use DEA (phases 2 and 3) to determine relative efficiencies for all members in our sample, and thereby determine which cluster (determined earlier by CA) is on average more efficient than the others (i.e., which cluster should be designated as the "Leaders"). However, at this time, we do not know whether the differences in the scores for relative efficiency obtained in phases 2 and 3 are due to differences in the levels of inputs and outputs, or due to differences in the efficiency of the processes by which inputs are converted into outputs. Using NN (phase 4) and DEA again (phase 5), we can determine whether the relative inefficiencies are resource- or process-driven, and consequently suggest an appropriate course of action. NN in phase 4 determines the "transfer functions," i.e., the processes by which the two clusters, the *Leaders* and the *Followers*, convert inputs (investments in telecoms) into outputs (revenues from telecoms). Switching these transfer functions between the two clusters then allows us to obtain simulated results for each cluster showing outputs for the *Leaders* using the conversion process of the *Followers*, and vice versa. Finally, DEA in phase 5 allows us to compare these simulated results (inputs and outputs) to determine the best course of action for the Followers to take, in order to improve their production of revenues.

The five phases are summarized in Table 20.4 and described in more detail in the following subsections.

Phase 1: CA

The purpose of this phase (phase 1 in Figure 20.2) is to test the homogeneity assumption of DEA and to identify subgroups (clusters) within the set of TEs. A previous study by Samoilenko and Osei-Bryson (2007) found two clusters with homogenous members, but that differed substantially from each other (see Table 20.1).

344 ■ *Theoretical Research Frameworks Using Multiple Methods*

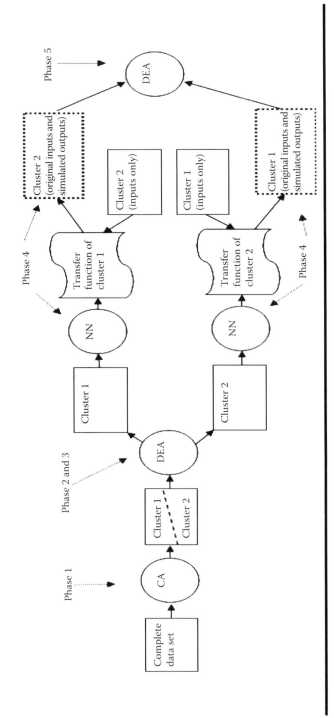

Figure 20.2 Illustration of the research approach.

Improving the Efficiency of Revenue Generation from ICT in TEs ■ 345

Table 20.4 Outcomes of the Phases of the Proposed Methodology

Phase	Method Used	Outcome
Phase 1	CA	One or more clusters of TEs (two in our study)
Phase 2	Output-oriented DEA	Scores of averaged relative efficiency for each cluster
Phase 3	Input-oriented DEA	Scores of averaged relative efficiency for each cluster
Phase 4	NNs	(1) A transfer function representing the process of converting inputs (investments in telecoms) into outputs (revenues) for each cluster (2) Simulated outputs for each cluster based on the transfer function derived from the other cluster
Phase 5	Output-oriented DEA	Scores of averaged relative efficiency for each cluster based on the original inputs and simulated outputs

Phase 2: Output-Oriented DEA

The purpose of this phase (phase 2 in Figure 20.2) is the identification of inefficiencies in the process of revenue production. The results of the output-oriented DEA allow us to calculate average scores for the relative efficiencies of the TEs in each cluster; we do this for CRS, VRS, and NIRS output-oriented DEA models. Consequently, the results of phase 2 allow us to determine the subset of TEs with the highest average relative efficiency, which we will call the *Leaders* group of the sample. The other groups are labeled as *Followers* (in our study, we are dealing only with two clusters, as obtained in phase 1, one cluster being the Leaders group and the other the Followers group).

Once the average relative efficiencies of the *Leaders* and *Followers* groups have been obtained, we are able to determine which assumption regarding the return to scale (viz., CRS, VRS, or NIRS) yields the greatest difference. Keeping in mind the S-shape of the PLC *Sales* curve, and the fact that such curves contain regions of increasing, constant, and decreasing returns to scale, we are now able to map the relative efficiencies of the *Leaders* and *Followers* groups obtained under conditions of CRS, VRS, and NIRS to the appropriate areas on the S-curve. We do so by parsing the general PLC curve into segments corresponding to constant, variable, and non-increasing returns to scale. This allows us to identify the area on the PLC sales curve where the relative inefficiency of the *Followers* is most pronounced in comparison to the *Leaders* group. Consequently, the identified area is labeled as the area of intervention, for this is the area of the greatest inefficiency of revenue production on the PLC curve.

Phase 3: Input-Oriented DEA

The purpose of this phase (phase 3 in Figure 20.2) is the isolation of the area of the production of revenue where an intervention will be applied. Input-oriented DEA is employed to determine the area of the least relative inefficiency (or greatest relative efficiency) on the PLC curve in terms of the utilization of resources. In order to do so, we follow a similar sequence of steps as outlined in phase 2, that is, we determine under what assumption of return to scale the difference between the three pairs of averaged scores is the smallest. By mapping this area of the smallest difference to the PLC curve, we can determine the area on the PLC curve where additional resources can be utilized with the greatest efficiency.

Phase 4: NN Simulation

In this phase (phase 4 in Figure 20.2), we determine NN models for the process of converting inputs (e.g., investments in telecoms) into outputs (e.g., revenues from telecoms) for each of the clusters identified in phase 1. Input variables from the DEA model of the two previous phases will serve as input nodes, and output variables as the output nodes of the NN models. In the case of the two-cluster solution in our study, we end up with two NN models, one for the *Followers* and one for the *Leaders*. The resultant NN models represent "black-box" models for the process of converting inputs into outputs for each cluster. The overall goal of this phase is to obtain simulated outputs for each of the clusters by switching inputs and conversion models, i.e., by using the inputs from one cluster with the conversion model of the other cluster.

Thus, this approach allows us to use the NN model created for the *Followers*, and the input values of the *Leaders* to simulate what the outputs would have been for the *Followers*, if they had the input resources of the *Leaders* but used their own, less efficient process of conversion. Similarly, we can use the NN model created for the *Leaders* and the input values of the *Followers* to simulate what the outputs would have been for the *Leaders*, if they had the input resources of the *Followers* but used their own, more efficient process of conversion (see Figure 20.3). Thus, we end up with three sets of outputs: (1) the original data, (2) the hypothetical outputs of the *Leaders* based on the conversion process of the *Followers*, and (3) the hypothetical outputs of the *Followers* based on the conversion process of the *Leaders*.

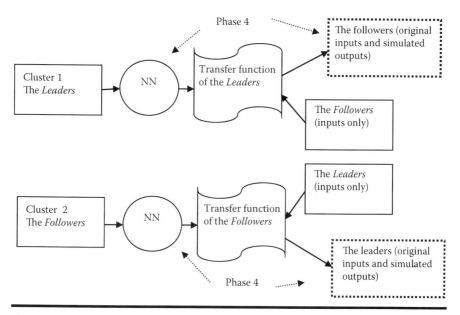

Figure 20.3 Illustration of the NN simulation.

Phase 5: Output-Oriented DEA of the Simulated Data

The purpose of the last phase of our methodology (phase 5 in Figure 20.2) is the identification of the appropriate type of intervention that should be applied to the production of revenues in the area of the PLC curve that was isolated in phase 3. In the fifth phase of our methodology, we conduct DEA on the data sets obtained in phase 4. By comparing the resultant averages of the scores of the relative efficiencies, we can identify the simulated set that produces the smallest difference in terms of the levels of relative efficiencies. This allows us to identify whether the TEs in the less efficient cluster (e.g., the *Followers*) will benefit more from an increase in the level of inputs (e.g., investments in telecoms), or whether they benefit more from improving the process by which inputs are converted into outputs (e.g., revenues from telecoms). We summarize the results from each phase in our approach in Table 20.5.

Discussion of Results of the Data Analysis

Results of Phase 1

Based on the findings of a previous study (Samoilenko & Osei-Bryson 2007), we divided our data into two subsets, the *Leaders*, a group of TEs that have a higher

Table 20.5 Summary of the Results from the Proposed Approach

Phase	Data Analytic Method	Purpose
Phase 1	CA	Identification of the heterogeneous subsets (clusters) of the sample
Phase 2	Output-oriented DEA	Identification of the area of the greatest inefficiency of revenue production on the PLC curve
Phase 3	Input-oriented DEA	Identification of the area of the smallest inefficiency of resources utilization on the PLC curve
Phase 4	NN simulation	Simulated outputs based on, first, increased level of inputs, and, second, increased level of relative efficiency of conversion of inputs into outputs
Phase 5	Output-oriented DEA	Identification of the changes associated with, first, increase in the level of inputs and, second, increase in the level of relative efficiency of conversion of inputs into outputs

level of investments and revenues from telecoms, and the *Followers*, a group of TEs characterized by much lower levels of investments and revenues from telecoms (see Table 20.1). The results of CA offered evidence that the sample of TEs is not homogenous in regard to the levels of capital investments in telecoms, i.e., the two clusters differed substantially from each other. This finding is important, as it has been suggested that the differences in capital stock and levels of investments impact the macroeconomic outcomes of investments in ICT (Piatkowski 2002). However, at this point, it is not clear whether the TEs with different levels of inputs and outputs also differ in terms of the relative efficiency of the conversion of inputs into outputs. This is a question important to the national development of any country, for prior to a significant increase in the level of investments, policy makers must possess information regarding the efficiency of utilization of investments and the production of revenues, in order to being able to gauge the possible impact of the additional investments. For example, the results of the CA indicate that Moldova and Hungary belong to two different subgroups of TEs. Thus, should policy makers in Moldova advocate an increase in investments in ICT in order to benchmark Hungary's level of revenues, or should they invest in improving their processes?

Results of Phase 2

Phase 2 of our approach utilized output-oriented DEA and served the purpose of answering the first research question of the study:

Which part of the revenue production process is characterized by minimum efficiency?

Results of phase 2, summarized in Table 20.6, provide evidence that the *Leaders* have a higher level of relative efficiency in the production of revenues from telecoms than the *Followers* under any assumption of the returns to scale (viz., CRS, VRS, or NIRS). (The actual results of the DEA are provided in Tables 20A.1 through 20A.3 in the Appendix of the paper.) The data show that for every dollar that the *Leaders* make in revenues from telecoms, the *Followers* on average only make about 80 cents, i.e., 20% less. However, at this point, we still cannot advise a policy maker regarding the appropriate course of action directed at improving the process of production of revenues, for it is not clear what phase of the process is the most inefficient and should be targeted first.

As discussed earlier, we assume that the PLC *Sales* curve is S-shaped, and that revenues from investments in telecoms follow a general PLC curve with different returns to scale during different times of the PLC. By parsing the PLC curve into three separate areas, one for variable returns to scale (VRS), one for constant returns to scale (CRS), and one for non-increasing returns to scale (NIRS), we can now use our results to identify the area of the greatest relative inefficiency in the production of revenues by the *Followers*. To justify this partitioning, we note that it is commonly accepted that the sales of new products do not transition smoothly from *Introduction* phase to *Maturity* phase; instead, there is a *takeoff point* that signifies a transition from a "long introduction period when sales linger at low levels" to the period of "rapid growth, associated with a huge jump in sales" (Tellis et al. 2003).

Using the data from Table 20.6, we note that the percentage scores for relative efficiency of the *Followers* versus the *Leaders* (assuming 100% benchmark of the *Leaders*) are 76.33% for CRS, 83.99% for VRS, and 78.33% for NIRS. By placing these percentage scores on the corresponding areas of the curve (see Figure 20.4), we see that the area of greatest relative inefficiency, which in our case corresponds to the CRS part of the curve, spans from somewhere in the middle of the *Growth* stage to somewhere in the middle of the *Maturity* stage of the *Sales* curve.

We argue that the part of the *Growth* phase after *takeoff point*, where the rate of growth may be higher than 400% (Golder & Tellis 1997), is better suited to be modeled under the assumption of constant, rather than variable, return to scale. We are not making, however, any assumptions regarding the length of the period over which assumption of CRS might hold. Based on the results depicted in Figure 20.4, we suggest that the area of the greatest relative inefficiency of the *Followers* is after the takeoff

Table 20.6 Comparison of the *Leaders* and *Followers* Based on Output-Oriented DEA

Criterion for Comparison	Leaders Cluster	Followers Cluster	Difference	Difference %	Efficiency of Followers vs. Leaders
Average efficiency score, CRS	1.94	2.54	0.60	23.67%	76.33%
Average efficiency score, VRS	1.89	2.25	0.36	16.01%	83.99%
Average efficiency score, NIRS	1.89	2.41	0.52	21.67%	78.33%

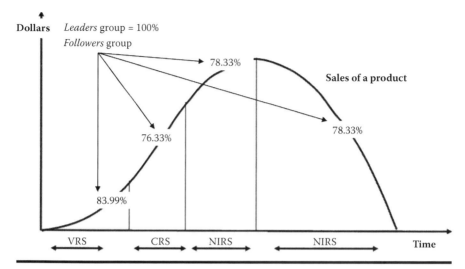

Figure 20.4 Identification of greatest relative inefficiency in revenue production.

point, where "rapid growth requires extensive resources in terms of advertising, sales staff, manufacturing, distribution, and inventory support" (Tellis et al. 2003).

This insight is important for economic development of TEs, for it offers a policy maker a vantage point for assessing the nature of the possible intervention directed at the improvement of the production of revenues. If Belarus, for example, benchmarks Poland in terms of production of revenues, then policy makers must know what set of action are needed in order to improve Belarus' stream of revenues. That is, where along the *Sales* curve of the PLC model should the intervention take place? Taking into consideration insights offered by our CA and DEA at this point, we can suggest that policy makers in Belarus should direct their attention to sustaining an effective and efficient growth of the market for the product at the time of rapid increase in consumer demand, as this is the phase where Belarus lags Poland the most. Overall, it appears that while the *Followers* are most effective and efficient during the *Introduction* phase, possibly due to the demands of the internal markets for new offerings, they fail to effectively and efficiently develop economies of scale (see Tables 20A.7 and 20A.8, discussed later in the chapter), and sustain the momentum during the development and the maintenance of the market.

Results of Phase 3

Phase 3 of our approach involved conducting input-oriented DEA and served the purpose of answering the second research question:

> *Which part of the revenue production process exhibits maximum efficiency with respect to utilization of additional resources?*

The results of the input-oriented DEA provide evidence that the *Leaders* have a higher level of relative efficiency of the utilization of investments in telecoms than the *Followers* under any assumption of returns to scale (viz., CRS, VRS, or NIRS). The disparity in the levels of efficiency of utilization of investments, however, is significantly less than the disparity in the case of the production of revenues. Nevertheless, the data show that for every dollar invested, the *Followers* lose up to 15 cents due to input inefficiencies.

Based on the results summarized in Table 20.7, we note that the percentage scores for relative efficiency of the *Followers* versus the *Leaders* (again, assuming 100% benchmark of the *Leaders*) are 84.97% for CRS, 99.79% for VRS, and 84.93% for NIRS. Following a similar process as described for phase 2 and placing these scores on the *Sales* curve (see Figure 20.5), we see that the area of greatest relative efficiency of utilization of resources is the VRS part of the curve, which includes the *Introduction* stage and part of the *Growth* stage of the *Sales* curve.

Consequently, we suggest that the *Followers* are most efficient in terms of utilization of resources during the penetration of the market by a product and up to the take-off point. It is reasonable to suggest that the *Followers* lack the capability of efficiently building economies of scale, for their level of relative efficiency of utilization of resources drops significantly upon the entry into the CRS region. We decided to investigate this issue further and conducted input- and output-oriented DEA with the purpose of obtaining the averaged values of scale efficiency scores for the *Leaders* and the *Followers*. The summary of the results is presented in Table 20.8; the results seem to support our suggestion.

The information obtained in this phase is very useful in the case that a policy maker must advise on the appropriate course of action of improving the generation of revenues, while keeping the cost of intervention down. If Kazakhstan, for example, benchmarks Estonia in terms of the production of revenues, then according to phase 2, *Growth* is the most appropriate area of intervention. However, in the case of limited resources available for this intervention to Kazakhstan, the relative impact of the allocated resources becomes more important. Resultantly, the resources should be dedicated to improvements in the *Introduction* phase of the PLC, because during this period, additional investments can be utilized most efficiently; it seems reasonable to advise against allocating the limited intervention resources to improvements in the *Growth* phase, as during that period, every 15 cents out of a dollar is wasted due to inefficiencies.

Results of Phases 4 and 5

The purpose of phases 4 and 5 of our approach was to address the third research question of our study and to identify the appropriate type of intervention directed

Table 20.7 Comparison of the *Leaders* and *Followers* Based on Input-Oriented DEA

Criterion for Comparison	Leaders Cluster	Follcwers Cluster	Difference	Difference %	Efficiency of Followers vs. Leaders
Average efficiency score, CRS	0.62	0.54	0.08	15.03%	84.97%
Average efficiency score, VRS	0.73	0.73	0.00	0.21%	99.79%
Average efficiency score, NIRS	0.62	0.54	0.08	15.07%	84.93%

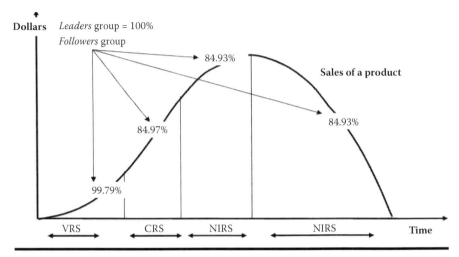

Figure 20.5 Identification of greatest relative efficiency of resource utilization.

Table 20.8 Comparison of the *Leaders* and the *Followers* in Terms of the Scale Efficiency

Criterion for Comparison	Leaders	Followers	Difference
Averaged scale efficiency score, input-oriented DEA	0.94	0.89	4.96%
Averaged scale efficiency score, output-oriented DEA	1.09	1.24	12.15%

at increasing the level of the relative efficiency of the production of revenues. The third research question was stated as

> *Given the current level of investments in ICT, should additional resources be allocated as additional investments in ICT, or should these resources be used to increase the efficiency of the current investments?*

We would like to elaborate on the importance of the results of phases 4 and 5. Based on the results of CA and DEA, we determined that the *Followers* cluster is characterized by lower levels of investments and revenues from telecoms, as well as by lower levels of averaged relative efficiencies of utilization of investments and production of revenues. However, this information, by itself, is not sufficient to determine the path that will allow the *Followers* to increase the level of revenues from telecoms, for it does not indicate whether the lower levels of revenues from telecoms by the *Followers* are caused by the insufficient levels of investments in telecoms, or whether they are caused by the inefficient processes of conversion of investments into revenues. The results provided in phases 4 and 5, however, allow us to discern which of the factors is affecting the level of the production of revenues the most.

Let us place this issue within the context of our sample more specifically. Consider a case of the less efficient Kyrgyzstan vs. the more efficient Slovenia: Kyrgyzstan invests less in telecoms than Slovenia, and Kyrgyzstan receives less revenue from telecoms than Slovenia. If Kyrgyzstan aims to benchmark Slovenia and to increase the level of revenues from telecoms, then what course of action should a policy maker in Kyrgyzstan advocate: allocate additional resources in the form of additional investments in telecoms, or allocate additional resources to improve the processes by which investments are converted into revenues? This is an important question for economic development in any country: In order to improve an economic bottom line, should one simply invest more, with the expectation that the increase in investments will translate into an increase in revenues, or should one address the existing inefficiencies first, and then perhaps increase the level of investments?

Let us again turn our attention to the empirical results of DEA that are provided in the Appendix of the paper. By themselves, the results are important if we want to simply compare the relative efficiencies of our TEs for a given year, or for a 10-year period, or even if we want to compare the averaged scores for the two groups of TEs. However, DEA alone does not allow us to obtain any insights as to why the averaged scores differ. Moreover, even after conducting input- and output-oriented DEA, all that we get is the information that inefficient TEs should be able to produce a higher level of revenues utilizing the same level of inputs (see Tables 20A.5 through 20A.7), or, conversely, to achieve the same level of revenues while utilizing a lower level of inputs (see Tables 20A.2 through 20A.4). Similarly, the differences in the scores of relative efficiency under constant (see Tables 20A.2 and 20A.5), variable (see Tables 20A.3 and 20A.6), and non-increasing (see Tables 20A.4 and 20A.7) returns to scale provide us with important, but insufficient information for decision-making and policy formulating purposes. The multitechnique, multistep methodology followed in the investigation reported on in this paper allows us to build on the results of DEA and to gain a greater insight into the nature of the inefficiencies (see Tables 20A.8 and 20A.9).

The summary of the results of phase 4 (NN simulation) and phase 5 (DEA) is provided in Tables 20.9 and 20.10. The values from Table 20.9 refer to the utilization of additional investments by the *Followers* (keeping their current processes in place), whereas the values from Table 20.10 refer to the efficiency of improving the processes of the *Followers* (keeping their current levels of investments).

We now place the percentage differences of the *Followers* taken from Tables 20.9 and 20.10, on the corresponding areas of the PLC *Sales* curve (see Figure 20.6). By comparing the magnitudes of the percentage differences, we conclude that the *Followers* can derive greater benefit from allocating additional resources not toward additional investments in ICT, but rather toward increasing the level of efficiency of the process of converting investments into revenues.

Table 20.9 Comparison of Clusters Based on Output-Oriented DEA: Simulated Outputs of *Leaders* Based on the Conversion Processes of the *Followers*

Criterion for Comparison	Leaders (Simulated)	Followers (Actual)	Difference	Difference %
Average efficiency score, CRS	2.09	2.30	−0.21	−9.20%
Average efficiency score, VRS	1.38	2.00	−0.62	−30.87%
Average efficiency score, NIRS	1.38	2.17	−0.79	−36.26%

Table 20.10 Comparison of Clusters Based on Output-Oriented DEA: Simulated Outputs of *Followers* Based on the Conversion Processes of the *Leaders*

Criterion for Comparison	Leaders (Simulated)	Followers (Actual)	Difference	Difference %
Average efficiency score, CRS	2.04	1.62	0.42	25.62%
Average efficiency score, VRS	1.79	1.14	0.65	57.32%
Average efficiency score, NIRS	1.80	1.14	0.65	57.45%

The implications of these findings are significant as they suggest that prior to increasing the level of investments in telecoms, the *Followers* should turn their attention to improving the efficiency of the processes of utilization of investments and production of revenues. Moreover, the results of the data analyses conducted in phases 4 and 5 suggest that the *Followers* should concentrate specifically on improving the revenue generation processes during the *Maturity*, *Saturation*, and *Decline* stages of the PLC.

Within the context of our illustrative example comparing Kyrgyzstan and Slovenia, the results indicate that in order to improve the production of revenues, Kyrgyzstan should not allocate any additional resources to simply increase the level of investments; instead, a policy maker of Kyrgyzstan should promote a policy of improving the quality of the processes by which investments are converted into revenues first. Furthermore, the insights provided by our data analysis suggest that the most effective allocation of resources is to improve the processes for the period after the *Introduction* phase of the PLC.

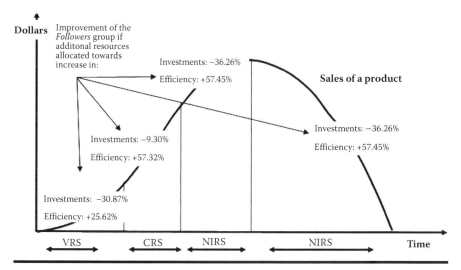

Figure 20.6 Identification of the Appropriate Factor for Efficiency of Revenue Production.

Conclusions

Summary of Results

Overall, the insights provided by our approach and our data analyses show that

- The 18 TEs in our sample form two distinct groups that differ substantially in terms of investments in telecoms and revenues from telecoms.
- The group of TEs with the higher levels of investments and revenues, i.e., the *Leaders* group, also is relatively more efficient in utilizing these investments and producing revenues than the *Followers* group.
- The *Followers* are least efficient in revenue production during the PLC period associated with the development of economies of scale.
- In order to improve the production of revenues, the *Followers* should concentrate on improving the efficiency of the revenue production processes during the *Maturity*, *Saturation*, and *Decline* stages of the PLC.

Thus, the results of our investigation allow us to summarize the answers to our research questions as follows:

> *While the Followers utilize the resources most efficiently during the Introduction phase of the PLC, it is during the Maturity, Saturation, and Decline phases that they generate revenues most inefficiently; these inefficiencies should be addressed not by means of additional investments but by means of improving the process of conversion of investments into revenues.*

Contribution

In this study, we presented an innovative approach to determine a strategy for TEs to improve their relative efficiency in the production of revenues from investments in ICT. We demonstrated our approach using an example of investments in telecoms and revenues from telecoms in the context of 18 TEs.

We approached the research questions posed in the introduction of this paper from a neoclassical growth accounting perspective, while incorporating into our study another established model, that of the PLC. Thus, one contribution of our study is in demonstrating how a general theoretical model can be used in a specific, context-dependent undertaking for strategy formulation. Another contribution of this study is methodological. We outlined a novel approach based on several well-known, established, and reliable data analytic methods, which can easily be employed by researchers and practitioners in the field alike.

From a practical perspective, the results of our analysis indicate that TEs that lag behind in their utilization of ICT may benefit more from dedicating additional resources toward improving the processes that accompany their current technology investments, rather than investing in more technology. Furthermore, the improvement of processes seems especially important in the later stages, i.e., after the introduction stage, of the technology investments. This insight may be valuable for macroeconomic planning in TEs that have difficulty in catching up. Though our analysis was limited to TEs, it is quite likely that similar conclusions apply to other developing economies that do not seem to be able to reap the expected benefits out of their ICT investments.

Limitations

One of the limitations of this study lies in the subjectivity of the concept of homogeneity. There is no hard and fast rule for separating a sample into heterogeneous subsets similar to the *Leaders* and *Followers* of this study. Consequently, this responsibility resides under the purview of the policy maker. Moreover, the issue of identifying an appropriate peer group used for evaluating the relative efficiency can become problematic as well, especially in the context of TEs. Another limitation of our study is that we cannot statistically validate the analytical findings of our research. This limitation is due primarily to the nature of our data set (a sample of convenience) and the use of nonparametric data analytic techniques in conducting our investigation.

We also made some assumptions that, though not unreasonable, may not always hold. For example, we are using telecoms as a surrogate for ICT in general, assuming that results for telecoms carry over to other ICT. Also, the use of the PLC model and the S-shaped *Sales* curve is based on the assumption that investments in telecoms behave like sales of products. And of course, there may be other factors, which we did not recognize and consider, that may have skewed the data we used in our study. Nevertheless, our findings should be of interest and should be useful to strategists in TEs, concerned with allocating limited resources and getting maximum returns from their investments.

Future Research

We suggest that future inquiries in this area look at the issues associated with the complementarity of the investments in telecoms and the role of the telecom labor force. When results of investments in ICT in the context of developed countries manifest themselves at the macroeconomic level, such investments are accompanied by complementary investments. At this point, it is not clear what types of factors are complementary to investments in telecoms, especially in the context of TEs. Regarding the telecom-related labor force, it is important to determine whether the increase in the level of investments in telecoms should be accompanied by an increase in size of the labor force, or rather an increase in the quality of telecom workers. Moreover, it is important to examine whether there exists an optimal level of "productivity" of telecom workers (which could be expressed as a ratio of revenues to investments in telecoms), which, once achieved, would signal the necessity of increasing the size of the telecom-related workforce.

Other avenues for future research may be to look at different technologies other than telecoms. Also, the analysis need not be restricted to TEs but may be expanded to look at other developing and emerging economies.

Acknowledgment

Material in this chapter previously appeared in "Improving the relative efficiency of revenue generation from ICT in transition economies: A product life cycle approach," *Journal of Information Technology for Development* 16:4, 279–303.

References

Anderson, C. R. & Zeithaml, C. P. (1984). Stage of the product life cycle, business strategy and business performance. *Academy of Management Journal*, 27(1), 5–24.

Bescherer, F. (2005). Established life cycle concepts in the business environment—Introduction and terminology. Laboratory of Industrial Management Report Series, Report 1/2005, Helsinki University of Technology.

Bollou, F. & Ngwenyama, O. (2008). Are ICT investments paying off in Africa? An analysis of total factor productivity in six West African countries from 1995 to 2002. *Information Technology for Development*, 14(4), 294–307.

Çelebi, D. & Bayraktar, D. (2008). An integrated neural network and data envelopment analysis for supplier evaluation under incomplete information. *Expert Systems with Applications*, 35(4), 1698–1710.

Colecchia, A. & Schreyer, P. (2001). The impact of information communications technology on output growth, STI Working Paper 2001/7, OECD, Paris.

Colecchia, A. & Schreyer, P. (2002). ICT investment and economic growth in the 1990s: Is the United States a unique case? A comparative study of nine OECD countries. *Review of Economic Dynamics*, 5(2), 408–442.

Cooper, W. W. & Tone, K. (1997). Measures of inefficiency in data envelopment analysis and stochastic frontier estimation. *European Journal of Operational Research*, 99(1), 72–88.

Daveri, F. (2002). The new economy in Europe, 1992–2001. *Oxford Review of Economic Policy*, 18(3), 345–362.

Emrouznejad, A. & Shale E. (2009). A combined neural network and DEA for measuring efficiency of large scale datasets. *Computers and Industrial Engineering*, 56(1), 249–254.

Gielens, K. & Steenkamp, J. B. E. M. (2004). What drives new product success? An investigation across products and countries. Marketing Science Institute Report 04-108.

Golder, P. N. & Tellis, G. J. (1997). Will it ever fly? Modeling the takeoff of really new consumer durables. *Marketing Science*, 16(3), 256–270.

Gruber, H. & Verboven, F. (2001a). The evolution of markets under entry and standards regulation—The case of global mobile telecommunications. *International Journal of Industrial Organization*, 19(7), 1189–1212.

Gruber, H. & Verboven, F. (2001b). The diffusion of mobile telecommunications services in the European Union. *European Economic Review*, 45(3), 577–588.

Hall, W. K. (1980). Survival strategies in a hostile environment. *Harvard Business Review*, 58(5), 75–85.

Hauser, J., Tellis, G. J., & Griffin, A. (2006). Research on innovation: A review and agenda for marketing science. *Marketing Science*, 25(6), 687–717.

Hirschberg, J. G. & Lye, J. N. (2001). Clustering in a data envelopment analysis using bootstrapped efficiency scores. Department of Economics—Working Paper Series 800, The University of Melbourne.

Holsheimer, M. & Siebes, A. P. J. M. (1994). Data mining: The search for knowledge in databases, Report CS-R9406, Centre for Mathematics and Computer Science, Amsterdam, The Netherlands.

Hoskisson, R., Eden, L., Lau, C., & Wright, M. (2000). Strategy in emerging economies. *Academy of Management Journal*, 43(3), 249–267.

Jalava, J. & Pohjola, M. (2002). Economic growth in the new economy: Evidence from advanced economies. *Information Economics and Policy*, 14(2), 189–210.

Jorgenson, D. W. & Stiroh, K. J. (2000). Raising the speed limit: US economic growth in the information age. *Brookings Papers on Economic Activity*, (1), 125–211.

Jorgenson, D. W. (2001). Information technology and the US economy. *The American Economic Review*, 91(1), 1–32.

Khouja, M. (1995). The use of data envelopment analysis for technology selection. *Computers and Industrial Engineering*, 28(1), 123–132.

Lemos, C. A. A., Lins, M. P. E., & Ebecken, N. F. F. (2005). DEA implementation and clustering analysis using the K-means algorithm. In *Data Mining VI—Data Mining, Text Mining and Their Business Applications*. Transactions of the Wessex Institute, Southampton, UK.

Mahajan, V., Muller, E., & Bass, F. (1995). Diffusion of new products: Empirical generalizations and managerial uses. *Marketing Science*, 14(3), G79–G88.

Magnan, G. M., Fawcett, S. E., & Birou, L. M. (1999). Benchmarking manufacturing practice using the product life cycle. *Benchmarking: An International Journal*, 6(3), 239–253.

Meffert, H. (2000). *Marketing—Grundlagen marktorientierter Unternehmensführung, Konzepte—Instrumente—Praxisbeispiele*, 9th edition, Gabler-Verlag, Wiesebaden.

Mostafa, M. M. (2009). A probabilistic neural network approach for modelling and classifying efficiency of GCC banks. *International Journal of Business Performance Management*, 11(3), 236–258.

OECD (2001). *The New Economy: Beyond the Hype*. OECD, Paris.

OECD (2005). Good practice paper on ICTs for economic growth and poverty reduction. *OECD Journal on Development*, 6(3), 27–95.

Oliner, S. D. & Sichel, D. E. (2002). Information technology and productivity: Where are we now and where are we going? *Economic Review*, 3(3), 15–41.

Parthasarathy, S. & Anbazhagan, N. (2008). Evaluating ERP projects using DEA and regression analysis. *International Journal of Business Information Systems*, 3(2), 140–157.

Piatkowski, M. (2002). The "New Economy" and Economic Growth in Transition Economies. WIDER Discussion Paper No. 2002/63, WIDER, Helsinki.

Piatkowski, M. (2003). Does ICT investment matter for growth and labor productivity in transition economies? TIGER Working Paper Series 47.

Rogers, E. M. (2003). *Diffusion of Innovations*. Simon and Schuster, New York, USA.

Roztocki, N. & Weistroffer, H. R. (2008). Information technology in transition economies. *Journal of Global Information Technology Management*, 11(4), 2–9.

Samoilenko, S. & Osei-Bryson, K. M. (2007). Increasing the discriminatory power of DEA in the presence of the sample heterogeneity with cluster analysis and decision trees. *Expert Systems with Applications*, 34(2), 1568–1581.

Samoilenko, S. (2008). Contributing factors to information technology investment utilization in transition economies: An empirical investigation. *Information Technology for Development*, 14(1), 52–75.

Samoilenko, S. & Green, L. (2008). Convergence and productive efficiency in the context of 18 transition economies: Empirical investigation using DEA. Proceedings of the Southern Association for Information Systems Conference, Richmond, VA, USA March 13–15.

Samoilenko, S. & Osei-Bryson, K.-M. (2007). Increasing the discriminatory power of DEA in the presence of the sample heterogeneity with cluster analysis and decision trees. *Expert Systems with Applications*, 34(2), 1568–1581.

Samoilenko, S. & Osei-Bryson, K.-M. (2008a). Strategies for telecoms to improve efficiency in the production of revenues: An empirical investigation in the context of transition economies. *Journal of Global Information Technology Management*, 11(4), 56–75.

Samoilenko, S. & Osei-Bryson, K.-M. (2008b). An exploration of the effects of the interaction between ICT and labor force on economic growth in transitional economies. *International Journal of Production Economics*, 115(2), 471–481.

Shao, B. & Lin, W. (2001). Measuring the value of information technology in technical efficiency with stochastic production frontiers. *Information and Software Technology*, 43(7), 447–456.

Shin, H. W. & Sohn S. Y. (2004). Multi-attribute scoring method for mobile telecommunication subscribers. *Expert Systems with Applications*, 26(3), 363–368.

Stiroh, K. (2002). Information technology and the U.S. productivity revival: What do the industry data say? *American Economic Review*, 92(5), 1559–1576.

Stremersch, S. & Tellis, G. J. (2004). Understanding and managing international growth of new products. *International Journal of Research in Marketing*, 21(4), 421–438.

Sultan, F., Farley, J. U., & Lehmann, D. R. (1990). A meta-analysis of applications of diffusion models. *Journal of Marketing Research*, 27(1), 70–77.

Talukdar, D., Sudhir, K., & Ainslie, A. (2002). Investigating new product diffusion across products and countries. *Marketing Science*, 21(1), 97–114.

Thorelli, H. B. & Burnett, S. C. (1981). The nature of product life cycles for industrial goods. *Journal of Marketing*, 45(4), 97–108.

Tellis, G. J., Stremersch, S. & Yin, E. (2003). The international takeoff of new products: The role of economics, culture, and country innovativeness. *Marketing Science*, 22(2), 188–208.

UN ICT Task Force Report (2005). Innovation and investment: Information and communication technologies and the millennium development goals. Report prepared for the United Nations ICT Task Force in support of the Science, Technologies and Innovation Task Force of the United Nations Millennium Project. Available online at http://www.unctad.info/upload/STDEV/docs/icttf.pdf

Van Ark, B., Melka, J., Mulder, N., Timmer, M., & Ypma, G. (2002). ICT investments and growth accounts for the European Union, 1980-2000. Research Memorandum GD-56, Groningen Growth and Development Centre, Groningen.

Van den Bulte, C. & Stremersch, S. (2004). Social contagion and income heterogeneity in new product diffusion: A meta-analytic test. *Marketing Science*, 23(4), 530–544.

WT/ICT Development Report (2006). Measuring ICT for social and economic development. International Telecommunication Union's World Telecommunication/ICT Development Report, 8th edition. Available online at http://www.itu.int/dms_pub/itu.../D-IND-WTDR-2006-SUM-PDF-E.pdf

Wu, D. (2009). Supplier selection: A hybrid model using DEA, decision tree and neural network. *Expert Systems with Applications*, 36(5), 9105–9112.

Yearbook of Statistics (2004). Telecommunication services chronological time series 1993–2002. ITU Telecommunication Development Bureau (BDT), International Telecommunication Union. Available online at http://www.itu.int/ITU-D/ict/publications/yb/1993-2002.html

Appendix

Table 20A.1 Variables Used in Cluster Analysis

Variables Representing Investments in Telecoms	Variables Representing Revenues from Telecoms
Annual telecom investment per capita (current US$)	Total telecom services revenue (current US$)
Annual telecom investment (% of GDP)	Total telecom services revenue per capita (current US$)
Annual telecom investment per worker (current US$)	Total telecom services revenue per worker (current US$)
Annual telecom investment per telecom worker (current US$)	Total telecom services revenue per telecom worker (current US$)

Source: Samoilenko, S. & Osei-Bryson, K.-M. (2007). *Expert Systems with Applications*, 34(2), 1568–1581.

364 ■ *Theoretical Research Frameworks Using Multiple Methods*

Table 20A.2   DEA, Output-Oriented Model, CRS

Country	DMU #	1993	1994	1995	1996	1997	1998	1999	2000	2001	2002
Albania	DMU1	1	1	1	1	1	1	1	1.25	1	1
Armenia	DMU2	1.27	1	1	1	1	1.13	1.24	1.37	1.16	1.25
Azerbaijan	DMU3	2.02	1.68	1	1	1	1.14	1.98	1.19	2.04	2.07
Belarus	DMU4	1.77	3.2	1.47	1.56	1.8	2.14	1.75	2.29	2.54	1.9
Bulgaria	DMU5	1	1	1	1.13	1.17	1.21	1.08	1	1.39	1
Czech Republic	DMU6	1	1	1.28	1.67	1.23	1.03	1.02	1.23	1.48	1.12
Estonia	DMU7	1.24	1.26	1	1	1	1	1	1	1	1
Hungary	DMU8	1	1	1	1	1	1	1	1	1	1
Kazakhstan	DMU9	1.25	2.5	1.67	1.49	1.33	1.53	1.97	1.86	1.96	1
Kyrgyz Republic	DMU10	1	1	1	1.4	1.77	1	1	1	1	1
Latvia	DMU11	2.47	1.01	1.06	1.06	1	1.18	1.44	1.5	1.83	1.94
Lithuania	DMU12	1.22	1.22	1.36	1.18	2.24	2.17	2.01	2.15	1.83	1.04
Moldova	DMU13	4.61	1	1	1	1	1	1	1	1	1
Romania	DMU14	1.81	1.92	1.62	1.69	2.1	2.16	1.81	1.93	1.59	1.11
Slovak Republic	DMU15	1.22	1.26	1.27	1.15	1.81	1.95	1.42	1	1	1
Slovenia	DMU16	1	1	1	1	1	1	1	1.41	1.23	1
Ukraine	DMU17	1	1.44	1.43	1	1.09	1	1	1	1.17	1.14
Poland	DMU18	1.08	1	1.11	1.32	1.18	1	1.19	1	1	1

Improving the Efficiency of Revenue Generation from ICT in TEs ■ 365

Table 20A.3 DEA, Output-Oriented Model, VRS

Country	DMU #	1993	1994	1995	1996	1997	1998	1999	2000	2001	2002
Albania	DMU1	1	1	1	1	1	1	1	1	1.13	1
Armenia	DMU2	1	1	1	1	1	1.09	1.17	1.1	1.32	1.16
Azerbaijan	DMU3	1.89	1.6	1	1	1	1.14	1	1	1	1
Belarus	DMU4	1.72	2.36	1.46	1.55	1.57	1.93	1.7	1.86	2.02	2.33
Bulgaria	DMU5	1	1	1	1.04	1.02	1.11	1	1	1	1.37
Czech Republic	DMU6	1	1	1.24	1.43	1.14	1.03	1	1.1	1.15	1.44
Estonia	DMU7	1.22	1.26	1	1	1	1	1	1	1	1
Hungary	DMU8	1	1	1	1	1	1	1	1	1	1
Kazakhstan	DMU9	1.16	1.96	1.48	1.34	1	1.23	1.93	1	1.73	1.93
Kyrgyz Republic	DMU10	1	1	1	1	1	1	1	1	1	1
Latvia	DMU11	2.47	1	1	1.06	1	1.12	1.41	1.93	1.23	1.82
Lithuania	DMU12	1.21	1.04	1.31	1.13	2.1	2.1	1.96	1	1.56	1.83
Moldova	DMU13	1	1	1	1	1	1	1	1	1	1
Romania	DMU14	1.8	1.89	1.54	1.6	1.8	1.96	1.78	1.1	1.92	1.58
Slovak Republic	DMU15	1.18	1.26	1.25	1.12	1.74	1.92	1.25	1	1	1
Slovenia	DMU16	1	1	1	1	1	1	1	1	1.17	1.22
Ukraine	DMU17	1	1.17	1.35	1	1	1	1	1	1	1.17
Poland	DMU18	1.06	1	1.1	1.3	1.17	1	1.18	1	1	1

Table 20A.4 DEA, Output-Oriented Model, NIRS

Country	DMU #	1993	1994	1995	1996	1997	1998	1999	2000	2001	2002
Albania	DMU1	1	1	1	1	1	1	1	1.25	1	1
Armenia	DMU2	1.27	1	1	1	1	1.09	1.17	1.32	1.16	1.1
Azerbaijan	DMU3	1.89	1.6	1	1	1	1.14	1.98	1.19	2.04	2.07
Belarus	DMU4	1.72	2.36	1.46	1.55	1.57	1.93	1.7	2.02	2.33	1.86
Bulgaria	DMU5	1	1	1	1.04	1.02	1.11	1	1	1.37	1
Czech Republic	DMU6	1	1	1.24	1.43	1.14	1.03	1	1.23	1.44	1.12
Estonia	DMU7	1.24	1.26	1	1	1	1	1	1	1	1
Hungary	DMU8	1	1	1	1	1	1	1	1	1	1
Kazakhstan	DMU9	1.25	1.96	1.48	1.34	1	1.23	1.93	1.73	1.93	1
Kyrgyz Republic	DMU10	1	1	1	1.4	1.77	1	1	1	1	1
Latvia	DMU11	2.47	1	1	1.06	1	1.18	1.44	1.5	1.82	1.94
Lithuania	DMU12	1.21	1.04	1.36	1.13	2.1	2.1	1.96	2.15	1.83	1.04
Moldova	DMU13	4.61	1	1	1	1	1	1	1	1	1
Romania	DMU14	1.81	1.92	1.54	1.6	1.8	1.96	1.78	1.93	1.58	1.1
Slovak Republic	DMU15	1.18	1.26	1.25	1.12	1.74	1.92	1.25	1	1	1
Slovenia	DMU16	1	1	1	1	1	1	1	1.41	1.23	1
Ukraine	DMU17	1	1.17	1.35	1	1	1	1	1	1.17	1
Poland	DMU18	1.06	1	1.1	1.3	1.17	1	1.19	1	1	1

Table 20A.5 DEA, Input-Oriented Model, CRS

Country	DMU #	1993	1994	1995	1996	1997	1998	1999	2000	2001	2002
Albania	DMU1	1	1	1	1	1	1	1	0.8	1	1
Armenia	DMU2	0.79	1	1	1	1	0.89	0.81	0.73	0.86	0.8
Azerbaijan	DMU3	0.49	0.6	1	1	1	0.88	0.51	0.84	0.49	0.48
Belarus	DMU4	0.56	0.31	0.68	0.64	0.56	0.47	0.57	0.44	0.39	0.52
Bulgaria	DMU5	1	1	1	0.88	0.86	0.83	0.92	1	0.72	1
Czech Republic	DMU6	1	1	0.78	0.6	0.81	0.97	0.98	0.81	0.68	0.9
Estonia	DMU7	0.81	0.79	1	1	1	1	1	1	1	1
Hungary	DMU8	1	1	1	1	1	1	1	1	1	1
Kazakhstan	DMU9	0.8	0.4	0.6	0.67	0.75	0.65	0.51	0.54	0.51	1
Kyrgyz Republic	DMU10	1	1	1	0.72	0.56	1	1	1	1	1
Latvia	DMU11	0.41	0.99	0.94	0.94	1	0.85	0.69	0.67	0.55	0.52
Lithuania	DMU12	0.82	0.82	0.74	0.85	0.45	0.46	0.5	0.46	0.55	0.96
Moldova	DMU13	0.22	1	1	1	1	1	1	1	1	1
Romania	DMU14	0.55	0.52	0.62	0.59	0.48	0.46	0.55	0.52	0.63	0.9
Slovak Republic	DMU15	0.82	0.8	0.79	0.87	0.55	0.51	0.71	1	1	1
Slovenia	DMU16	1	1	1	1	1	1	1	0.71	0.81	1
Ukraine	DMU17	1	0.7	0.7	1	0.92	1	1	1	0.85	0.88
Poland	DMU18	0.93	1	0.9	0.76	0.85	1	0.84	1	1	1

Table 20A.6 DEA, Input-Oriented Model, VRS

Country	DMU #	1993	1994	1995	1996	1997	1998	1999	2000	2001	2002
Albania	DMU1	1	1	1	1	1	1	1	0.95	1	1
Armenia	DMU2	1	1	1	1	1	0.89	0.83	0.75	0.87	0.8
Azerbaijan	DMU3	0.82	0.92	1	1	1	0.95	1	1	1	1
Belarus	DMU4	0.75	0.59	0.77	0.72	0.64	0.6	0.64	0.67	0.66	0.77
Bulgaria	DMU5	1	1	1	0.93	0.95	0.83	1	1	0.72	1
Czech Republic	DMU6	1	1	0.8	0.75	0.87	0.97	1	0.95	0.74	0.94
Estonia	DMU7	0.86	0.83	1	1	1	1	1	1	1	1
Hungary	DMU8	1	1	1	1	1	1	1	1	1	1
Kazakhstan	DMU9	1	0.48	0.62	0.7	1	0.65	0.7	0.78	0.75	1
Kyrgyz Republic	DMU10	1	1	1	1	1	1	1	1	1	1
Latvia	DMU11	0.57	1	1	0.95	1	0.93	0.9	0.94	0.83	0.76
Lithuania	DMU12	0.87	0.93	0.82	0.87	0.6	0.63	0.81	0.91	0.86	1
Moldova	DMU13	1	1	1	1	1	1	1	1	1	1
Romania	DMU14	0.61	0.63	0.67	0.66	0.63	0.65	0.73	0.75	0.72	0.91
Slovak Republic	DMU15	0.84	0.82	0.8	0.87	0.68	0.66	0.74	1	1	1
Slovenia	DMU16	1	1	1	1	1	1	1	0.96	0.87	1
Ukraine	DMU17	1	0.7	0.75	1	1	1	1	1	0.86	1
Poland	DMU18	0.93	1	0.91	0.85	0.93	1	0.96	1	1	1

Table 20A.7 DEA, Input-Oriented Model, NIRS

Country	DMU #	1993	1994	1995	1996	1997	1998	1999	2000	2001	2002
Albania	DMU1	1	1	1	1	1	1	1	0.8	1	1
Armenia	DMU2	0.79	1	1	1	1	0.89	0.81	0.73	0.86	0.8
Azerbaijan	DMU3	0.49	0.6	1	1	1	0.88	0.51	0.84	0.49	0.48
Belarus	DMU4	0.56	0.31	0.68	0.64	0.56	0.47	0.57	0.44	0.39	0.52
Bulgaria	DMU5	1	1	1	0.93	0.95	0.83	1	1	0.72	1
Czech Republic	DMU6	1	1	0.78	0.6	0.81	0.97	1	0.81	0.68	0.9
Estonia	DMU7	0.81	0.79	1	1	1	1	1	1	1	1
Hungary	DMU8	1	1	1	1	1	1	1	1	1	1
Kazakhstan	DMU9	0.8	0.4	0.6	0.67	1	0.65	0.51	0.54	0.51	1
Kyrgyz Republic	DMU10	1	1	1	0.72	0.56	1	1	1	1	1
Latvia	DMU11	0.41	1	1	0.94	1	0.85	0.69	0.67	0.55	0.52
Lithuania	DMU12	0.82	0.93	0.74	0.87	0.45	0.46	0.5	0.46	0.55	0.96
Moldova	DMU13	0.22	1	1	1	1	1	1	1	1	1
Romania	DMU14	0.55	0.52	0.62	0.59	0.48	0.46	0.55	0.52	0.63	0.9
Slovak Republic	DMU15	0.82	0.8	0.79	0.87	0.55	0.51	0.74	1	1	1
Slovenia	DMU16	1	1	1	1	1	1	1	0.71	0.81	1
Ukraine	DMU17	1	0.7	0.7	1	1	1	1	1	0.85	1
Poland	DMU18	0.93	1	0.9	0.76	0.85	1	0.84	1	1	1

Table 20A.8 DEA, Output-Oriented Model, Scale Efficiency Scores

Country	DMU #	1993	1994	1995	1996	1997	1998	1999	2000	2001	2002
Albania	DMU1	1	1	1	1	1	1	1	1.11	1	1
Armenia	DMU2	1.27	1	1	1	1	1.04	1.06	1.04	1	1.13
Azerbaijan	DMU3	1.07	1.05	1	1	1	1	1.98	1.19	2.04	2.07
Belarus	DMU4	1.03	1.35	1	1.01	1.14	1.11	1.03	1.13	1.09	1.02
Bulgaria	DMU5	1	1	1	1.09	1.15	1.09	1.08	1	1.01	1
Czech Republic	DMU6	1	1	1.04	1.17	1.09	1.01	1.02	1.07	1.02	1.01
Estonia	DMU7	1.01	1	1	1	1	1	1	1	1	1
Hungary	DMU8	1	1	1	1	1	1	1	1	1	1
Kazakhstan	DMU9	1.07	1.28	1.13	1.11	1.33	1.25	1.02	1.07	1.02	1
Kyrgyz Republic	DMU10	1	1	1	1.40	1.77	1	1	1	1	1
Latvia	DMU11	1	1.01	1.06	1	1	1.05	1.02	1.22	1.01	1
Lithuania	DMU12	1.01	1.18	1.03	1.05	1.07	1.03	1.03	1.38	1	1.04
Moldova	DMU13	4.61	1	1	1	1	1	1	1	1	1
Romania	DMU14	1.01	1.02	1.05	1.05	1.17	1.10	1.02	1	1.01	1
Slovak Republic	DMU15	1.04	1	1.02	1.03	1.04	1.02	1.14	1	1	1
Slovenia	DMU16	1	1	1	1	1	1	1	1.20	1.01	1
Ukraine	DMU17	1	1.22	1.06	1	1.09	1	1	1	1	1.14
Poland	DMU18	1.02	1	1.01	1.01	1	1	1	1	1	1

Table 20A.9 DEA, Input-Oriented Model, Scale Efficiency Scores

Country	DMU #	1993	1994	1995	1996	1997	1998	1999	2000	2001	2002
Albania	DMU1	1	1	1	1	1	1	1	0.84	1	1
Armenia	DMU2	0.79	1	1	1	1	1	1	0.98	0.99	1
Azerbaijan	DMU3	0.60	0.65	1	1	1	0.93	0.98	0.84	0.49	0.48
Belarus	DMU4	0.75	0.53	0.88	0.89	0.86	0.79	0.51	0.65	0.59	0.68
Bulgaria	DMU5	1	1	1	0.94	0.90	1	0.89	1	1	1
Czech Republic	DMU6	1	1	0.98	0.80	0.94	0.99	0.92	0.86	0.91	0.95
Estonia	DMU7	0.94	0.96	1	1	1	1	0.98	1	1	1
Hungary	DMU8	1	1	1	1	1	1	1	1	1	1
Kazakhstan	DMU9	0.80	0.84	0.96	0.96	0.75	1	0.72	0.69	0.68	1
Kyrgyz Republic	DMU10	1	1	1	0.72	0.56	1	1	1	1	1
Latvia	DMU11	0.71	0.99	0.94	1	1	0.91	0.77	0.71	0.66	0.68
Lithuania	DMU12	0.95	0.87	0.90	0.97	0.74	0.74	0.62	0.51	0.64	0.96
Moldova	DMU13	0.22	1	1	1	1	1	1	1	1	1
Romania	DMU14	0.91	0.82	0.92	0.90	0.75	0.71	0.75	0.69	0.88	1
Slovak Republic	DMU15	0.98	0.97	0.98	0.99	0.81	0.77	0.96	1	1	1
Slovenia	DMU16	1	1	1	1	1	1	1	0.74	0.93	1
Ukraine	DMU17	1	1	0.93	1	0.92	1	1	1	0.99	0.88
Poland	DMU18	1	1	0.99	0.90	0.91	1	0.88	1	1	1

Chapter 21

Determining Strategies for Telecoms to Improve Efficiency in the Production of Revenues

Introduction

Revenue generation is one of the ways by which investments in information and communication technologies (ICT) could affect the macroeconomic bottom-line (UN ICT Task Force Report 2005; WT/ICT Development Report 2006). Thus, it is reasonable to expect that improvements in the effectiveness and efficiency of production of revenue would benefit the overall macroeconomic impact of investments in ICT. Multiple studies conducted in the context of developed economies have demonstrated that such investments can offer robust returns and contribute to overall economic growth (Jorgenson & Stiroh 2000; Colecchia & Schreyer 2001; Jorgenson 2001; Daveri 2002; Jalava & Pohjola 2002; Oliner & Sichel 2002; Stiroh 2002; van Ark et al. 2002; OECD 2005a,b). Similar studies in the context of developing countries and transition economies (TEs), however, identified significantly lower levels of returns on investments in ICT (Dewan & Kraemer 2000; Pohjola 2001; Piatkowski 2003a). The term *transition economy* characterizes a country transitioning from a centrally planned economy to a market-driven economy (Ollman 1997; Myers 2004). Despite sharing a common label, TEs do not constitute a homogenous group in terms of the level of economic development. The World Bank, for example, may group some of them with the developed countries, and some with the developing countries, depending on the level of industrialization (World Bank 2004). The heterogeneous context of TEs is quite unusual, for it has been acknowledged that TEs share economic characteristics with both developed and less developed economic regions (OECD 2004). As a result, the context of TEs offers to investigators a unique perspective from which the relationship between investments in and macroeconomic outcomes from ICT can be studied and results generalized.

The general subject of our inquiry is the relationship between investments in telecoms and revenues from telecoms. To justify that relationship theoretically, we conduct this investigation based on the framework of neoclassical growth accounting, which allows for relating investments in telecoms to revenues from telecoms in a theoretically rigorous manner. We chose to study investments in telecoms primarily for two reasons. First, it represents a subset of investments in ICT. Second, it is also a common type of investment regularly made by almost any economy of the world. Overall, our inquiry fits into the established stream of research investigating the relationship between investments in ICT and their macroeconomic outcomes (Avgerou 1998; Morales-Gomez & Melesse 1998; Colecchia & Schreyer 2001, 2002; IMF 2001; Kraemer & Dedrick 2001; Pohjola 2001, 2002; Lee & Khatri 2003; OECD 2003a,b,c, 2004, 2005a,b,c; Piatkowski 2003a,b, 2004). Despite the ample evidence of the positive impact of ICT investments on the economies of developed countries, the research concerning the effects of ICT on developing and transition economies is scarce. The consensus is that more research is needed because hardly any relevant studies have been done outside 30 democracies that are members of the Organisation for Economic Co-operation and Development (OECD 2004). Consequently, the justification of our study is that "substantive research is urgently required if investment commitments are to be made—by the private sector or development agencies—with any real understanding of likely outcomes" (OECD 2004, p. 4).

How would the results of our investigation possibly differ from the results of the inquiry conducted in the context of fully developed market economies? A study by OECD (2004) concluded that a number of complementary factors in the business environment influence the extent of the impact of investments in ICT on business performance, and consequently, on the production of revenues. In the context of developed countries, the following five factors are considered particularly important. The first factor refers to the nature of the business, i.e., whether or not the business can make more extensive use of ICTs to change processes and their relationships with customers and suppliers. The second factor refers to the extent of competition and the nature of the regulatory environment. In this regard, OECD research concluded that more competitive and less regulated business environments could take greater advantage of innovation provided by ICT, thus contributing to macroeconomic performance. The third factor reflects the relative costs of ICT deployment, which takes into consideration both direct and indirect costs. The fourth factor considers the amount and quality of human capital available, where a better-skilled and better-equipped workforce is more likely to achieve higher rates of ICT-related innovation and increased productivity. Finally, the fifth factor reflects the importance of the flexibility of the business environment, where capability for restructuring and reorganization allows for better utilization of the opportunities offered by ICT (OECD 2004). Due to the obvious differences in the business environment of developed market economies and TEs, we have sound reason to expect that these differences make the findings of our inquiry specific to the context of TEs. Discussion about the peculiarities and a description of the additional characteristics specific to TEs can also be found in the work of Slay (2005), Sasse (2005), Sarychev (2005), and Schelkle (2005).

There seems to be a consensus (Dewan & Kraemer 1998, 2000; Murakami, 1997; Piatkowski 2002) that the size of the ICT investment and overall accumulated ICT capital are some of the factors that affect the relationship between investments in ICT and economic outcomes. This means that the level of investments should be above a certain threshold to manifest itself at the detectable level (Jorgenson & Stiroh 2000; Oliner & Sichel 2000; Council of Economic Advisors 2001; Jorgenson 2001). Unfortunately, there is no direct linear relationship between the level of investments and level of revenues (Murakami 1997; Dewan & Kraemer 2000; Piatkowski 2002; OECD 2004); an increase in the level of investments does not directly result in a proportional increase in the level of revenues. Instead, multiple factors, internal as well as external, mediate the relationship. One such factor is efficiency of the process of conversion of investments into revenues. Clearly, the level of efficiency of this process is instrumental to revenue production and is likely to be relative to the size of investments. As a result, it is possible to conceive that a policy of increasing levels of investments without regard to the efficiency of the processes by which those investments are converted into revenues can actually backfire and be detrimental to the efficient production of revenues.

The following scenario demonstrates the importance of this problem. Let us suppose there are two economies: a relatively less efficient economy A, with a lower level of investments in and revenues from telecoms, and a more efficient economy B, with a higher level of investments in and revenues from telecoms. The problem can be formulated as follows: how should a policy maker go about choosing and promoting a strategy of increasing the level of relative efficiency of economy A? Regarding the choice of a strategy, should a policy maker disregard the differences in the context and consider adopting and following the strategy of economy B? Regardless of the choice, strategy implementation should take into consideration two variables, namely, the level of investments and the expected level of revenues. Consequently, shall a policy maker consider promoting the strategy of investing more and more in telecoms, hoping that such an increase will manifest itself in the form of increased efficiency of production of revenues? Alternatively, should a policy maker consider increasing the efficiency of the processes by which revenue production takes place first? We will find the answers to these questions by addressing the research problems of our inquiry.

In this chapter, we present an empirical investigation conducted in the context of the 18 TEs classified by International Monetary Fund (IMF) as members of the group of *transition economies in Europe and the former Soviet Union* (IMF 2000). We would like to provide a background reference for the current study (we offer a more detailed overview in the section "Previous Findings"). During our previous inquiries (Samoilenko & Osei-Bryson 2007), we determined that over the 10-year period from 1993 to 2002, 18 TEs were not homogenous in terms of the levels of yearly investments and revenues from telecoms. In fact, we identified two subsets, *leaders* and *followers*, which clearly differ in their levels of investments and revenues from telecoms. Furthermore, we also determined that leaders differ significantly from followers in terms of the efficiency of utilization of investments in telecoms (Samoilenko 2008a). Subsequently, we established that leaders differ from followers in many aspects, such as those related to militarization of the economy, quality of human resources (i.e., education and health), level of sociotechnical development, and others. In this study, we build on the findings of our previous inquiries in order to explore two issues:

1. The identification of strategies by which two groups of economies, leaders and followers, characterized by different levels of performance in terms of the relative efficiency of production of revenues, can increase their respective levels of performance. Justification for exploration of this issue lies in the relativity of the concept of *level of performance*, which in the context of this research denotes the level of relative efficiency of production of revenues. Firstly, it is a relative measure from an internal perspective, where it is relative to the levels of inputs and outputs. Secondly, it is a relative measure from an external perspective, where it is relative to the peer group against which the level of performance is measured. As a result, we expect the level of relative performance to be influenced by three factors: level of inputs, the efficiency of conversion of inputs into outputs, and the peer group that is used for comparison. Consequently, economies belonging to the different peer groups, and having different levels of inputs and outputs, are likely to have different strategies for increasing the level of their performance.
2. The identification of the routes by which performance-increasing strategies should be implemented. Justification of this research question is intuitive; once the possible strategy has been identified, it could be implemented by means of altering such internal parameters as level of inputs and level of efficiency of conversion of inputs into outputs. However, while the answer to the first research question allows us to determine the *external* reason for relative inefficiency, dependent on the peer group, the *internal* reason for relative inefficiency has not been yet identified. According to the assumption of this study, the level of relative efficiency is dependent on the level of inputs and the level of efficiency with which inputs are converted into outputs. The

question becomes, then, which factor must be targeted first? Let us clarify that this is not an issue of using a breakeven analysis comparing two factors; rather, it is an issue of determining which factor at a given point in time can affect the level of revenues the most. Consequently, the second research question aims to determine whether the nature of the differences between leaders and followers is due to the differences in their levels of inputs and outputs, or whether it is due to the differences in processes by which investments are converted into revenues.

To explore the research questions of this study, we employ a three-phase methodology utilizing decision trees (DTs), neural networks (NNs), and data envelopment analysis (DEA). The rest of the chapter is structured as follows: The next section of the paper offers a description of the data used in our inquiry. Chapters 4, 6, and 9 of the book contain an overview of the data analytic methods used in this study. Section "Previous Findings" offers a summary of the previous investigations that this study builds on. Section "Methodology" presents our methodology used for inquiry. Section "Result of the Data Analysis" presents the results of the inquiry. Section "Interpretation of the Results" offers interpretation of the results of the data analysis. A discussion of the findings and the conclusion are presented in the last section of the paper.

Overview of the Data

We collected the data for our inquiry from two sources. The first source was the *World Development Indicators* database (http://web.worldbank.org/WBSITE/EXTERNAL/DATASTATISTICS), the *World Bank's* (http://web.worldbank.org) comprehensive database containing development data. The second source was the *Yearbook of Statistics* (2004; http://www.itu.int/ITU-D/ict/publications), an annual publication of the *International Telecommunication Union* (ITU; http://www.itu.int). To minimize the heterogeneity of our sample, we wanted to use TEs that belong to the same group and started transition at about the same time. We have chosen 25 countries classified as *transition economies in Europe and the former Soviet Union* by IMF (2000).

After the overview and analysis of the data, we concentrated on the following 18 TEs: Albania, Armenia, Azerbaijan, Belarus, Bulgaria, Czech Republic, Estonia, Hungary, Kazakhstan, Kyrgyzstan, Latvia, Lithuania, Moldova, Poland, Romania, Slovakia, Slovenia, and Ukraine. Unfortunately, data on the other seven TEs that we wanted to include in our analysis, namely, Croatia, the former Yugoslav Republic of Macedonia, Georgia, Russia, Tajikistan, Turkmenistan, and Uzbekistan, were not available, or contained too many missing data points to be useful in the analysis. Overall, for our 18 TEs, we were able to construct the data set covering the period from 1993 to 2002.

Previous Findings

In a previous inquiry (Samoilenko & Osei-Bryson 2008), we used cluster analysis (CA) to determine that a 10-year data set on 18 TEs, spanning a period from 1993 to 2002, is not homogenous in terms of the investments in and revenues from telecoms.

The variables that were used to conduct CA are listed in Table 21.1.

The intent in presenting the chosen variables in terms of ratios, instead of absolute dollar terms, was to present the differences between TEs in terms of their size, population, and level of wealth, while representing the investments and revenues more broadly (i.e., relative to the whole population, labor force of a country, and the telecom industry). We decided that such relative representation provides a more objective depiction of not only the investments and revenues themselves, but also the environment within which the investments were converted into revenues. By using CA, we obtained a solution that partitions our data set into two clusters. The membership of each cluster is provided in Table 21.2.

Our results confirm that even within the same geographic region, emerging market economies are not homogenous (Hoskisson et al. 2000). Also, results of the CA corroborate the findings of Piatkowski (2003b) that in the period between 1995 and 2000, the Czech Republic, Hungary, Poland, and Slovenia have benefited from investments in ICT the most.

Once the results of CA were obtained, the data set was partitioned into two subsets accordingly. Then input- and output-oriented DEA of the complete data set was conducted to calculate the scores of the averaged relative efficiency for each cluster. The variables used to perform DEA are listed in Table 21.3. For justification of the variables constituting our DEA model, as well as an overview of the theoretical framework used in the study, see the work of Samoilenko (2008b).

Table 21.1 Variables Selected for the CA

1. Total telecom services revenue (current US$)
2. Total telecom services revenue per capita (current US$)
3. Total telecom services revenue per worker (current US$)
4. Total telecom services revenue per telecom worker (current US$)
5. Annual telecom investment per capita (current US$)
6. Annual telecom investment [% of gross domestic product (GDP)]
7. Annual telecom investment per worker (current US$)
8. Annual telecom investment per telecom worker (current US$)

Table 21.2 Membership of the Two-Cluster Solution

Cluster 1 (Followers)	Cluster 2 (Leaders)
Albania (1993–2002)	Bulgaria (2002)
Armenia (1993–2002)	Czech Rep (1993–2002)
Azerbaijan (1993–2002)	Estonia (1994–2002)
Belarus (1993–2002)	Hungary (1993–2002)
Bulgaria (1993–2001)	Latvia (1994, 1995, 1997–2002)
Estonia (1993)	Lithuania (1999–2002)
Kazakhstan (1993–2002)	Poland (1993–2002)
Kyrgyzstan (1993–2002)	Slovenia (1993–2002)
Latvia (1993, 1996)	Slovakia (1995–1998, 2000–2002)
Lithuania (1993–1998)	
Moldova (1993–2002)	
Romania (1993–2002)	
Slovakia (1993, 1994, 1999)	
Ukraine (1993–2002)	

Table 21.3 List of Variables for DEA Models

Role	Subset of Variables
Input	GDP per capita (in current US$)
	Full-time telecommunication staff (% of total labor force)
	Annual telecom investment per telecom worker
	Annual telecom investment (% of GDP in current US$)
	Annual telecom investment per capita
	Annual telecom investment per worker
Output	Total telecom services revenue per telecom worker
	Total telecom services revenue (% of GDP in current US$)
	Total telecom services revenue per worker
	Total telecom services revenue per capita

According to the results of DEA, one of the clusters, members of which include the Czech Republic, Hungary, Poland, and Slovenia, has higher averaged relative efficiency scores than the cluster containing Albania, Armenia, Azerbaijan, Belarus, Kazakhstan, Kyrgyzstan, Moldova, Romania, and Ukraine. Subsequently, we call the first group the leaders and the second group the followers. Side-by-side comparison of the two clusters based on the results of the input-oriented DEA (where the focus is on the minimization of the use of the investments for achieving a given level of revenues) and the output-oriented DEA (where the focus is on the maximization of the level of revenues per given level of investments) is presented in Table 21.4.

Table 21.4 Results of the DEA

DEA Model	Criterion for Comparison, Return to Scale	Leaders Cluster	Followers Cluster	Difference	Difference, %
Input oriented	Avg. efficiency score, CRS	0.89	0.79	0.10	12.54%
Input oriented	Avg. efficiency score, VRS	0.95	0.88	0.07	7.48%
Input oriented	Avg. efficiency score, NIRS	0.89	0.80	0.09	11.63%
Output oriented	Avg. efficiency score, CRS	1.21	1.44	0.22	15.58%
Output oriented	Avg. efficiency score, VRS	1.18	1.30	0.12	8.88%
Output oriented	Avg. efficiency score, NIRS	1.21	1.38	0.18	12.78%

Note: CRS, constant returns to scale; NIRS, non-increasing returns to scale; VRS, variable returns to scale.

The results of our previous inquiries can be summarized as follows. The data set representing 18 TEs over the period from 1993 to 2002 contains two clusters that are clearly different in terms of the levels of investments in telecoms, revenues from telecoms, as well as the levels of efficiency of utilization of resources and effectiveness of the production of revenues. At this point, it is not clear what strategies leaders and followers should adopt in order to improve their respective levels of effectiveness and how these strategies should be implemented. In the next section of the chapter, we present a methodology that aims at answering the research questions of this study.

Methodology

In this part of the chapter, we describe in a systematic fashion a three-phase methodology that we used to conduct our inquiry. First, we would like to provide some justifications for the chosen approach. The cornerstone of our inquiry is DEA; consequently, we would like to justify the use of DT induction and NNs with DEA. All three techniques are nonparametric methods without conflicting assumptions, which require no alteration or modification of the data to be used together.

Samoilenko and Osei-Bryson (2008) previously successfully argued for the use of DTs after clustering and DEA as an approach allowing for increasing the discriminatory power of DEA. While the *purpose* of DT analysis in the current study is different from that of Samoilenko and Osei-Bryson (2008), the *application* of the technique follows the accepted methodology.

Samoilenko (2008b) also previously argued for the use of NN with DEA as an appropriate methodology allowing for determining the effects of the scale heterogeneity of the sample on scores of the relative efficiency of decision-making units (DMUs). The black-box approach used by NN is not unlike that of DEA and fits the purpose of our study well, for we are not trying to determine *how* investments in telecoms are converted into revenues.

Phase 1: DT Induction

In the first phase of our study, we use DT induction to generate rules that describe the relative efficiency categories in terms of the input and output variables of the DEA model. In order to do that, we created a target variable *EfficiencyCategory*, which identifies the efficiency category of each data point decision-making units (DMUs) in our sample. Because we identified that our data set consists of two clusters, we represent the domain of values for our target variable as follows:

"10"—relatively inefficient DMU with membership in cluster 1 ("inefficient followers")
"11"—relatively efficient DMU with membership in cluster 1 ("efficient followers")
"20"—relatively inefficient DMU with membership in cluster 2 ("inefficient leaders")
"21"—relatively efficient DMU with membership in the cluster 2 ("efficient leaders")

We have created for the purposes of this analysis a derived variable, *productivity ratio per telecom worker*, as the ratio of the variable *telecom revenue per telecom worker* to *annual telecom investment per telecom worker*. However, because the total number of telecom workers was considered to be unchanging in the period of 1 year, the variable *productivity* represents, in fact, a ratio of the *total telecom revenue* to *annual telecom investment* for each year. It should be noted that this variable does not represent the productivity of telecom workers in terms of conversion of investments into revenues, but rather, the purpose of this variable is to give an intuitive representation of the fiscal situation within each TE in terms of the levels of yearly investments, labor, and revenues from telecoms.

The purpose of the first phase of our data analysis is to identify the variables involved in the top-level splits. Keeping in mind that DT utilizes greedy splits, we consider these variables to be the most important in differentiating the followers from the leaders. Similarly, we also determined the set of variables that differentiate the relatively efficient members of each cluster from the relatively inefficient ones.

Phase 2: NN Simulation

According to the assumption of this study, two factors could be responsible for the differences in the levels of relative efficiency of the TEs in our sample. The first factor is the level of investments in telecoms, and the second factor is the level of the relative efficiency and effectiveness of the process by which investments are transformed into revenues; we named this process *transformative capacity*. Consequently, we need to find a way of determining the effects of the level of investments and the transformative capacity on the level of relative efficiency separately. The separation of the effects of the two factors is the purpose of the second phase of our methodology, which proceeds in a two-step fashion.

First, we use the set of input variables for each cluster as the input nodes, and the set of output variables as the output nodes of the NN. Then we train the NN and obtain the cluster-specific *model of transformative capacity* of a given cluster. Consequently, we end up generating two models of transformative capacity, one corresponding to the leaders and another corresponding to the followers. At this point, we have successfully isolated one of the factors influencing the relative efficiency of each DMU in the sample, namely, the effect of the transformative capacity.

The purpose of the next step is to obtain, based on the NN models generated previously, the sets of simulated outputs for each cluster. This process allows us to determine the hypothetical outputs of the *Followers* that they could produce using their own inputs while utilizing the "borrowed" transformative capacity of the leaders, as well as to determine the hypothetical outputs of the leaders that they could produce using their own inputs while utilizing the borrowed transformative capacity of the followers. Conversely, this could be interpreted as generating the set of outputs of the followers based on the inputs of the leaders.

Phase 3: DEA

We perform DEA using the same model as that used by Samoilenko (2008b). Input and output variables comprising the model are listed in Table 21.4. During the third phase of our methodology, we subject the original inputs and the simulated outputs of the followers obtained in the phase 2, as well as the original inputs and the simulated outputs of the leaders, to DEA. Then we calculate the averages of the relative efficiency of both groups again and determine whether the averages of the followers have improved. If this is the case, and the average relative efficiencies of the followers have gone up, we have a reason to suggest that the disparity between the relative efficiencies of the leaders and the followers is due to the differences in their transformative capacities.

However, we still need to determine whether the level of the inputs plays a role in the disparity of the levels of the relative efficiencies of the leaders and the followers. In order to do so, we conduct DEA using the data set consisting of the original inputs and the outputs of the followers, and the original inputs and simulated outputs, based on the transformative capacity of the followers, of the leaders. Once the scores of relative efficiency have been obtained, we group them according to cluster membership (i.e., followers and leaders) and average the relative efficiency scores for each group. If, after the comparison of the average relative efficiencies, the leaders still have a higher average relative efficiency score, then we have a reason to suggest that the disparity between the relative efficiencies of the two groups is due, in part, to the differences in their levels of inputs.

Results of the Data Analysis

In the following two parts of this section, we describe the DT and NN analysis that we have conducted using *Enterprise Miner* by SAS Institute. The third phase involved the use of DEA and was performed using software application *OnFront* by Lund Corporation.

Decision Tree Analysis

The numbers of DMUs in each of our four categories are as follows: 38 *efficient leaders*, 31 *inefficient leaders*, 43 *efficient followers*, and 68 *inefficient followers*.

We generated our DT model, which enabled us to identify conditions associated with our four categories. The variables that turned out to be the most important in separating the four categories are provided in Table 21.5. Figure 21A.1 depicts the complete DT of phase 1 and is included the Appendix of this chapter.

Table 21.5 Variables That Separate Efficiency Categories within the Sample of 18 TEs

	Efficiency Category	
Variable(s)	*Group 1*	*Group 2*
Annual telecom investment per worker	Leaders (overall)	Followers (overall)
GDP per capita Total telecom services revenue Productivity ratio per telecom worker	Inefficient leaders	Efficient leaders
Annual telecom investment per worker Productivity ratio per telecom worker	Inefficient followers	Efficient followers

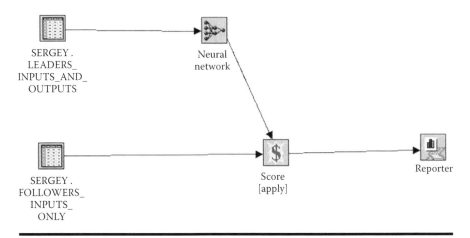

Figure 21.1 Process flow diagram of NN simulation.

NN Simulation

In this part of the data analysis, we used SAS Enterprise Miner to obtain a set of simulated outputs. The process flow diagram, depicting the simulation model created for this purpose, is provided in Figure 21.1. The two leftmost nodes on the diagram represent the data sources. The purpose of the *NN* node is to create a model according to which inputs of the leaders are converted into outputs. The created model then is stored in the *score* node, and the predicted values of the outputs for the majority, simulated by the saved NN model, then could be retrieved from the *distribution explorer* node. We started by using NN to simulate the outputs of the followers based on the transformative capacity of the leaders, and then created a similar model for the purposes of simulating the outputs of the leaders based on the transformative capacity of the majority.

Results of DEA

To perform DEA, we used the software application OnFront, version 2.02, produced by Lund Corporation (http://www.emq.com). Once the simulated outputs were obtained, they were substituted for the real outputs. After that, we conducted DEA again and obtained the new values of relative efficiencies for the followers and the leaders. The results are summarized in Tables 21.6 and 21.7.

Interpretation of the Results

In this section of the chapter, we offer the interpretation of the results of the study. DT induction analysis, described in the section "Decision Tree Analysis",

Table 21.6 Comparison of Simulated Outputs of Followers Based on the Transformative Capacity of Leaders

Criterion for Comparison	Leaders Cluster	Followers Cluster	Difference	Difference, %
Average efficiency score, CRS	2.04	1.62	0.42	25.62%
Average efficiency score, VRS	1.79	1.14	0.65	57.32%
Average efficiency score, NIRS	1.80	1.14	0.65	57.45%

Table 21.7 Comparison of Simulated Outputs of Leaders Based on the Transformative Capacity of Followers

Criterion for Comparison	Leaders Cluster	Followers Cluster	Difference	Difference, %
Average efficiency score, CRS	2.09	2.30	0.21	9.20%
Average efficiency score, VRS	1.38	2.00	0.62	30.87%
Average efficiency score, NIRS	1.38	2.17	0.79	36.26%

allows us to suggest that the level of relative efficiency of the followers can be increased by means of a gradual increase in the level of investments in telecoms. Simply put, the followers do not invest enough in telecoms to compete with the leaders in terms of the relative efficiency. One of the possible explanations for the link between investments and levels of efficiency refers to economies of scale; it is plausible that an insufficient level of investments precludes followers from developing economies of scale in the area of telecoms, and this prevents them from reaping the benefits in the form of increased efficiency.

Based on the result of phase 1, we have enough evidence to suggest that in order to reduce the level of relative inefficiency, the *external strategy*, or *intercluster* strategy, of the followers should be directed at an increase in the level of investments in telecoms. However, the *internal strategy*, or *intracluster* strategy, of the followers should aim at an increase in the levels of revenue from telecoms, as well as at an increase in the levels of revenue from telecoms *relative* to the level of investments.

The suggested strategies, however, are based on the assumption of homogeneity of the transformative capacities of the leaders and followers. Unfortunately, at this point in our investigation we cannot determine whether this assumption holds. Consequently, we cannot determine whether the followers can handle an increased level of investments with the same level of efficiency as the leaders.

After NN simulation conducted in phase 2, we performed DEA in phase 3. Based on the results of the DEA, we were able to establish that the followers are capable of becoming relatively more efficient than the leaders in the case where they improve their level of transformative capacity. Thus, at this point, we can state that the lower level of the averaged relative efficiency of the followers is, at least partially, a result of inefficient processes of revenue production.

We also determined, based on the results summarized in Table 21.7, that a simple increase in the level of the inputs does not improve the level of relative efficiency of the followers. Consequently, prior to increasing the level of investments in telecoms, the followers must improve the effectiveness of revenue production associated with the current level of investments.

Discussion and Conclusion

The main purpose of this study was to answer two research questions regarding routes for identifying and implementing strategies allowing for increasing the level of relative efficiency of production of revenues from investments in telecoms in the context of 18 TEs. We offered a solution approach that takes into consideration both the external and internal circumstances of a given TE. The peer group of countries relative to which the level of efficiency of the production of revenues is determined reflects external circumstances. The level of investments and the level of efficiency of conversion of investments into revenues represent the internal circumstances of a TE. Incorporation of the two perspectives, internal and external, allows us to identify not only the appropriate strategy leading to an increase in relative efficiency of the production of revenues, but also internal strategies suggesting how such an increase can be obtained. In the case of our illustrative example, the followers must begin by improving their internal capabilities of conversion of investments into revenues, and only then proceed with increasing the level of investments in telecoms.

How do the results of our inquiry contribute to an understanding of the economics of ICT in the context of TEs? First, the findings of our study suggest that, in the context of TEs, a simple increase in the level of inputs is not likely to result in a proportional increase in the levels of outputs. Rather, the evidence suggests that the increase in inputs must be gauged in accordance with the internal capabilities of TEs to efficiently utilize investments. Second, our investigation demonstrated that the notion of the *peer group* in the context of TEs is a rather complex one; despite having a common denomination of TEs, this group of countries is not homogenous. Consequently, the process of selection of a reference group for a given TE must be conducted with caution. Our results, for example, suggest that

Hungary and Poland should not be directly compared to Armenia and Albania; consequently, these four TEs should not be assigned to the same peer group.

These findings are different from what one would expect to find in the context of developed market economies. First, the relative homogeneity of developed economies (Arcelus & Arocena 2000; Barro & Sala-i-Martin 1995; Sala-i-Martin 1996) allows for easier sharing of successful strategies. This means that if a developed economy pursues a goal of increasing its level of efficiency of production of revenues, then it is possible that such a goal can be accomplished by adopting a strategy of a better-performing peer. Unlike developed economies, TEs do not represent a homogenous group (Hoskisson et al. 2000); thus, the adoption of a strategy of a better-performing TE may not work. As the results of our inquiry demonstrate, for a given TE, not every-better performing TE is a *better-performing peer*. Second, homogeneity of developed economies also suggests that implementation paths of the adopted strategies can be adopted also. This refers to the internal processes by which investments are converted into revenues. While benchmarking and adoption of the best practices have been and are common among developed countries, it might not be a viable option in the context of TEs.

We suggest that our study makes several contributions to the research in this area of inquiry. First, we use the well-established framework of neoclassical growth accounting as a theoretical foundation of our study. Hoskisson et al. (2000) state that one of the difficulties associated with research on strategies in emerging market economies is that the "theories promulgated for developed market economies may not be appropriate for emerging economies" (p. 257). Our study demonstrated that the framework of neoclassical growth accounting extensively applied in the context of developed countries could also be applied to the context of TEs. Second, in our research, we outlined and tested a novel multitechnique methodology that utilizes well-established and reliable data analytic tools. Third, Hoskisson et al. (2000) warn that the clear differences between emerging economies may create problems in cases where "for example, samples from two distinct emerging economies are grouped together to overcome small-sample problems" (p. 259). The approach taken in our inquiry allows for explicitly dealing with the issue of heterogeneity in small samples.

Our inquiry contributes to practice as well. First, our approach allows for easy identification of the appropriate peer group; thus, a decision maker should be able to identify TEs that should, and conversely, should not, be used for a direct comparison in terms of performance. Second, the outlined approach allows a decision maker to take explicitly into consideration external and internal factors affecting performance. Finally, the methodology introduced in this chapter can be used as a simulation tool allowing a decision maker to forecast possible consequences of increased investments or changes in the relative efficiency of conversion of investments into revenues. Our study raises some new research questions as well. First, if the level of investments in telecoms should be increased with regard to the level of efficiency of revenue production, then what is the role of telecom employees in the process? Should we consider the number of full-time telecom employees as complementary to investments in telecoms? This is an important question to answer because if a full-time telecom labor force serves as a caretaker of investments in telecoms, then it is possible that these two factors should not be considered in isolation. Second, as the level of investments increases, it is likely to bring about an increased level of the workforce. Should the increase in the number of employees involve hiring of full-time employees or consultants? Third, if additional investments are made, should they be accompanied by an increase in productivity of existing employees or by an increase in the total number of employees? Intuitively, it makes sense to think that there exists a threshold of *productivity*, represented as a ratio of revenues to investments per full-time employee, which signals the need for expansion of the current workforce.

Our study is not without its limitations. The first limitation is associated with the data used in this study. In our inquiry, we were limited to 18 TEs represented by a limited number of variables; this could only allow us to render an incomplete picture of a complex context. However, this limitation seems to be typical of the research associated with emerging market economies, where "on empirical side, researchers face sampling and data collection problems" (Hoskisson et al. 2000, p. 257). The second limitation is associated with the data analytic methods used in our analysis. For example, in DT analysis, determination of what constitutes heterogeneity of a sample ultimately resides with the decision maker, and black-box approach of NN to the process of conversion of inputs into outputs does not offer any insights to the decision maker regarding how the process can be improved. Despite these shortcomings, however, we hope that the contributions of this study outweigh its limitations.

Acknowledgment

Material in this chapter previously appeared in "Strategies for Telecoms to Improve Efficiency in the Production of Revenues: An Empirical Investigation in the Context of Transition Economies," *Journal of Global Information Technology Management* 11:4, 56–75.

References

Arcelus, F. J. & Arocena, P. (2000). Convergence and productive efficiency in fourteen OECD countries: A non-parametric frontier approach. *International Journal of Production Economics*, 66, 105–117.

Avgerou, C. (1998). How can IT enable economic growth in developing countries? *Information Technology for Development*, 8, 15–28.

Barro, R. & Sala-i-Martin, X. (1995). *Economic Growth*. McGraw-Hill, Boston.

Bishop, C. M. (1995). *Neural Networks for Pattern Recognition*. Oxford University Press, Oxford.

Charnes, A., Cooper, W. W., Lewin, A. Y., & Seiford, L. M. (eds.). (1994). *Data Envelopment Analysis: Theory, Methodology, and Applications*. Kluwer, Boston.

Colecchia, A. & Schreyer, P. (2001). The Impact of Information Communications Technology on Output Growth. STI Working Paper 2001/7, OECD, Paris.

Colecchia, A. & Schreyer, P. (2002). ICT investment and economic growth in the 1990s: Is the United States a unique case? A comparative study of nine OECD countries. *Review of Economic Dynamics*, 5, 408–442.

Council of Economic Advisors (2001). The annual report of the council of economic advisors. In *The Economics of the President*. U.S. Government Printing Office, Washington, DC.

Daveri, F. (2000). Is Growth an Information Technology Story in Europe too? Working Paper, University of Parma, Parma, Italy.

Daveri, F. (2002). The new economy in Europe, 1992–2001. *Oxford Review of Economic Policy*, 18(3), 345–362.

Dewan, S. & Kraemer, K. (1998). International dimensions of the productivity paradox. *Communications of the ACM*, 41(8), 56–62.

Dewan, S. & Kraemer, K. (2000). Information technology and productivity: Evidence from country level data. *Management Science (Special Issue on the Information Industries)*, 46(4), 548–562.

Hoskisson, R., Eden, L., Lau, C., & Wright, M. (2000). Strategy in emerging economies. *Academy of Management Journal*, 43(3), 249–267. Available online at http://www.itu.int/ITU-D/ict/publications/wtdr_06/index.html, retrieved January 20, 2008.

IMF (2000). Transition Economies: An IMF Perspective on Progress and Prospects. Available online at http://www.imf.org/external/np/exr/ib/2000/110300.htm#I, retrieved January 20, 2008.

IMF (2001). *International Financial Statistics*. IMF, Washington, DC.

Jalava, J. & Pohjola, M. (2002). Economic growth in the new economy: Evidence from advanced economies. *Information Economics and Policy*, 14(2), 189–210.

Jorgenson, D. W. (2001). Information technology and the US economy. *American Economic Review*, 91(March), 1–32.

Jorgenson, D. W. & Stiroh, K. J. (2000). Raising the speed limit: US economic growth in the information age. *Brookings Papers on Economic Activity*, 2(1), 125–211.

Kraemer, K. L. & Dedrick, J. (2001). Information technology and economic development: Results and policy implications of cross-country studies. In *Information Technology, Productivity, and Economic Growth*, ed. M. Pohjola, Oxford University Press, Oxford.

Lee, M. I. H., & Khatri, M. Y. (2003). Information technology and productivity growth in Asia (No. 3–15). International Monetary Fund.

Morales-Gomez, D. & Melesse, M. (1998). Utilizing information and communication technologies for development: The social dimensions. *Information Technology for Development*, 8, 3–13.

Murakami, T. (1997). The Impact of ICT on Economic Growth and the Productivity Paradox. Available online at http://www.tcf.or.jp/data/19971011_Takeshi_Murakami_2.pdf, retrieved January 20, 2008.

Myers, D. (2004). *Construction Economics*. Spon Press, UK.

OECD (2003a). *The Sources of Economic Growth in OECD Countries*. OECD, Paris.

OECD (2003b). *ICT and Economic Growth: Evidence from OECD Countries, Industries and Firms.* OECD, Paris.
OECD (2003c). *Integrating Information and Communication Technologies in Development Programmes, Policy* Brief, OECD, Paris.
OECD (2004). *DAC Network on Poverty Reduction: ICTs and Economic Growth in Developing Countries.* OECD, Paris. Available online at http://www.oecd.org/dataoecd/15/54/34663175.pdf, retrieved January 20, 2008.
OECD (2005a). Background paper: The contribution of ICTs to pro-poor growth: No. 384. *OECD Papers*, 5(2), 15–52.
OECD (2005b). The contribution of ICTs to pro-poor growth: No. 379. *OECD Papers*, 5(1), 59–72.
Oliner, S. D. & Sichel, D. E. (2000). The resurgence of growth in the late 1990s: Is information technology story? *Journal of Economic Perspectives*, 14(4), 3–22.
Oliner, S. D. & Sichel, D. E. (2002). Information technology and productivity: Where are we now and where are we going? *Economic Review*, 3(3), 15–41.
Ollman, B. (1997). *Market Socialism: The Debate among Socialists.* Routledge, UK.
Piatkowski, M. (2002). The "New Economy" and Economic Growth in Transition Economies. WIDER Discussion Paper No. 2002/63, Helsinki: WIDER.
Piatkowski, M. (2003a). The Contribution of ICT Investment to Economic Growth and Labor Productivity in Poland 1995–2000. *TIGER Working Paper Series* 43. July. Warsaw. Available online at http://www.tiger.edu.pl, retrieved January 20, 2008.
Piatkowski, M. (2003b). Does ICT Investment Matter for Output Growth and Labor Productivity in Transition Economies? *TIGER Working Paper Series*, No. 47. December. Warsaw. Available online at http://www.tiger.edu.pl, retrieved January 20, 2008.
Piatkowski, M. (2004). The Impact of ICT on Growth in Transition Economies, *TIGER Working Paper Series*, No. 59. July. Warsaw. Available online at http://www.tiger.edu.pl/publikacje/TWPNo59.pdf, retrieved January 20, 2008.
Pohjola, M. (ed.). (2001). *Information Technology, Productivity and Economic Growth: International Evidence and Implications for Economic Development, WIDER Studies in Development Economics.* Oxford University Press, Oxford.
Pohjola, M. (2002). New Economy in Growth and Development, WIDER Discussion Paper 2002/67, United Nations University World Institute for Development Economics Research (UNU/WIDER). Helsinki, Finland.
Sala-i-Martin, X. (1996). The classical approach to convergence analysis. *Economic Journal*, 106(4), 1019–1036.
Samoilenko, S. & Osei-Bryson, K.-M. (2007). Increasing the discriminatory power of DEA in the presence of the sample heterogeneity with cluster analysis and decision trees. *Expert Systems with Applications*, 34(2), 1568–1581.
Samoilenko, S. (2008a). Contributing factors to information technology investment utilization in transition economies: An empirical investigation. *Information Technology for Development*, 14(1), 52–75.
Samoilenko, S. (2008b). Determining Sources of Relative Inefficiency in Heterogeneous Samples: Methodology Using Cluster Analysis, DEA and Neural Networks. Working paper.
Sarychev, A. (2005). Productivity, prosperity, and development. *Development & Transition*, 1(July), 15–16. Available online at http://www.developmentandtransition.net, retrieved January 20, 2008.

Sasse, G. (2005). Lost in transition: When is transition over? *Development & Transition*, 1(July), 10–11. Available online at http://www.developmentandtransition.net, retrieved January 20, 2008.

Schelkle, W. (2005). Linking economic development and transition. *Development & Transition*, 1(July), 11–13. Available online at http://www.developmentandtransition.net, retrieved January 20, 2008.

Slay, B. (2005). Development versus transition. *Development & Transition*, 1(July), 7–9. Available online at http://www.developmentandtransition.net, retrieved January 20, 2008.

Smith, A. (1776). *The Wealth of Nations*. Available online at http://www.marxists.org/reference/archive/smith-adam/works/wealth-of-nations/, retrieved January 20, 2008.

Stiroh, K. (2002). Information technology and the U.S. productivity revival: What do the industry data say? *American Economic Review*, 92(5), 1559–1576.

UN ICT Task Force Report (2005). Innovation and Investment: Information and Communication Technologies and the Millennium Development Goals. Report Prepared for the United Nations ICT Task Force in Support of the Science, Technologies and Innovation Task Force of the United Nations Millennium Project. Available online at http://www.unicttaskforce.org/, retrieved January 20, 2008.

Van Ark, B., Melka, J., Mulder, N., Timmer, M., & Ypma, G. (2002). ICT Investments and Growth Accounts for the European Union, 1980–2000. Research Memorandum GD-56, Groningen Growth and Development Centre, Groningen. Available online at http://www.eco.rug.nl/ggdc/homeggdc.html, retrieved January 20, 2008.

World Bank (2004). Beyond Economic Growth: An Introduction to Sustainable Development. The World Bank Institute, 2nd edition. Available online at http://www.worldbank.org/depweb, retrieved January 20, 2008.

WT/ICT Development Report (2006). Measuring ICT for social and economic development. International Telecommunication Union's World Telecommunication/ICT Development Report, 8th edition. Available online at http://www.itu.int/ITU-D/ict/publications/wtdr_06/index.html, retrieved January 20, 2008.

Yearbook of Statistics (2004). Telecommunication Services Chronological Time Series 1993–2002. ITU Telecommunication Development Bureau (BDT), International Telecommunication Union. Available online at http://www.itu.int/ITU-D/ict/publications, retrieved January 20, 2008.

392 ■ *Theoretical Research Frameworks Using Multiple Methods*

Appendix

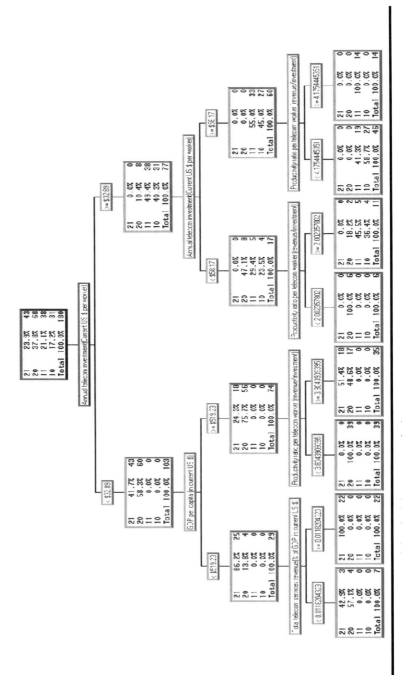

Figure 21A.1 Results of the decision tree analysis.

Index

A

ARM, *see* Association rule mining
Artificial neural networks (NNs), 91–105
    backpropagation approach, 93
    error functions, 97, 98
    forward-propagation approach, 93
    incremental mode, 94
    induction process for directed learning, 93–97
    inputs to the nodes of the hidden and output layers, computing of, 94
    layers, 92
    multilayer perceptron, 92
    network architectures, 95
    outputs of the nodes of the hidden and output layers, computing of, 95
    software implementation, 97–102
    termination test, 93
    weight adjustment for the backpropagation algorithm for multilayer perceptron, 95
Association rule mining (ARM), 139, 317, 321–322
Average Variance Extracted (AVE) measure, 203

B

Bartlett test of sphericity, 86, 199
BCC model, 76
Black-box model, 72

C

CA, *see* Cluster analysis
CCR model, 76
CFA, *see* Confirmatory factor analysis

Chaos theory (CT), 25–27, 108; *see also* Information systems development, chaos and complex systems theories and
    control parameter, 26
    Feigenbaum number, 27
    fixed-point attractor, 26
    limit-cycle attractor, 26
    Lorenz attractor, 26
    period doubling, 26
    phase transition, 26
    strange attractor, 26
    torus attractor, 26
Classification trees (CTs), 39, 40, 41
Cluster analysis (CA), 55–70, 269, 290
    clustering algorithms, 60–61, 61–64
    cluster node parameter settings, 67
    cluster validity, 62
    goals for clustering, 62–63
    internal standardization, 67
    investment utilization in transition economies, 294–295
    knowledge discovery via data mining, 63, 64
    output of clustering, 57
    process model for clustering, 63–64
    revenue generation from ICT, 337
    seed initialization, 67
    similarity metrics, 61
    software implementation, 65–67
Cobb–Douglas production function, 15
Complex systems (CSs), 130
Complex systems theory (CST), 19–25, 108, 315; *see also* Information and communication technology capabilities in sub-Saharan economies, socioeconomic impact of

emergent behaviors, 22
functional perspective, 24–25
overview, 19
structural perspective, 20–23
system elements, 20
Confirmatory factor analysis (CFA), 87
Constant return to scale (CRS), 233, 349
Cross industry standard procedure for data mining (CRISP-DM), 63
CSs, *see* Complex systems
CST, *see* Complex systems theory
CT, *see* Chaos theory
CTs, *see* Classification trees
Cybernetics, 134, 136

D

Data envelopment analysis (DEA), 71–79, 145, 189; *see also* Revenue production, strategies for telecoms to improve efficiency in
   additive model, 76
   approach and efficiency types, 73–74
   base-oriented model, 75
   BCC model, 76
   black-box model, 72
   capital investment, 73
   CCR model, 76
   changes in technology, 78
   clusters of TEs and, 269
   DEA model, 72
   input-oriented model, 75
   investment utilization in transition economies, 292–294
   labor force, 73
   linear programming, 73
   Malmquist index, 77–78
   model guidelines and assumptions, 72–73
   model orientations, 74–75
   model types, 76–77
   output-oriented model, 75
   price efficiency, 74
   revenue generation from ICT, 337
   revenue production and, 377
   test grade, 72
Data mining (DM), 55
DEA, *see* Data envelopment analysis
Decision-making units (DMUs), 71, 147, 189, 320
Decision tree (DT); *see also* Revenue production, strategies for telecoms to improve efficiency in
   analysis, 269
   building of, 57
   investment utilization in transition economies, 295–296
   revenue production and, 377
   telecom investment and, 217
Decision tree (DT) induction, 39–54
   accuracy, example of splitting methods on, 47–49
   classification tree, 39, 40, 41
   DT generation process, 43–53
   postpruning, 46
   prepruning, 46
   recursive splitting, 45
   recursive splitting example, 50–51
   regression tree, 41–43
   selection of splitting method, 46
   sibling rules, 41
   software implementation, 51–53
   unpruned DT, 45
DM, *see* Data mining
DMUs, *see* Decision-making units
DT, *see* Decision tree

E

E-commerce, 20
Emerging economies, 188
Endogenous growth models, 12–13
Exogenous growth models, 12–13
Exploratory factor analysis (EFA), 87

F

Factor analysis (FA), 86
Factor loading, 87
Feigenbaum number, 27
First-order cybernetics, 134
Fixed-point attractor, 26

G

Gain-ratio (GR) method, 46
General least squares (GLS), 85
Gross domestic product (GDP), 73, 144, 266
Growth accounting, 290–291

H

Health of Economy, 195
Heckscher-Ohlin (H-O) model, 20
High-income country, 287

Human Assets Index, 287
Human development and macroeconomic
    returns (investments in telecoms),
    169–185
  background, 172–176
  discussion, 178–182
  followers group, 182
  groups sample, 172
  Human Development Index (HDI), 170
  ICT capitalization, 178
  leaders group, 178
  overview of data, 177–178
  previous research results, 171
  results of the data analysis, 178
  theoretical framework and research
      questions, 176–177
  total factor productivity, 172

I

ICT Capitalization, 178, 194, 226, 247
Information and communication technologies
    for development (ICT4D), research
    on, 1–9
  context (countries and communities), 3
  data analysis methods, 4
  developing world, 3
  ICT issues and focus areas, 2–3
  options for configuring ICT4D research
      project, 2–4
  research methodologies, 3
  theoretical frameworks, 3–4
Information and communication technology
    (ICT) capabilities in sub-Saharan
    economies, socioeconomic impact of,
    315–333
  association rule mining, 317, 321–322
  data, 318
  DEA, 320–321
  DEA results, 325–326
  decision-making units, 320
  discussion of the results, 324–332
  low-income economies, 330–331
  market basket analysis, 322, 327–329
  methodology, 320–322
  Networked Readiness Index, 316
  overview, 315
  research framework, 318
  research questions and null hypotheses,
      322–323
  results of data analysis, 323–324
  variable return to scale DEA model, 320

Information and communication technology
    (ICT) and economic growth in
    transition economies, human capital
    dimension of, 247–264
  analysis procedure, 257–260
  background, 249
  complementarity, using MR to determine
      the presence of, 253
  data analysis, 257–260
  determining appropriate policy options,
      251–252
  discussion, 260–261
  proposed approach to policy analysis,
      253–256
  relative inefficiency, using DEA to
      determine the sources of, 253–256
  research problem, 250–251
  sample and panel data, 256
Information and communication technology
    and labor force on economic growth,
    interaction in transition economies,
    265–286
  annual investments in telecoms, 285, 286
  annual revenues from telecoms, 284
  caretakers of investments in telecoms, 275
  cluster analysis, 269
  complementarity and translog production
      function, theory of, 270–271
  data and background of study, 267–269
  data source node, 274
  decision tree analysis, 269
  formal definition of research problem,
      272–274
  gross domestic product, 266
  leaders and followers, 279
  partial correlation, 272
  Political Constraint index, 278
  reporter node, 275
  research problem, importance of, 266–267
  results of data analysis, 274–279
  telecom staff, 285
  theoretical framework, 269–274
  World Development Indicators data set, 268
Information systems development (ISD), chaos
    and complex systems theories and,
    107–127
  conceptual foundations, 109–112
  CT and ISD, 113–114
  emergent properties, 109
  environment, diversity and complexity of, 124
  fitness landscapes, 118–120
  fitness landscapes and ISD, 120–123

information system fitness, definition of, 119–120
insights and implications for ISD, 114–117
IS and ISD from the perspective of CST, 118–120
reporting structure, 122
rugged landscapes, 118
rule trajectory, 111
scenarios, 121–123
self-organization of the CSs, 118
unstable behavior, study of, 111
Information technology investment utilization in transition economies, contributing factors to, 287–314
CA, 294–295
cluster analysis, 290
contribution of the study, 303–305
DEA, 292–294
DT, 295–296
emerging economies, 288
growth accounting, 290–291
Human Assets Index, 287
input-saving measure of efficiency, 311–312, 313, 314
methodology, 292–296
overview of the data, 291–292
results, 296–303
theoretical framework, 290–291
International Monetary Fund (IMF), 315, 376
International Telecommunication Union (ITU), 146, 177, 292, 377
International trade
Heckscher-Ohlin model of, 20
new product, 31
participation of an economy in, 194, 210
ISD, *see* Information systems development, chaos and complex systems theories and
ITU, *see* International Telecommunication Union

K

Kaiser–Meyer–Olkin (KMO) test of sampling adequacy, 86, 199
Knowledge discovery via data mining (KDDM), 63, 64

L

Labor force, *see* Information and communication technology and labor force on economic growth, interaction in transition economies

Least-squares (LS) method, 90
Leontief paradox, 30
Limit-cycle attractor, 26
Linear programming (LP), 73
Lorenz attractor, 26
Low-income country, 287
LP, *see* Linear programming
LS method, *see* Least-squares method

M

Macroeconomic returns, *see* Human development and macroeconomic returns (investments in telecoms)
Malmquist index (MI), 77, 170, 189, 269
Maximum simulated likelihood estimation (MSL), 85
Middle-income country, 287
Militarization of Economy, 195
MR, *see* Multivariate regression
MSL, *see* Maximum simulated likelihood estimation
Multifactor productivity, 12
Multilayer perceptron (MLP), 92
Multitheoretical support, design of the research workbench for investigations relying on, 129–141
cybernetics-based analytic support system, 133–134
general principles of cybernetic systems for designing the PAS, implications of, 135, 137
performance analysis for complex systems, 130–133
performance analysis system, 132
structural components of PAS, 135–140
Multivariate regression (MR), 145, 253

N

Neoclassical growth accounting, framework of, 11–18
exogenous and endogenous growth models, 12–13
formulation of neoclassical production function, 13–15
multifactor productivity, 12
Solow residual, 12
transcendental logarithmic (translog) production function, 15–17
weaknesses and criticisms, 17–18
Networked Readiness Index (NRI), 316

Index ■ 397

Neural networks (NN), 139; *see also* Artificial neural networks; Revenue production, strategies for telecoms to improve efficiency in
   based clustering algorithm, 60
   revenue production and, 337, 377
   simulation, 269, 346
Non-increasing return to scale (NIRS), 233, 349
Nonrelaxed LP, 73
NRI, *see* Networked Readiness Index

O

Ordinary least squares (OLS), 85, 145
Output of clustering, 57
Output-oriented DEA, 345–346

P

Partial correlation, 272
Partial least squares (PLS), 85, 195
PCA, *see* Principal component analysis
PCT, *see* Product cycle theory and the product life cycle model
Performance analysis system (PAS), 132
PFA, *see* Principal factor analysis
PLC, *see* Product cycle theory and the product life cycle model
PLS, *see* Partial least squares
Political Constraint index, 278
Price efficiency, 74
Principal component analysis (PCA), 81, 199
Principal factor analysis (PFA), 87
Product cycle theory (PCT), 129
Product cycle theory (PCT) and the product life cycle (PLC) model, 29–37
   Leontief paradox, 30
   PLC model, criticisms and applications of, 36–37
   PLC model decline, 35–36
   PLC model growth, 34–35
   PLC model introduction, 34
   PLC model maturity, 35
   product cycle theory, 29–33
   product life cycle model, 33–37
Productivity, Total factor productivity, investigation of determinants of
Product life cycle (PLC), 337; *see also* Revenue generation from ICT, improving the relative efficiency of
   model, 129, 338–339
   telecom investment and, 233

R

Regression analysis (RA), 139
Regression trees (RTs), 39, 41–43
Relaxed LP, 73
Revenue generation from ICT, improving the relative efficiency of, 335–371
   CA, 343
   contribution, 358
   DEA, input-oriented model, CRS, 367
   DEA, input-oriented model, NIRS, 369
   DEA, input-oriented model, scale efficiency scores, 371
   DEA, input-oriented model, VRS, 368
   DEA, output-oriented model, CRS, 364
   DEA, output-oriented model, NIRS, 366
   DEA, output-oriented model, scale efficiency scores, 370
   DEA, output-oriented model, VRS, 365
   description of data and background of the study, 339–341
   discussion of results of the data analysis, 347–356
   Followers, 348, 352, 353
   future research, 359
   ICT product, 336
   input-oriented DEA, 346
   Leaders, 347, 352, 353
   limitations, 358–359
   methodological approach, 343–347
   methods and techniques used in the study, 341–343
   model stages, 338
   NN simulation, 346
   output-oriented DEA, 345–346
   output-oriented DEA of the simulated data, 347
   PLC model, 338–339
   summary of results, 357
   theoretical framework, 338–339
   two-cluster solution, 340
   variables used in cluster analysis, 363
Revenue production, strategies for telecoms to improve efficiency in, 373–392
   DEA, 382–383
   DEA results, 384
   decision tree analysis, results of, 392
   discussion, 386–388
   DT analysis, 383
   DT induction, 381
   interpretation of the results, 384–386
   level of performance, 376

methodology, 380–383
NN simulation, 382, 384
overview of data, 377
previous findings, 378–380
results of the data analysis, 383–384
transition economy, 373
two-cluster solution, 379
RTs, *see* Regression trees

S

SAS Enterprise Miner, 229, 274
Second-order cybernetics, 134
SEM, *see* Structural equation modeling
Sibling rules, 41
Socioeconomic impact (ICT capabilities), *see* Information and communication technology capabilities in sub-Saharan economies, socioeconomic impact of
Software implementation
    clustering process, 65–67
    DT induction, 51–53
    NN induction process, 97–102
Solow residual, 12, 146
Spillover effects, *see* Telecom investment, spillover effects of, investigating factors associated with
Strange attractor, 26
Structural equation modeling (SEM), 81–90,195
    assessment of measurement model, 88–89
    assessment of structural model, 89–90
    Bartlett test of sphericity, 86
    common approaches, 85–86
    component based method, 86
    confirmatory factor analysis, 87
    endogenous variable, 82
    exogenous variable, 82
    exploratory factor analysis, 87
    factor loading, 87
    full information method, 89
    Kaiser–Meyer–Olkin test of sampling adequacy, 86
    latent variables, 82
    measurement model, 81
    mediational effect, 83
    model specification, 84–85
    preliminary data analysis and factor analysis, 86–88
    reflective and formative measurement models, 83–84
    structural model, 81
    unidimensionality of the model, 88
    validity of measurement model, 88
Sub-Saharan Africa (SSA), 316; *see also* Information and communication technology capabilities in sub-Saharan economies, socioeconomic impact of

T

Telecom investment, spillover effects of, 187–214
    appendix, 214
    Average Variance Extracted measure, 203
    convergent validity of measures, 201, 202
    discussion of the results, 203–209
    emerging economies, 188
    Followers subgroup, 188, 196
    Health of Economy, 195
    ICT Capitalization, 194
    Leaders subgroup, 188, 196
    limitations of the study, 209–210
    measures of research model, 197–198
    Militarization of Economy, 195
    overview of the data, 195–199
    partial least squares method, 195
    PLS analysis, 199–203
    preliminary data analysis, 199
    previous research results, 190
    research problem, 191–195
    results of data analysis, 199–203
    structural equation modeling, 195
    transition economies, 214
Telecom investment, spillover effects of, investigating factors associated with, 215–245
    background of the study, 218–220
    balanced investment policy, 236
    change in efficiency, 219
    change in technology, 219
    DEA results, scores of relative efficiencies, 245
    decision trees, 217
    discussion of the results, 235–239
    endogeneity problem, 221
    ICT Capitalization, 226
    limitations of the study and future research, 239–240
    measures of current research model, 225
    overview of the data, 223–224
    research methodology, 224–229

results of the data analysis, 229–235
theoretical foundation and research
 questions, 220–223
Telecoms, investments in
 human development and macroeconomic
  returns, 169–185
 total factor productivity, investigation of
  determinants of, 143–168
TEs, *see* Transition economies
Torus attractor, 26
Total factor productivity (TFP), 12
 human development and, 170
 yearly changes in, 269
Total factor productivity (TFP), investigation of
  determinants of, 143–168
 annual telecom investment, 149
 comparison of sources of growth in
  productivity, 163
 conclusion, 164
 conversion efficiency, 165
 data availability, 166
 decision-making unit, 147
 description of the data, 146
 full-time telecommunication staff, 149
 macroeconomic impact of investments and,
  172
 null hypotheses of the study, 152–154
 OnFront, version 2.02, 154
 research methodology, 146–152
 results and discussion, 154–163
 SAS Enterprise Miner, 154
 Solow residual, 146
Transcendental logarithmic (translog)
 production function, 15–17
Transition economies (TEs), 169, 288
 human development and macroeconomic
  returns (investments in telecoms),
  169–185

information and communication technology
 and economic growth in transition
 economies, human capital dimension
 of, 247–264
information and communication technology
 and labor force on economic growth,
 interaction in transition economies,
 265–286
information technology investment
 utilization in transition economies,
 contributing factors to, 287–314
leaders and followers, 214
revenue generation from ICT, improving the
 relative efficiency of, 335–371
revenue production, strategies for telecoms
 to improve efficiency in, 373–392
telecom investment, spillover effects of,
 187–214
total factor productivity, investigation of
 determinants of, 143–168

U

Unpruned DT, 45
Unstable behavior, study of, 111

V

Validity of measurement model, 88
Variable return to scale (VRS), 233, 349, 320

W

World Development Indicators (WDI), 224, 268

Y

Yearbook of Statistics, 268